WITHDRAWN
UTSA Libraries

In Search of the True Universe

Astrophysicist and scholar Martin Harwit examines how our understanding of the Cosmos advanced rapidly during the twentieth century and identifies the factors contributing to this progress. Astronomy, whose tools were largely imported from physics and engineering, benefited mid-century from the U.S. policy of coupling basic research with practical national priorities. This strategy, initially developed for military and industrial purposes, provided astronomy with powerful tools yielding access – at virtually no cost – to radio, infrared, X-ray, and gamma-ray observations. Today, astronomers are investigating the new frontiers of dark matter and dark energy, critical to understanding the Cosmos but of indeterminate socio-economic promise. Harwit addresses these current challenges in view of competing national priorities and proposes alternative new approaches in search of the true Universe. This is an engaging read for astrophysicists, policy makers, historians, and sociologists of science looking to learn and apply lessons from the past in gaining deeper cosmological insight.

MARTIN HARWIT is an astrophysicist at the Center for Radiophysics and Space Research and Professor Emeritus of Astronomy at Cornell University. For many years he also served as Director of the National Air and Space Museum in Washington, D.C. For much of his astrophysical career he built instruments and made pioneering observations in infrared astronomy. His advanced textbook, *Astrophysical Concepts*, has taught several generations of astronomers through its four editions. Harwit has had an abiding interest in how science advances or is constrained by factors beyond the control of scientists. His book *Cosmic Discovery* first raised these questions. The present volume explores how philosophical outlook, historical precedents, industrial progress, economic factors, and national priorities have affected our understanding of the Cosmos. Harwit is a recipient of the Astronomical Society of the Pacific's highest honor, the Bruce Medal, which commends "his original ideas, scholarship, and thoughtful advocacy."

In Search of the True Universe

The Tools, Shaping, and Cost of Cosmological Thought

MARTIN HARWIT

CAMBRIDGE
UNIVERSITY PRESS

CAMBRIDGE
UNIVERSITY PRESS

32 Avenue of the Americas, New York, NY 10013-2473, USA

Cambridge University Press is part of the University of Cambridge.

It furthers the University's mission by disseminating knowledge in the pursuit of education, learning, and research at the highest international levels of excellence.

www.cambridge.org
Information on this title: www.cambridge.org/9781107044067

© Martin Harwit 2013

This publication is in copyright. Subject to statutory exception and to the provisions of relevant collective licensing agreements, no reproduction of any part may take place without the written permission of Cambridge University Press.

First published 2013

Printed in the United States of America

A catalog record for this publication is available from the British Library.

Library of Congress Cataloging in Publication data
Harwit, Martin, 1931– author.
In search of the true universe : the tools, shaping, and cost of cosmological thought / Martin Harwit.
    pages   cm
Includes bibliographical references and index.
ISBN 978-1-107-04406-7 (hardback)
1. Astronomy–History–20th century.   2. Science and state–History–20th century.
3. Astronomy–Social aspects–History–20th century.   I. Title.
QB32.H27   2013
523.1–dc23   2013015856

ISBN 978-1-107-04406-7 Hardback

Cambridge University Press has no responsibility for the persistence or accuracy of URLs for external or third-party Internet Web sites referred to in this publication and does not guarantee that any content on such Web sites is, or will remain, accurate or appropriate.

**Library
University of Texas
at San Antonio**

Contents

Preface

When I was a student in college, Albert Einstein was still alive. My friends and I grew up with the myth of this great man who, the story went, while a clerk, third class, at the Swiss Patent Office in Bern, had emerged from nowhere at age 26 to set down the laws of relativity in a paper so fresh, so novel that not a single other scholar needed to be cited as a source of at least partial inspiration for this monumental paper. Einstein's paper contained no list of references; none had existed or could even be found!

We all aspired to emulate Einstein and write a paper as great as his. To this end it seemed we would need only to foster self-reliance, reject outside influences, and rely solely on an inner intellect.

Of course, this could not happen.

Indeed, it had not happened!

But in the first half of the twentieth century, the myth could not be dispelled. Historians of science active at the time preferred to write about a distant past. And, as a young man, Einstein himself may have quietly enjoyed the mystique that surrounded his work. Only three times, as far as I am aware – twice late in life – did he describe the road he had traveled, the difficulties with which he had wrestled, and the inspiration the work of others had provided.

The first occasion came on December 14, 1922, possibly one of the headiest days of his life. Four days earlier, he might have been in Sweden to receive the Nobel prize in physics from the King of Sweden. Instead he was now in Kyoto.

Einstein had not been at home in Berlin when the telegram from Stockholm arrived on November 10.[1] By then, he and his wife had already embarked on the long voyage to the other side of the globe to accept an invitation from Japan.

Einstein didn't seem worried, this December day, that he had missed the trip to Stockholm.[a] Responding to an impromptu request, he talked spontaneously without written notes to students and faculty at Kyoto University. He spoke in German, the language in which he best expressed himself. The professor of physics at Tohoku University, Dr. Jun Ishiwara, who had studied in Munich under the

---

[a] On Einstein's return to Berlin, the Swedish ambassador dropped by to hand him the award.

Figure i. 'The Nose as Reservoir for Ideas.' A drawing by Ippei Okamoto sketched in December 1922 during Einstein's visit to Japan. (*From the Japanese newspaper* Asahi Shimbun. *Courtesy AIP Emilio Segre Visual Archives.*)

theoretical physicist Arnold Sommerfeld, took careful notes and gave a running translation to the Japanese audience.

Ishiwara's notes were published the following year in the Japanese monthly *Kaizo*, but a complete translation did not become available in English until 60 years later, when it appeared in the August 1982 issue of *Physics Today* with the title 'How I created the theory of relativity.'[2]

In this Kyoto talk, Einstein spoke of the early influences on his thinking that the experiments on the speed of light by the American physicist Albert Michelson and the French physicist Hippolyte Fizeau had played. Einstein had also read the 1895 monograph on electrodynamics by Hendrik Lorentz[3] in the Netherlands and was fully familiar with the *electromagnetic theory* that James Clerk Maxwell had produced a generation earlier in Scotland. He recalled the "enormous influence" on his thoughts by the Austrian philosopher and physicist Ernst Mach.

Einstein may not have referred to the work of others in his ground-breaking paper on relativity, but he had certainly been thoroughly familiar with their efforts.

I wish I had known as a student that the young Albert Einstein had spent most of his undergraduate years reading the great works of Boltzmann, Helmholtz, Hertz, Lorentz, Maxwell, and others. Often he preferred this way of learning to attending lectures at the Eidgenössische Polytechnikum in Zürich, today's Eidgenössische Technische Hochschule (ETH).[4] It might have disabused me of concentrating on my own strivings instead of more carefully consulting the work of others to learn how they had gone about their contributions to science. For, science is a craft, and craftsmanship can best be acquired by learning from great masters. Their work is fascinating not only in what it tells us about the *Universe*. It is intriguing also in what it teaches us about the roles we humans play in discovering truths.

The scientific method asserts that its use inexorably leads to a faithful portrayal of Nature. A closer look, however, reveals a picture fashioned by individual talents and preferences; by the availability of new tools; by common mistakes and misapprehensions; by an inability to wrest ourselves free from established beliefs adopted and defended over millennia; and by the continual clash of political, military, and economic realities.

My aim in writing the present book has been to show how strongly these factors have shaped and continue to shape our understanding of the Universe. To keep the investigation in bounds, I have largely restricted it to advances made just in the twentieth century and on just one set of intertwined fundamentals – the origin and evolution of matter, of *stars*, and of the Universe, from earliest cosmic times down to the present. These three topics were actively pursued throughout the twentieth century. Their developments were so tightly interwoven that they cannot well be separated.

The book's early chapters show how varying personal choices helped different astrophysicists to arrive at new insights. During the first half of the twentieth century most of these researchers pursued their investigations alone or sometimes with a single student or colleague. Many of these efforts involved importing new theoretical tools from physics or mathematics to enable a novel approach and a convincing advance in our understanding.

* * *

World War II dramatically changed all this. The war had shown that strength in science and technology can determine a nation's future. In the United States, this realization led to a visionary report, 'Science – The Endless Frontier,' commissioned by President Franklin Delano Roosevelt at the suggestion of Vannevar Bush, throughout the war Director of the Office of Scientific Research and Development (OSRD).[5] Widely praised for its persuasive promotion of increased governmental support of science, the report issued in July 1945 advocated a seamlessly integrated program of basic and applied research, dedicated to the nation's health, welfare, and security.

The resulting close cooperation between basic research and practical efforts geared to national priorities soon provided U.S. astronomers powerful new *radio*, *infrared*, *X-ray*, and *gamma-ray* instrumentation. Balloons, aircraft, rockets, and *satellites* for transporting these exceptional tools to high altitudes or into space provided the clearest view of the Universe ever attained. With these technologies, often developed for national security at enormous cost but made available to astronomers free of charge, success followed success in astronomical discoveries and theoretical advances.

* * *

Starting in the early 1980s and gaining speed ever since, the field largely reorganized itself into big science projects shouldered by hundreds if not thousands of scientists, engineers, and managers.

Publications might then be co-authored by scientists working at dozens of separate institutions. In the United States, major research directions no longer were defined by individual investigators or by their observatory directors, but rather through a community-wide consensus established at 10-year intervals, through a Decadal Survey conducted under the auspices of the U.S. National Academy of Sciences. Communal agreement had become necessary because the support of astronomy had become costly, often bordering on the extremes the nation could afford.

Observatories under construction today, whether designed to operate on the ground or in space, have become so expensive that only intercontinental consortia can raise the required means. Europe and the United States already are collaborating on many space missions. Ground-based observatories similarly require the collaboration of nations on separate continents.

Future facilities may soon reach construction costs requiring even larger international consortia and outlays affordable only if budgeted across many decades or longer, though we do not yet have experience in designing scientific enterprises spread over such long periods.

* * *

These are some of the problems our quest to understand the Universe are beginning to encounter as the really hard cosmological problems we are about to face may become unaffordable unless we find new ways for astronomy to advance.

Why unaffordable?

Over the past decade, our investigations have revealed that atomic matter in the Universe – *galaxies*, stars, interstellar gases, *planets*, and life – constitutes a mere 4% of all the mass in the Universe. The far larger 96% is due to a mysterious *dark matter* in galaxies and an equally mysterious *dark energy* pervading the entire Universe. Apart from the gravitational forces these components exert, we know little about either.

This lack of insight is largely due to our current dependence on tools and capabilities originally developed for industrial or defense purposes before their adaption for astronomical *observations*. They enable investigations mainly limited to observations of atomic matter – a constraint that biases our perception of the Universe. We know this because a similar bias misled us once before.

In the years leading up to World War II, astronomy was driven by just one set of tools, those of optical astronomy: The Universe perceived then was quite different from the Cosmos we survey today. Astronomers of the first half of the century described the Universe in serene majestic terms. Today we view with awe the violence of galaxies crashing into each other, forming massive new stars that soon explode to tear apart much of the colliding matter. Yet, our current understanding still comes from piecing together impressions delivered by a limited toolkit. It may be far more versatile than the capabilities available early in the twentieth century, but it still restricts us to dealing primarily with observations of just the 4% of the mass attributed to atoms.

We will require a new set of tools to study the other 96% of the cosmic inventory. These are unlikely to have practical applications and will no longer be conveniently available through simple adoption and adaption of industrial or defense-related technologies. Their development may require support beyond reasonable governmental means unless stretched out over many decades or longer.

More than six decades since first adopting the economic model for astronomy that 'Science – The Endless Frontier' bequeathed, we may accordingly need to review our options for successfully completing our cosmological quest.

I am concerned because, with limited instrumental means, we may reveal only those aspects of the Cosmos the instruments are able to unveil. We might then arrive, just as astronomers working before World War II did, at a false sense of the Cosmos. This is why I place such importance on realistically planning the acquisition of the potentially costly tools we may require to reliably shape our cosmological views and help us succeed in our search for the true Universe.

Contributions from Publishers and Archives

I am indebted to the many publishers and archives that permitted me to include material to which they own copyrights. These figures and quotes have vividly enriched the book. Although the contributions are also acknowledged elsewhere in figure captions and bibliographic citations, I wish to explicitly thank the copyright holders here as well:

I thank the American Astronomical Society (AAS) and its publishers, the Institute of Optics, for permission to quote and include a figure from the *Astrophysical Journal* and to cite a passage from 'The American Astronomical Society's First Century,' published jointly by the AAS and the AIP; The American Institute of Physics (AIP) for the use of nine images from the Emilio Segre Visual Archives; The American Physical Society for figures from four of its journals and quotes from several of these journals as well as others; *The Annual Reviews of Astronomy and Astrophysics* for permission to cite the recollections of Hans Bethe in 'My Life in Astrophysics' and Edwin E. Salpeter in 'A Generalist Looks Back'; Basic Books, a division of Perseus Books, gave me permission to quote material from my book 'Cosmic Discovery'. I thank Cornell University's Carl A. Kroch Library Division of Rare and Manuscript Collections for permission to publish a photograph of Hans Bethe; Dover Publications for permitting me to include a passage from Henri Poincaré's 'Science and Hypothesis.' I thank Elsevier for permission to quote from an article by Yakov Borisovich Zel'dovich and Maxim Yurievich Khlopov in *Physics Letters B*; Europa Verlag, Zürich provided permission to quote several passages from an autobiographical sketch of Albert Einstein in 'Helle Zeit – Dunkle Zeit' edited by Carl Seelig; The George Gamow Estate granted permission to quote several recollections from Gamow's book 'My World Line'; The Harvard University John G. Wolbach Library gave me permission to quote parts of the PhD thesis of Cecilia Payne; I thank *The Journal for the History of Astronomy* – Science History Publications Ltd. for allowing me to quote passages from Sir Bernard Lovell's 'The Effects of Defense Science on the Advance of Astronomy'; The National Academy of Sciences of the USA allowed me to reproduce Edwin Hubble's original diagram relating the redshifts of nebulae to their distances, as well as a quote from an article by Duncan J. Watts on 'A simple model of global cascades on random networks'; I appreciate the permission of Nature Publishing Group to make use of several quotes and figures published in *Nature* over the course of the twentieth century; The Open Court Publishing Company provided permission to cite Albert Einstein in 'Albert Einstein: Philosopher-Scientist,' edited by Paul Arthur Schilpp. I thank Princeton University Press for its permission to use excerpts from 'Henry Norris Russell' by David DeVorkin; The Royal Astronomical Society (London) gave me permission to cite some words of Arthur Stanley Eddington in *The Observatory* and to quote from several articles, respectively by Hermann Bondi and Thomas Gold, by Fred Hoyle, and by Roger Blandford and Roman Znajek in *The Monthly Notices of the Royal Astronomical Society (London)*. I thank the Royal Society of London for permission to cite some of the writings of Meghnad Saha and of Paul A. M. Dirac, respectively published in the *Proceedings of the Royal Society A* in 1921 and 1931; St. John's College Cambridge provided photographs of Paul A. M. Dirac and Fred Hoyle with permission to reproduce them; Trinity College Cambridge similarly provided photographs of Arthur Stanley Eddington, Ralph Howard Fowler, and Subrahmanyan Chandrasekhar, with permission to include them. The University of Chicago Press permitted my quoting from 'The Structure of Scientific Revolutions' by Thomas S. Kuhn and 'How Experiments End' by Peter Galison, as well as use of a quote from the *Publications of the Astronomical Society of the Pacific*. The University of Illinois, permitted me to cite Claude Shannon in an excerpt from its republication of his 'The Mathematical Theory of Communication'; The University of Louvain Archives kindly provided and gave me permission to use a photograph of Georges Lemaître; and W. W. Norton publishers allowed me to cite a passage from 'Six Degrees – The Science of a Connected Age' by Duncan J. Watts.

Acknowledgments

I thank the many academic institutions and people who made 'In Search of the True Universe' possible.

I started work on the book in the spring of 1983 as a visiting Fellow at the Smithsonian Institution's National Air and Space Museum. I meant the book to be a companion volume to 'Cosmic Discovery,' published two years earlier and dealing largely with the way astronomical discoveries have come about; but I did not get very far on my newer efforts before other priorities intervened.[6] In the summer of 2002, the Kapteyn Institute at the University of Groningen invited me to spend three months in Groningen as its Adrian Blaauw Professor. Again, I started writing, but again I did not get as far as I had hoped. Finally, in the fall term of 2007, the Institute of Advanced Study at Durham University in Britain extended its hospitality. There, two other visiting Fellows introduced me to work of which I had been unaware. The sociologist David Stark of Columbia University introduced me to recent studies of social networks, and the Australian theoretical anthropologist Roland Fletcher, an expert on the growth of human settlements, familiarized me with his work. Finally, these tools, which I had been looking for without knowing whether they existed, were at hand to complete the present book satisfactorily. I particularly thank the directors of the Institute of Advanced Study at Durham University for their hospitality.

For nearly half a century the U.S. National Aeronautics and Space Administration has supported my astronomical research, most recently on joint efforts on which NASA partnered with the European Space Agency, initially on its Infrared Space Observatory and later on the Herschel Space Observatory. These efforts have colored many of the approaches pursued in the book's later chapters.

The theoretical astrophysicist Ira Wasserman, my long-term colleague and friend at Cornell, was the first person I asked to read and critique an early version of the completed book. His insight and recommendations were of immense value. A number of colleagues and family members then read and commented on successive drafts of the book: I thank our sons Alex and Eric Harwit and our daughter Emily and her husband Stephen A. Harwit-Whewell. The astronomer Peter Shaver and the historian of astronomy Robert W. Smith both read the entire book draft as well. Two other historians, Karl Hufbauer and David DeVorkin, kindly read and critiqued specific chapters in which I had heavily drawn on their informed scholarship. Karl also provided me with a number of informative unpublished articles.

The MIT astrophysicist and writer Alan Lightman gave me permission to quote sections of interviews he had conducted with leading astrophysicists. David DeVorkin similarly permitted my quoting some of his insights in his biography of Henry Norris Russell. The theoretical physicist Mark E. J. Newman of the University of Michigan gave me permission to include and adapt figures and tables from several of his publications. A number of colleagues, among them David Clark,

Bruce Margon, Vera Rubin, and Alexander Wolszczan generously answered questions about their personal contributions to astronomy, when I was unsure about how their discoveries had come about. Others, including Bernhard Brandl, Harold Reitsema, Jill Tarter, and Kip Thorne, provided information on issues I had not been able to resolve. Members of the staff at the Cornell University Library, in Ithaca, NY, none of whom I ever had the opportunity to meet and thank as I now live and work in Washington, DC, over the years found and e-mailed me copies of innumerable early scientific articles from many countries. I would not have had access to these on my own. And Peter Hirtle, also at Cornell, provided invaluable expert advice on the complex restrictions of the copyright law.

Working with Vince Higgs, Editor for Astronomy and Physics, and Joshua Penney, Production Editor, at Cambridge University Press, has been a pleasure. I also thank Abidha Sulaiman, Project Manager at Newgen Knowledge Works in Chennai, India, who coordinated the production of the book, and Theresa Kornak whose good judgment as a copy editor helped to improve the book. The book's cover design was produced by James F. Brisson.[b] I thank Brien O'Brien, with whom I first worked nearly thirty years ago, for his design of Figure 11.1 reproduced in gray tone from a 1985 NASA brochure, as well as for the fragment from the figure reproduced in its original colors on the book's back cover.

I owe a debt to my wife Marianne. She consistently encouraged me as I spent innumerable days writing and rewriting. It is clear to me that the book would not have been written without her support.

Notes

1. *Subtle Is the Lord – The Science and the Life of Albert Einstein*, Abraham Pais, Oxford, 1982, pp. 503–504.
2. How I created the theory of relativity, Albert Einstein, Translated from the German into Japanese by the physicist J. Ishiwara, and published in the Japanese periodical *Kaizo* in 1923; then translated from the Japanese into English by Yoshimasa A. Ono and published in *Physics Today*, 35, 45–47, 1982.
3. Versuch Einer Theorie der Elektrischen und Optischen Erscheinungen in Bewegten Körpern, H. A. Lorentz, Leiden, The Netherlands: E. J. Brill, 1895; reprinted unaltered by B. G. Teubner, Leipzig, 1906.
4. Erinnerungen eines Kommilitonen, Louis Kollros, in *Helle Zeit – Dunkle Zeit*. Zürich, (Carl Seelig, editor), Europa Verlag, 1956, p. 22.
5. *Science – The Endless Frontier, A Report to the President on a Program for Postwar Scientific Research*, Vannevar Bush. Reprinted on the 40th Anniversary of the National Science Foundation, 1990.
6. *Cosmic Discovery – The Search, Scope and Heritage of Astronomy*, Martin Harwit. New York: Basic Books, 1981.

---

[b] The book's cover by James F. Brisson shows an hourglass displaying, at top, the early Universe crisscrossed by myriad colliding subatomic particles. Over time, today's Universe emerges, below, as an intricate web of galaxies pervading all space.

Notes on Usage

Chapter 1 begins by portraying where the physical sciences stood at the end of the nineteenth century. The chapters that follow deal almost entirely with advances in our understanding of the Universe during the twentieth century and at the turn of the millennium.

Astronomy is an observational science. It differs from the experimental sciences in its inability to manipulate the objects it studies. These difficulties are endemic, and so several of the book's chapters, notably Chapters 3 to 6, 8, and 10, aim at depicting the principal problems twentieth-century scientists faced in delving ever more deeply into the nature of the Universe. Chapters 2, 7, 9, and 11 to 16, in contrast, describe how interactions among the scientists themselves, and their increasing exposure to changing societal forces, affected the ways astrophysics came to be conducted as the century progressed. The scientific questions and the societal factors, although presented in these largely separate chapters, are part and parcel of a more integrated set of processes that the book's final chapters aim to depict.

Because the book is aimed, at least in part, at young researchers as well as scholars from the humanities and social sciences, I hope that unfamiliarity with the few astrophysical formulae and equations I have inserted will not dissuade newcomers from reading on. The text usually seeks to augment the meanings of mathematical expressions with explanations in everyday English. Where I insert symbolic statements, they are there primarily for the benefit of astrophysicists for whom they provide a familiar short-hand way of quickly perceiving the thrust I seek to convey. Wherever possible mathematical expressions are placed in footnotes, so as not to intrude on readers happier to read on without them.

The main impediment to reading a book that may involve a large number of unfamiliar concepts can be a lack of adequate explanation. I approach this difficulty by including a comprehensive glossary. On the first occasion at which I use a term that many readers might not immediately recognize, I *italicize* it. This is meant to indicate that if I do not explain its meaning on the spot, its definition can usually be found in the glossary following the book's epilogue.

* * *

Since the beginning of the twentieth century, scientific notation has steadily evolved. In the 1920s the oxygen isotope weighing 16 *atomic mass units* was written as O^{16}, where as today we write ^{16}O. In the 1930s, expressions representing nuclear transformations were written as though they were equations, with an equality sign (=) inserted between initial and final constituents. Today, the direction of the transformation, from its initial to its final state is indicated by an arrow ($\rightarrow$).

Highly energetic helium nuclei, electrons, and X-rays, respectively are named α-particles, β-particles and γ-rays, using the Greek letters, alpha, beta, and gamma.

Temperatures measured in degrees Kelvin that earlier were written as $^{\circ}K$ now are contracted to K. In astrophysics centimeter-gram-second (cgs) units remain in common use, even though officially the meter-kilogram-second (mks) units are supposed to have been universally adopted. The unit of time, the second, used to be abbreviated as *sec*. Today, physicists and astrophysicists instead write *s*.

In citing the literature written in different years I have opted to uniformly employ current usage, based largely on the cgs set of units, even where I quote earlier scientific texts verbatim. The one exception is that I have kept the abbreviation *sec* for seconds, because *s* in isolation might sometimes be ambiguous. I hope that adoption of this uniform notation throughout will reduce demands on the reader.

Aside from such changes in scientific notation, quoted material always appears in its original form. If a quotation contains words in *italics*, it is because the original text included them that way. Parenthetic remarks of my own that I may insert in direct quotes always appear in [square brackets].

I expect that at least some readers will wish to know how conclusions that astrophysicists reached at various epochs in the twentieth century might be viewed today. In order not to interrupt the flow of the text, but still satisfy curiosity, I occasionally insert such comparisons in footnotes. Bibliographic information appears in endnotes appended to each chapter.

Where the precise expression used by a scientist is worth citing and originally appeared in German, the language in common scientific use in continental Europe up to World War II, I have provided my own translation. Often this is not a verbatim translation which, a century later, might no longer convey its original sense, but rather expresses the intent of the statement as closely as possible in current English. The original text then also appears as a footnote for interested readers.

I should clarify five words I use throughout the book:

I use *Universe* and *Cosmos* interchangeably to encompass all of Nature.

In speaking about *astronomers* and *astrophysicists*, I distinguish colleagues who, respectively, concentrate on *observing*, as distinct from *explaining* the Universe. But the two functions are seldom separable; many colleagues think of themselves as both astronomers and astrophysicists. Accordingly, I use the two designations almost interchangeably.

In using the word *we*, which consistently appears throughout the text, I try to convey the sense of what *we* – the community of astronomers and astrophysicists – would say. In writing this way, I may occasionally misrepresent my colleagues; but astronomy is a rapidly changing field, and *we* may not always fully agree.

The appendix included at the end of the book defines scientific expressions, lists the meanings of symbols, and elucidates the relations between different units that may be unfamiliar to a reader.

A final index locates topics cross-referenced in the text.

1

The Nineteenth Century's Last Five Years

In the fall of 1895, the 16-year-old Albert Einstein traveled to Zürich to seek admission to the engineering division of the Eidgenössische Polytechnikum. He had prepared for the entrance examinations on his own, in Italy, where his family had recently moved. Most students took these exams at age 18, and the Polytechnikum's Rector, instead of admitting the youngster straight away, recommended him to the Swiss canton school in Aarau, from which Einstein graduated the following year.

Graduation in Aarau led to acceptance at the Polytechnikum without further examinations. Albert began his studies there in October 1896 and graduated in late July 1900. Only, instead of pursuing engineering, he registered for studies in mathematics and physics.

Looking back at those five student years between the fall of 1895 and the summer of 1900, one can hardly imagine a more exciting era in science.

* * *

Late on the afternoon of Friday, November 8, 1895, Wilhelm Conrad Röntgen, professor of physics at Würzburg, noticed an odd shimmer. He had for some weeks been studying the emanations of different electrical discharge tubes, and had previously noted that a small piece of cardboard painted with barium platinocyanide fluoresced when brought up to one of these tubes. To understand better the cause of the fluorescence, he had now shrouded the tube with black cardboard so no light could escape. In the darkened room, he checked the opacity of the shroud. It looked sound, but there was a strange shimmer accompanying each discharge. Striking a match in the dark, he found the shimmer to come from the barium platinocyanide–coated cardboard he had set aside to use next.

Over the weekend, he repeated the procedure. Some novel effect appeared to be at work. In the weeks to follow he conducted a wide variety of experiments to study more closely the cause of the fluorescence. He found that whatever emanation was passing through the blackened cardboard shroud also darkened photographic plates.

The emanation was absorbed most strongly by lead. Unlike the electric discharge in the tube, it could not be deflected by a magnet. It traveled along straight lines but, unlike *ultraviolet radiation*, it was not reflected by metals. Not knowing precisely what to call these rays, he chose the nonprejudicial designation, *X-rays*.

On December 28, his first paper on his findings, titled 'Über Eine Neu Art von Strahlen' (On a New Kind of Rays) was published in the *Sitzungsberichte der Würzburger Physikalisch-Medizinischen Gesellschaft*. Six days earlier, Röntgen had asked his wife to place her hand over a photographic plate. His paper included the resulting picture, shown here as Figure 1.1. It caused an immediate sensation and may still be the most iconic scientific image of all time, showing the skeletal structure of her hand and, even more clearly, the ring she was wearing. At one glance it revealed both the discovery of a new, penetrating imaging technique, and clear indications of its vast promise for medical science and countless other investigations. In Britain, the journal *Nature* published

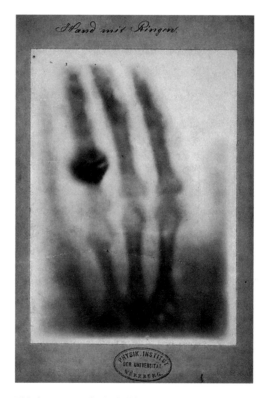

Figure 1.1. This image was included in the original paper Röntgen published, albeit without the stamp of the Physics Department of the University of Würzburg. An intensely private man, Röntgen captioned the published image simply 'Photograph of the bones in the fingers of a living human hand. The third finger has a ring upon it.' It was not in Röntgen's nature to publicize that the hand was his wife's. The German caption on the image reproduced here denotes even more sparingly 'Hand with rings.' (*Courtesy of the University of Würzburg*).

an English translation of the paper on January 23, 1896.[1] This, and the news-paper articles that had preceded it, led to instant world-wide attention. More than 1000 publications on X-rays, including books and pamphlets, appeared that same year, 1896.[2]

Röntgen was not sure precisely what kind of radiation he had discovered. He thought it was light-like or some variant of light. He knew about the medical applications X-rays would have and refused to take out a patent, insisting that all mankind should benefit from his work. But he could never have anticipated that a century later, Earth-circling X-ray telescopes would be studying signals emanating from cosmic *black holes* far out in the expanding Universe. Even uttering this last sentence would have made no sense; it conveys concepts unimaginable at the time!

* * *

The news of Röntgen's discovery spurred Antoine Henri Becquerel, professor of physics at the Polytechnic in Paris, to see whether all phosphorescent materials emitted similar rays. They did not. But Becquerel encountered another unexpected phenomenon, a spontaneous emission of radiation from uranium salts, which he announced in 1896.

The following year, 1897, Joseph John Thomson, Cavendish Professor of Physics at Cambridge, revealed the origin of the energetic cathode rays that had given rise to Röntgen's X-rays. The cathode rays, Thomson found, were streams of negatively charged particles. The magnitude of their charge was identical to that of the positively charged hydrogen atom in the electrolysis of dilute solutions. The mass of the particles he judged to be only 1/1700 of the mass of a charged hydrogen atom.[a] The careful measurements, and the completeness of Thomson's arguments led to the swift acceptance of the new particles we now call *electrons*.

By 1898, Becquerel's investigations had been pursued further by Pierre Curie, professor of physics at the Sorbonne in Paris, and by his young Polish wife, Marie Skłodowska Curie, who coined the word *radioactivity* for the phenomenon Becquerel had discovered. Among the many experiments they conducted, the Curies isolated two new radioactive materials from the uranium-containing ore *pitchblende*. Each was millions of times more radioactive than uranium. Both were previously unknown elements. The first, Marie Curie called *polonium*; the second, *radium*.

By late 1898, the Scottish chemist William Ramsay and co-workers in his laboratory had also isolated the inert elements helium, argon, neon, krypton, and xenon, in a succession of ingenious experiments conducted over the previous several years.

---

[a] Thomson's ratio of the mass of the *electron* to that of the *proton*, the nucleus of the hydrogen atom which Thomson called "the charged hydrogen atom," was remarkably close to today's accepted ratio $\sim 1/1836$.

As the nineteenth century was drawing to a close, physicists and chemists were beginning to understand the nature of the chemical elements. The periodic table the Russian chemist Dmitri Ivanovich Mendeleev of St. Petersburg had published in 1869 was serving as a roadmap whose details were emerging with increasing clarity. Full of confidence in the table he had devised, Mendeleev had provocatively left blank spaces in it. Now, chemists were rapidly isolating the new elements whose existence he had predicted and finding them to fall in line. Ramsay's new elements fit neatly into a new column added to the table, just as Marie Curie's new elements filled two of its previously blank spaces. Chemists were increasingly feeling they were on the right track.[3]

The atomic nature of the elements was gaining acceptance. Molecules were recognized as invariably composed of atoms. The structure of atoms was not yet known, but J. J. Thomson's experiments showed them to contain electrons, which could be removed from any number of different substances by strong electric fields.

In the course of the century, the nature of electricity and magnetism had become established through many experiments, particularly those that Michael Faraday had conducted in England. By 1865, James Clerk Maxwell in Scotland had extended Faraday's work and developed the electromagnetic theory of radiation as we know it today. This indicated that light was a wave oscillating transverse to its direction of propagation, transporting equal parts of electric and magnetic energy across space.[4] This prediction had been confirmed experimentally with radio frequency waves, by Heinrich Hertz in Germany in 1888.[5,6] Following Hertz's lead, the Italian inventor Guglielmo Marconi, by 1899, was demonstrating that radio waves could even be transmitted across the English Channel.[7]

* * *

For astronomers many of these results remained too new to find immediate application. Astrophysics was still too young a discipline. The first issue of the *Astrophysical Journal* had appeared only on January 1, 1895, and then was largely devoted to *spectroscopy*. Spectra of the Sun, stars, and laboratory sources were being pursued with vigor to discern parallels between chemical constituents found on Earth and those that might compose the atmospheres of planets, stars, and astronomical nebulae.

Although spectroscopic work was still largely devoted to gathering data and classification of stellar spectra, high-resolution spectroscopy was beginning to yield line-of-sight velocities of stars through observation of their *Doppler-shifted* – that is, velocity-shifted – spectra.[b] By the 1890s, Hermann Carl Vogel and Julius Scheiner at the Astrophysical Observatory at Potsdam had advanced spectroscopic techniques sufficiently to permit reliable determination of the line-of-sight velocities of stars

---

[b] The spectroscopic shift had been predicted half a century earlier, in 1842, by the mathematician Christian Doppler in Prague.[8]

relative to Earth and to provide a direct measurement of Earth's velocity during its annual orbit about the Sun.

By observing the Doppler shift of *spectral lines* over periods of months or years, the speeds at which close binary stars orbit each other could also be determined precisely, provided they happened to *eclipse* each other. Once their orbital velocities and periods were determined, Newton's laws of motion could be applied to derive the stars' masses.[9]

On several crucially important points, however, astronomy remained mute: We knew nothing about the size of the Universe. We knew nothing of its age – or whether it might be ageless, as most people believed. And how the Sun might have kept shining for as long as Earth appeared to have been warmed by sunlight also was quite uncertain.

The rate of erosion of terrestrial rocks by winds and rain, and the resulting salinity of the oceans into which the eroded matter was being swept, implied that Earth had been warmed by the Sun for hundreds of millions of years. The depth of stratigraphic deposits containing fossilized fauna led to similar conclusions. But how the Sun could have kept shining so long, nobody could explain. The required energy far exceeded any conceivable supplies!

The twentieth century would gradually answer this question.

Notes

1. On a New Kind of Rays, W. C. Röntgen, *Nature* 53, 274–76, 1896.
2. Wilhelm Conrad Röntgen, G. L'E. Turner, *Dictionary of Scientific Biography*, Vol.11. New York: Charles Scribner & Sons, 1981.
3. Ueber die Beziehungen der Eigenschaften zu den Atomgewichten der Elemente, D. Mendelejeff, *Zeitschrift für Chemie*, 12, 405–406, 1869.
4. A Dynamical Theory of the Electromagnetic Field, James Clerk Maxwell, *Philosophical Transactions of the Royal Society of London*, 155, 459–502, 1865.
5. Über die Entwicklung einer geradlinigen elektrischen Schwingung auf eine benachbarte Strombahn, H. Hertz, *Annalen der Physik*, 270, 155–70, 1888.
6. Über die Ausbreitungsgeschwindigkeit der electrodynamischen Wirkungen, H. Hertz, *Annalen de Physik*, 270, 551–69, 1888.
7. Wireless Telegraphic Communication, Guglielmo Marconi, *Nobel Lectures in Physics*, The Nobel Foundation, 1909.
8. Ueber das farbige Licht der Doppelsterne und einiger anderer Gestirne des Himmels, Christian Doppler, *Abhandlungen der königlich böhmischen Gesellschaft der Wissenschaften zu Prag*, Folge V, 2, 3–18, 1842.
9. *A History of Astronomy*, Anton Pannekoek, p. 451. London: George Allen & Unwin, 1961. Reprinted in Mineola, NY: Dover Publications, 1989.

Part I The Import of Theoretical Tools

2
———————

An Overview

The Universe We See Today

Twentieth century astrophysics has taught us that the origin and evolution of everyday matter, of the stars we see at night and the Universe we inhabit, share a coherent history dating back *billions* of years. Our knowledge remains fragmentary, but progress has been rapid and provides hope that our search will someday be complete, perhaps not in the sense that we will be all-knowing, but that we may have uncovered all that science can reveal.

One helpful feature in our search is that, as far back in time as we are able to probe, we find the known laws of physics holding firm. The *speed of light*, the properties of atoms, their constituent electrons and nuclei, and the mutual interactions of all these particles and radiation, appear unchanged ever since the first few seconds in the life of the Cosmos.

Also helpful has been that the Universe is expanding and that light, despite its high velocity, requires eons to cross cosmic distances. Using our most powerful telescopes we are able to directly view remote stars and galaxies, which emitted their light billions of years ago. We can compare how they appeared then and how stars and galaxies nearer to us in space appear now. And, as the Universe expands, light waves crisscrossing space expand with it. Short-wavelength light emitted by a distant galaxy reaches us with a longer *wavelength*; blue light is shifted toward the red. This *redshift* increases with the distance traversed and the time since the light was emitted. A galaxy's redshift then dates the epoch at which the galaxy emitted the light we observe.[a]

---

[a] We tend to think of telescopes as devices we use to look far out into space; but, in cosmology, they actually serve to look back in time. Galaxies coming into view today from the most remote early epochs are those that also are most distant. Time has been needed for their light to reach us. The nearest galaxies we see today can be seen only as they appeared just a few million years ago. We never will observe these as they might have appeared billions of years earlier. Light they emitted then bypassed us far too long ago.

9

Powerful telescopes can map a galaxy's appearance, discern its structure, and spectroscopically detect its internal motions and chemical constituents to determine the physical processes at work: Is the galaxy forming new stars? Just how much radiation is it emitting? Is it isolated, or possibly interacting with a neighboring galaxy?

As we probe ever more deeply in space and further back in time, large surveys provide a panoramic history of cosmic evolution. We see small galaxies shortly after their birth; we note their merging to form larger galaxies in an epoch during which the galaxies' nuclei everywhere briefly blazed more brightly than ever before or after. The early Universe appears almost devoid of chemical elements more massive than hydrogen and helium; at later times, we see the abundance of heavier chemical elements steadily rising.

The most difficult epochs to probe are the very earliest times. During the first few hundred thousand years after the birth of the Cosmos, a dense fog of electrons, *nucleons*, radiation, and *neutrinos* permeated space, degrading all data that visible light, radio waves, X-rays, or any other electromagnetic radiation could have transmitted about the birth of the Universe. We cannot yet tell precisely how much information about its creation the Universe may have eradicated this way. But we make up for this loss, with some success, by seeking other clues to cosmic history at early times.

One of these is provided by the ratio of hydrogen to helium atoms. The other is the temperature of a pervasive *microwave background radiation* bath that we now detect dating back to a time just after the initial fog had cleared. Between them these two features tell us how dense and extremely hot the Universe must have been when it was only a few minutes old, as primordial protons, *neutrons*, and electrons repeatedly collided to form the abundance of helium atoms that remain a major cosmic constituent to this day.

Further clues, which we do not yet know how to interpret, may be the high abundances of electrons and protons in the Universe and the virtually complete absence of their *antiparticles*, the *positrons* and *antiprotons*. At the high temperatures that existed at earliest times, collisions among particles and radiation should have formed protons and antiprotons, and electrons and positrons, in equal numbers. Could the laws of physics as we currently understand them have differed at those earliest epochs, after all? Or is our current knowledge of physics at extremely high energies merely incomplete? If so, experiments conducted at high-energy accelerators, or observations of high-energy *cosmic rays* naturally impinging on Earth from outer space, may some day tell us how these laws should be augmented.

Our search continues.

Discovery and Insight

Our understanding of the Cosmos is based on a confluence of discovery and insight. The two words can have many meanings. Here, I will consider *discovery* to be the recognition that a confirmed finding does not fit prevailing expectations. Observations, experiments, and exploration, can all lead to discovery. *Insight* enables us to place the discovery into a pattern shaped by everything else we believe we know. It is as though the discovery was a strangely shaped piece to be fitted into a larger jigsaw puzzle. To make the discovery fit, the shape of the puzzle may have to yield, or the entire puzzle may have to be disassembled and reconfigured through novel insight before the new piece can be accommodated.

The growth of understanding involves successive cycles of discovery and insight. But, unlike more orderly cyclic processes, discovery and insight generally do not follow neat periodic patterns. The sequence may be better described as times of accumulating discoveries, uncertainty, and doubt, followed by recognition that we will find clarity only by abandoning previously held views in favor of new perspectives.

Verification of the new perspectives, however, may require novel sets of tools. For astronomers this may mean construction and use of new kinds of telescopes and instrumentation. For theorists it may be the use of specially invented analytical approaches or mathematical concepts. For both, a lack of laboratory data may be a roadblock to be removed with yet a third set of tools. The proper tools are indispensable.

Due diligence alone cannot make up for a lack of tools. But the availability of the required tools, or even a knowledge of what those tools might be, often may be lacking. For decades, no astronomer could conceive that tools for detecting X-rays could revolutionize the field.

As we shall see later, when Einstein had struggled several years to understand the nature of space, time, and gravitation, he had to turn for help to his friend and former fellow-student, the mathematician Marcel Grossmann, who introduced him to an arcane differential geometry that the mathematician Bernhard Riemann and his successors had explored late in the nineteenth century. Advances are possible only when the right tools come to hand.

At times, theoretical tools abound and then astrophysical theory outstrips observations until theory threatens to detach from reality. At other times, observations are readily obtained, and our knowledge of the Universe becomes phenomenological. We may know what transpires, but cannot account for it with overarching principles that relate different phenomena to each other. Astronomers are most satisfied when their theories roughly match observations, both in depth and in the range of topics they cover. Such balance enables steady progress along a broad front where new problems can be tackled both with available observational and theoretical tools.

Ultimately, however, even discovery, insight, and the tools they require are not enough to ensure progress. Each astronomical advance leading to a new level of understanding must first become accepted before a succeeding advance can follow. For this, the persuasion of fellow astronomers is critical. Acceptance entails the consent of the community. An individual scientist may make a discovery, gain new insight, and find a journal willing to publish the findings – though even this is not assured if the finding does not find favor with the journal's editors and referees. But publication alone does not automatically lead to acceptance. Most published articles, particularly those announcing a novel advance, are largely ignored by the community, which prefers to deal with accepted knowledge rather than novelties. Unless a scientist can persuade critics that he or she has made a valuable contribution, the advance will remain buried in some publication – if it even gets that far. Most astronomers will tell you that it is easier to find acceptance for a finding that independently verifies a well-known truth than to persuade colleagues of the importance of an exceptional observation or a particularly daring insight.

Persuasion is so difficult because it normally proceeds not through a single advance but through an expanding list of convincing instances of how a new insight or novel discovery brings harmony to disparate findings that prevailing views fail to explain.

To perceive how our understanding of the Universe advances, the importance of each of the ingredients – discovery, insight, the existence of requisite tools, and persuasion of the astronomical community – must all be recognized. If we neglect or minimize the significance of any of these four, the growth in understanding becomes unfathomable and seemingly haphazard.

The Growth of Understanding

How new discoveries come about and lead to greater understanding of the Universe can depend on many factors. *Quasars* and *pulsars* were both discovered in the 1960s. But, whereas it took nearly 30 years to recognize fully what quasars are, the physical processes that account for pulsars were understood almost at once.

The first sign that quasars were quite unusual was that they emitted powerful radio signals from what appeared to be point-like sources in which even the largest radio telescopes could not resolve observable structure. Added to this, observations with optical telescopes showed that the radiation was strongly redshifted. All this was known by 1963. Wanting a crisp name to identify these newly found sources, but not wishing to prejudice the interpretation of what they entailed, we called them *quasars*, a contraction of *quasi-stellar* sources – a reference to their point-like appearance.

Einstein's general relativity offered two quite distinct interpretations of what we might be seeing. One potential explanation for the high redshift was that the quasars were at extreme distances where the Universe was expanding at ever greater speeds. An alternative possibility was that quasars were highly compact and massive, and that the observed redshift was due to the quasar's strong gravitational pull on light escaping its surface. Einstein's general theory of relativity had predicted such potential redshifts, and laboratory experiments had by then confirmed the prediction.

Determining the origin of the redshift thus became a priority. If the gravitational pull of quasars was very strong, they might be located quite nearby and perhaps not be particularly luminous. If their light was reaching us from far out in the Universe, quasars would have to be fantastically luminous to appear as bright as they did.

Not until the mid-1990s, 30 years after their initial discovery, were we quite sure what the quasars are. By then we had ascertained that they were highly luminous massive nuclei of distant galaxies. Because the quasars far outshone the starlight from their host galaxies, the very existence of these galaxies had long been questioned. But by the mid-1990s, the *Hubble Space Telescope's* powerful imaging capabilities began to show the faint radiation from the host galaxies' stars. The quasars and their host galaxies had to be at the same distance; chance superpositions of galaxies and quasars were easily ruled out.[1] This evidence confirmed the quasars as very distant, extremely powerful emitters of light. The gravitational contribution to the quasar redshift eventually showed itself to be quite minor compared to the redshift due to the cosmic expansion.

* * *

The discovery of pulsars followed a very different trend. First noted in radio observations published in 1968, these also were point-like radio sources. But pulsar radio emission took the form of short, intense pulses emitted on time scales of seconds or fractions of a second.[2] Again, in order not to prejudice the interpretation of what these pulsing sources might be, they were called *pulsars*, a contraction of *pulsating stars*. Recognition of what these were came swiftly. Within weeks, theorists proposed that they represented highly compact, rapidly rotating stars with an embedded *magnetic field* that dragged along an ionized atmosphere, a *magnetosphere*, as the star rotated. The radio bursts reaching us were emitted by this magnetosphere, whipping around, synchronized to the star's rotation. The interaction of the radio waves with interstellar gases along the line of sight also showed the initially observed pulsars to be located within our own Milky Way Galaxy. They were nowhere nearly as distant as quasars, many of which were a hundred thousand times further away, out in the Universe.

Theorists had long postulated the potential existence of stars so compact that electrons would be forced into their atomic nuclei, converting the stars into massive aggregates of neutrons. Such a collapse might be inevitable if the central portions of an aging star, which had used up all the nuclear energy that makes it shine, cooled off and no longer could prevent a catastrophic gravitational collapse. A star considerably more massive than our Sun might then implode to form a *neutron star* no larger in diameter than the city of New York. Any magnetic field the star might initially have had would be compressed in the collapse. And if the star had initially been rotating, it would rapidly spin up to conserve its angular momentum.

Once the pulsars were discovered, all these earlier hypotheses were available and could be explored as more detailed observations poured in. Soon there was little doubt. The pulsars were rapidly rotating neutron stars!

The discovery of these two distinct phenomena, quasars and pulsars, shows how one can *discover* a striking new astrophysical *phenomenon* strictly through its observational appearance, which differs so greatly from anything previously detected that intuition suggests it must be due to some novel process never identified before. *Insight* on what this discovery actually represents may take mere weeks, many decades, or conceivably much longer.

Our understanding of the Universe then emerges through two distinct steps. The first is the realization that an observation has defined an entirely new phenomenon. The second is the search for a physical process that accounts for the new finding. Each of these steps is essential to comprehending the world around us. Discovery calls a new observation to our attention. Insight on the discovery's significance may then provide an explanation, a theory, a deeper level of understanding. Alternatively, existing theory may predict a new phenomenon, which then is looked for and potentially discovered. Whether discovery precedes or follows insight depends largely on available tools.

The Chain of Reasoning

Science at the forefront almost inevitably requires judgment calls influenced by accumulated experience and one's understanding of basic principles. Because the ability to judge these correctly differs from one researcher to another, the conclusions they reach when confronted with the same data may vary greatly. This can lead to high-stakes disputes marked by impassioned debates that may take decades to resolve.

Even where the sifting of raw astronomical data is essential to remove *artifacts* or *noise*, respectively caused by instrumental defects or random environmental disturbances, personal judgments enter. The discovery of pulsars easily might have been missed if Jocelyn Bell, a graduate student at Cambridge University in the late 1960s, had decided to discard a half-inch "bit of scruff" on a 400-foot-long paper scroll recording radio signals.[3] But Bell's persistence in digging further through the data instead led to the discovery of these neutron stars. Later, other astronomers went back to look through their own earlier-obtained data, only to find that these contained similar signals but that they had ignored them as just some source of noise.

So, should we believe the striking astronomical pictures of the Universe conveyed by powerful telescopes we are now able to launch into space? These too require significant amounts of manipulation to correct for telescope and spacecraft characteristics, spurious signals produced by cosmic rays incident on detectors, or other defects. We have to trust that the scientists who *process* the data remove such defects solely to highlight the information the data convey.

Scientific results cannot be dissociated from the scientists who produce them. Science, like any significant undertaking, is very much a human and social enterprise. Where the interpretation of an astronomical finding involves a concatenation of independent steps, such as in the interpretation of the distance of a newly discovered galaxy, we become ever more dependent on the trust we have to place in the judgment of the many astronomers who provided the chain of observations and reasoning on which we base our conclusions.

Astronomical knowledge is a web that draws literally millions of independently conducted observations into a coherent picture based on physical principles that observers, experimenters, and theorists have worked out over the centuries. Because no individual scientist could ever repeat so many experiments and calculations to everyone's satisfaction, scientific understanding is built largely on trust and verification. Each important result is repeatedly checked by independent groups to weed out errors, misunderstandings or, occasionally, deliberate fraud.

The chain of reasoning that goes into answering even simple questions such as "Is the Milky Way a galaxy, similar in kind to the myriad spiral galaxies we see across the sky?" or "How far away is the Andromeda Nebula, the nearest sizable galaxy beyond the Milky Way?" is so complex that it is not amiss to ask exactly how we can feel so certain about anything we know about the world we inhabit.

Even simpler questions are, "What makes us think the stars we see at night are distant suns like our own? How do we know that our Sun was born some 4.6 billion years ago, or that primitive life took hold on Earth roughly two billion years later?"

Coming ever closer to the everyday world around us, we may ask an even simpler question, "Is the fossil dinosaur in the American Museum of Natural History in New York City real or a clever reconstruction?" "Were the cave drawings at Altamira put there by ancient cavemen or are they a hoax drawn more recently?"

These are not entirely idle questions. For decades Piltdown man was thought to be a forerunner of modern humans, but then new scientific techniques showed him to have been a clever fake.[4] The temptation to support or disprove theories bearing on the origins of life, the past history of Earth and its climate, or the origins of the Universe, in an attempt to promote or disprove passionately held personal, religious or political beliefs, is a factor that weighs heavily on resolving contentious scientific debates.

Fraud is most difficult to trace and thus especially abhorrent. A colleague discovered to have deliberately faked a result loses all trust, is of little further use to scientific inquiry, and generally is expelled from the profession. Fraud is the enemy of knowledge.

Even so, fraud is not the chief hindrance to scientific advance. Yet harder to recognize are communal prejudices or assumptions that shape the way we think, the way we sift scientific evidence, the extent to which we can imagine a universe whose behavior at times violates every conceivable experience of daily life. This was what made Einstein's special relativity so difficult to accept. How could light travel at identical speeds with respect to any and all observers, all moving around at quite distinct velocities? As we shall see in Chapter 3, the young Einstein long wrestled with this question before he fully understood the answer and felt sufficiently confident to publish his theory in 1905.

Similar preconceptions currently may be a primary hindrance to reconciling Einstein's general relativity with quantum mechanics – an essential step to overcoming some of the greatest astrophysical uncertainties we face today.

The Universe as Social Construct?

At some level, it is surprising that we seem to know so much about the Universe and so little about ourselves.

What we learn about the Cosmos is based on telescopes, instrumentation, and data processing systems that we calibrate, test under a wide range of operating conditions, and continually monitor to ensure that the data obtained are reliable and fully understood.

The process we understand least and calibrate most poorly is how the astrophysical community perceives the way the Cosmos functions. We are convinced that in our everyday work we faithfully reveal the true nature of the Universe. No intervention, no interpretation.

But historians and sociologists of science have shown, time and again, that communities of scholars tend to be trapped in thought conventions that shackle the way we view the world we study. We remain unaware of how restricting our most basic assumptions may be. We take them to be obvious truths and discard them only with greatest difficulty.

Can we find ways in which these limitations might be overcome? Answers to this question require us to look at how the scientific community is structured and interacts.

* * *

Three decades ago, the sociologist of science Andrew Pickering wrote 'Constructing Quarks – A Sociological History of Particle Physics,' a book that at the time greatly annoyed the high-energy physics community, perhaps because it contained a disquieting dose of truth.

Pickering argued that the theory of fundamental particles – the quarks and mesons, the baryons and leptons constituting all matter – was a construct that physicists had pieced together, through a process he termed a *communally congenial representation of reality*. The *standard theory* of particle physics, he claimed, was not an inherent description of Nature, but "deeply rooted in common-sense intuitions about the world and our knowledge of it."[5] In place of this theory, Pickering surmised, a better depiction of particle physics would eventually be found that would appear unrecognizably different from what had come to be the accepted way of viewing Nature's fundamental particles.

Today, many particle physicists would be more likely to agree with Pickering than they were then. Although the standard theory has successfully survived nearly half a century of testing, its scope is known to be limited. It fails to properly accommodate gravity. And the string theories, brane theories, and other attempts of particle physicists to produce a coherent theory of all the known forces of nature have so different a structure from the standard theory, topologically, as well as in terms of numbers of spatial dimensions, that they share little recognizable resemblance.

So, we may ask, was Pickering right? Are physicists and astronomers just constructing congenial representations that bear little relation to the inherent structure of the Universe we inhabit?

In astronomy, we have by now embraced what we term the *concordance model* based on general relativity, which we assert has led to tremendous strides in understanding the evolution of the Universe. But we find ourselves forced to postulate a new form of matter, *dark matter*, the existence of which is supported by little independent evidence, and we find ourselves forced to postulate the existence of a new form of energy, *dark energy*, for which there is similarly little independent evidence. Both are defined solely by gravitational effects for which we have no other explanations.

Together, dark matter and dark energy account for 96% of cosmic *mass–energy*. The atoms in galaxies, stars, planets, and people account for a mere 4%.

Perhaps, someday soon, the postulated dark matter and dark energy will be fully understood and certified as new forms of matter and energy that had mistakenly been overlooked. But we may equally well find a need for viewing the Universe in a totally different way that encompasses general relativity only as a limiting case and embraces dark matter and dark energy as a natural consequence of this new perspective, this new theory.

Such a depiction might then be just as mind-bogglingly different from what we conceive today as Einstein's postulates were when he first announced that the speed of light would always appear the same no matter how fast an observer was moving and that gravity represents the *curvature of space*. How could this be? Both violated common intuition!

* * *

One might argue that such drastic changes in how we view Nature are just a natural progression of science, whether physics or astronomy – that ultimately, after a long series of revision after drastic revision, we come ever closer to describing accurately the nature and evolution of the world around us. But when we look back at the history and sociology of science over past centuries, it is obvious that different cultures saw the world we inhabit in very different ways, judging the validity of their world views on grounds that we today dismiss as unscientific and therefore irrelevant.

Even within the established scientific community, heated arguments can arise because different factions dismiss one form of evidence while emphasizing another. The conclusions reached by antagonists can then lead to totally different world views, each the product of a very different cultural approach to science, as the medical researcher and thinker, Ludwik Fleck, first taught us in the mid-1930s in his seminal book 'Entstehung und Entwicklung einer wissenschaftlichen Tatsache' (Genesis and Development of a Scientific Fact). The book's German sub-title, 'Einführung in die Lehre vom Denkstil und Denkkollektiv' (Introduction to the Study of Thought Convention and Thought Collective), which is missing in the English translation, is particularly telling.

Fleck defined a *Denkstil*, which is more than a "style" in the English sense of the word: It is a *thought convention*, a mutually accepted way of reasoning, conversing, discussing, debating, persuading others, and viewing the world. The *Denkkollektiv* is a scientific leadership that delimits the bounds of acceptable scientific thought and discourse.[6]

Fleck demonstrated convincingly the extent to which a scientific community, in his case the medical community, conforms to cultural authority and resists new ideas and concepts, often remaining wedded to prevailing thought until long after it has outlived its usefulness.

A quarter century after Fleck, Thomas Kuhn echoed the same themes in his 1962 book 'The Structure of Scientific Revolutions.'[7] Fleck's *Denkstil* became Kuhn's *paradigm*, setting the tone for what he termed *normal science*.[b] So, if we wish to examine whether current cosmology and our concordance model are just *communal constructs*, we should begin by looking at whether or not Pickering, Fleck, and Kuhn, might be right – that we astronomers work under a tightly coordinated, culturally imposed authoritarian system – and, if so, whether warning signals could exist that might keep us from straying too far into self-indulgence.

This is what I will now try to do.

Perspective and Debate

The astronomical community's acceptance of new findings, in particular new theories or explanations, is unpredictable. Much depends on psychological preparedness. At times of mounting uncertainty, when an increasing number of problems appear intractable, new ideas indicating a way forward may be welcome. But when the work is humming along, making steady progress, novel proposals may be dismissed as annoying distractions.

Aside from psychological preparedness, however, is there a rational basis on which divergent astrophysical views are resolved? What are the kinds of arguments that eventually lead to general agreement and acceptance of an explanation,

---

[b] Many years later, Kuhn explicitly acknowledged his indebtedness to Fleck in his foreword to the English translation of Fleck's book published in 1979.[8]

independent of whether it may ultimately prove right or wrong? In short, how do astrophysicists reach at least provisory agreement on an issue? This is what we need to examine now.

To see why it can be almost impossible for some astrophysicists to reach agreement, one needs to appreciate that many scientists think in, and are persuaded by, only a few distinct forms of reasoning. This restriction to different thought patterns is one of the most difficult hurdles to overcome. Only a few astrophysicists are equally at ease with all such patterns. Those few can play invaluably decisive roles as translators in a heated discussion between colleagues who cannot reach agreement because their modes of thought differ so greatly that they do not realize they may be saying the same thing but are speaking past and misunderstanding each other.

To make this understandable, it is worth listing some of the more frequently encountered forms of reasoning.

1. One of the simplest aids to reasoning is the concept of *continuity*. Consider all the rains draining into the Black Sea from the surrounding land. Rather than accumulating there, the water drains through the straights of Bosporus into the Sea of Marmara and thence ultimately into the oceans. Where the Bosporus narrows, the currents flow faster than where it is wide; but the mass of water passing per minute through any segment along the straights is always the same. Nowhere along the straights does water accumulate or is its flow diminished. This observation represents the continuity of flow of an incompressible fluid. It may also be thought of as a steadiness, or *conservation*, of the flow-rate of water through any cross section of the channel.

In a related quest, I may place an ice cube in a sturdy, tightly sealed glass jar sitting on a balance. As I heat the ice, it first melts and then begins to evaporate. Throughout, the balance remains motionless. The mass of the water in the jar does not change, even as its *phase* changes from solid, to liquid, and then to vapor. The process exhibits a continuity, which we may speak of as the *conservation of mass*.

Mass and incompressible mass flow rates were just two of the quantities that scientists of the nineteenth century already recognized as being conserved. A third was energy. Yet a fourth was the momentum of a moving body. The expectation that these quantities must inevitably be preserved is a shorthand way of checking whether or not an observed phenomenon makes sense or whether we might be failing to recognize a missing component.

2. A second simple but equally incisive form of scientific reasoning entails arguments based on symmetries. To show both the power and limitations of symmetry arguments, let us consider a throw of dice shown in Figure 2.1. If we were to ask what the chances are that a single die will land with the pattern of five dots on its top surface, symmetry tells us that this will happen in one out of six throws. To reach this conclusion, we argue that the likelihood of the die landing with any face pointing up is the same for each. Because there are six faces, the chance of finding the top surface to have five dots is 1/6. We may generalize this result as a

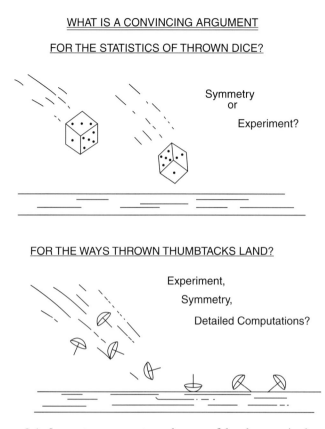

Figure 2.1. Symmetry arguments can be powerful and persuasive in astrophysics. As long as an astrophysical setting is not too complex, symmetry arguments tend to carry considerable weight. Once complexity sets in, most astronomers will be persuaded only by detailed calculations or experiment. Even the likely final resting position of thrown thumbtacks tends to be too complex to be predicted without detailed calculations or experiments.

conservation law: The likelihood of a die landing with a given face pointing up is identical for all its faces.

As it turns out, symmetry arguments and conservation laws are flip sides of the same coin, though most of us feel more at home with one or the other argument and may not realize their correspondence! Indeed, this insight was not always evident, and represents one of the triumphs of twentieth century mathematical physics.

In 1918, the 36-year-old Emmy Noether, widely considered the most brilliant woman ever to enter mathematics, proved one of the most powerful theorems of theoretical physics.[9] It showed that the conservation of energy, the conservation of angular momentum, and all such conservation laws correspond to particular symmetries of Nature. Initially the symmetries she considered were relatively simple. Her later work and that of others generalized her theorems to cover symmetries of increasing complexity. Quantum mechanics, theories of elementary particles, and the theory of general relativity, all are expressions of such symmetries.[c]

3. A different class of scientific argument arises when a skeptic mistrusts the symmetry explanation for the behavior of dice and insists that the hypothesis be tested experimentally. The skeptic tosses dice hundreds or thousands of times, only to find that the face with five dots lands facing upward, say, on average once in four times – not once in six.

Has the symmetry argument broken down? Evidently it has; the dice must be loaded.

But even when nothing untoward is going on, the symmetry argument tends to gain credence only with simple systems. If we replace the tossing of dice with randomly dropping thumbtacks onto a table from some height, the situation becomes more complex. Some of the thumbtacks will come to rest on their heads with the pin pointing straight up. Others will end up with both the rim of the head and the point of the pin touching the table top, as in Figure 2.1. The symmetry of the thumbtacks, though not especially complex, already makes a prediction of the way the tacks come to rest much more difficult and less intuitive.

4. For thumbtacks, a detailed theoretical approach might gain more general acceptance than the simple symmetry argument. It would involve a careful analysis of the shape of the head, whether completely flat or more rounded like a carpet tack. It would also take into account the mass of the head, compared to the mass and length of the pin. A comprehensive calculation based on the shape of the tacks and their mass distributions might then lead to reliable predictions about the probabilities of how the tacks will come to rest. A tack with a massive hemispherical head and a long, light weight pin inevitably settles with the pin pointing straight up. The tack acts like a stand-up doll.

An experimentalist still might distrust even these calculations and insist on throwing large numbers of thumbtacks to demonstrate their traits convincingly. This would leave little doubt about the behavior of a particular type of tack but might tell little about how other, differently shaped tacks would come to rest on the table.

5. A recent variant of theoretical explanations is computer modeling. The distinction between modeling and theoretical calculations may be considered

[c] Even before the publication of her eponymous theorem, her mathematical accomplishments in invariant theory had been so prodigious that the great mathematicians David Hilbert and Felix Klein brought her to the University of Göttingen in 1915. For a tradition-bound Göttingen faculty, Emmy Noether's achievements, however, were evidently not enough for her to gain even a modicum of recognition. For four years, members of the philosophical faculty denied her the right to *Habilitation*, a prerequisite to lecturing at the university under her own name. Women, they insisted, could not qualify. To no avail, Hilbert protested, "We are a university not a bathing establishment." Eventually, Habilitation was granted, but further years had to pass before she was appointed to a position that even paid a salary.[10]

merely one of scale, but when this difference becomes substantive, the need for distinction becomes urgent.

Massive astrophysical computer models generally invoke a large number of repetitive steps. A modeler wishing to analyze the interactions of stars in a giant complex may start out with a set of more-or-less randomly distributed stars all of which exert gravitational forces on each other. A first step is to see how the forces between these stars induce instantaneous accelerations and small changes in relative positions after a brief time interval.

This information is fed back into the model, and a new set of forces resulting from the induced displacements is calculated together with further displacements the forces produce. The computer model may include millions of stars all acting on each other in this fashion and may follow their displacements in the course of a million successive small steps. At the end of this massive computation, the assembly of stars may have arranged itself into what appears to be a galaxy, a *cluster of galaxies*, or a massive black hole, depending on the initial physical conditions assumed.

Computations of this type have become increasingly common in modeling processes believed to have taken place early in the evolution of the Universe. They indicate that, as the Cosmos cooled sufficiently, most of the *ionized hydrogen* – the protons and electrons – combined into neutral hydrogen atoms. With some delay for additional cooling of this neutral gas through cosmic expansion, atomic hydrogen initially distributed more or less evenly throughout the Universe will first have formed hydrogen molecules. The molecules, more efficient at radiating away energy, then enabled gaseous clouds to cool rapidly and collapse, through mutual gravitational attraction, to form stars and galaxies.

The computer modelers seek to show convincingly how all this happened. They may wish to determine also whether stars formed before galaxies, or vice versa, and whether quasars – the extremely luminous nuclei of galaxies now believed to be powered by *supermassive black holes* – formed before or after the first stars. This might indicate how these giant black holes, often as massive as a billion suns, may have formed in early epochs.

The difficulty of modeling with a sufficiently large number of iterative small steps is that the escalating number of iterations involved may permit the ascendance of anomalies that at first sight appear insignificant. But with so many iterations required, each potentially introducing the same small cumulative anomaly, the modeler needs to be particularly careful to take potential defects in the computations appropriately into account.

Even then, there is no guarantee that this has been done satisfactorily. A model gains credence only if it can first simulate a large number of different processes all of which are well understood experimentally, observationally, or through less complex calculations. It also helps to have a number of differently structured computer models reaching consistent conclusions, preferably if the models have been produced by independently working researchers employing different mathematical techniques.

A particularly pressing problem for astrophysics is the difficulty of checking massive computer-based simulations. Regrettably, many journals today do not require authors of papers publishing the results of massive computations to make their programs available for others to check their work. Verification of the computations is then difficult. A recent study by concerned experts argues persuasively for more open access to computer programs used in scientific computations.[11] They point out that programming errors by researchers typically exceed one error per thousand lines of code, and that even a perfectly coded program may not give identical results if run on different computers. All this necessitates insight on how a computation was coded and then executed before its results can be fully trusted. Unfortunately, computer programs that have commercial applications may be proprietary; journals then have no way of enforcing access to a computation. All this may leave much of the astrophysics community unable to check the validity of a computational finding – a potential breakdown of the ways science normally advances.

Scientific results are not supposed to be taken on faith. They should be accepted only after careful verification.

* * *

Each of the different types of objective evidence – astronomical observations, laboratory experiments, conservation principles, symmetry arguments, traditional calculations, and computer modeling – provides its own insights but may leave some unanswered questions. Where possible, astrophysical reasoning takes all these sources of evidence into account, as well as factors we accept as likely or even self-evident, because a trusted authority or everyone in the entire community thinks so. In Chapter 9 we will come across one self-evident belief that in mid-twentieth century proved to be unwarranted, namely the conviction that the Universe has no innate preference for left- over right-handed structures.

Astronomy is largely an observational science; but many astrophysical phenomena, such as explosions of *supernovae* sufficiently nearby to be studied in depth, are quite rare. This makes us heavily dependent on theoretical approaches to guide our understanding. Competing theories can help us define critical measurements to be carried out when the next opportunity for further observations presents itself. That way we may eventually come to distinguish between alternative explanations of, and models for, different types of supernova explosions and many other seldom-occurring phenomena. Observations and theory work hand in hand.

The next four chapters will illustrate, respectively, how the application of overarching symmetries and conservation principles, observations, and detailed calculations led to important advances in astrophysical understanding during the first half of the twentieth century. Computer simulations came into play only in the 1970s, and became progressively more detailed thereafter as technological advances provided increasingly powerful tools.

Notes

1. The Apparently Normal Galaxy Hosts for Two Luminous Quasars, John N. Bahcall, Sofia Kirhakos, & Donald P. Schneider, *Astrophysical Journal, 457,* 557–64, 1996.
2. Observation of a Rapidly Pulsating Radio Source, A. Hewish, S. J. Bell, J. D. H. Pilkington, P. F. Scott, & R. A. Collins, *Nature, 217,* 709–13, 1968.
3. Ibid., Observation of . . ., A. Hewish, et al., 1968.
4. The 100-year mystery of Piltdown Man, Chris Stringer, *Nature, 492,* 177–79, 2012.
5. *Constructing Quarks – A Sociological History of Particle Physics,* Andrew Pickering, University of Chicago Press, 1984, p. 413.
6. *Entstehung und Entwicklung einer wissenschaftlichen Tatsache – Einführung in die Lehre vom Denkstil und Denkkollektiv.* Basel: Ludwik Fleck, Benno Schwabe & Co., 1935; reprinted by Suhrkamp, Frankfurt am Main, 1980 (English translation: *Genesis and Development of a Scientific Fact,* Ludwik Fleck, translated by Fred Bradley & Thaddeus, J. Trenn, edited by Thaddeus J. Trenn & Robert K. Merton, University of Chicago Press, 1979).
7. *The Structure of Scientific Revolutions,* Thomas Kuhn. University of Chicago Press, 1962.
8. Ibid., *Genesis and Development,* Fleck pp. vii–xi.
9. Invariante Variationsprobleme, Emmy Noether, *Königliche Gesellschaft der Wissenschaften zu Göttingen, Nachrichten, Mathematisch-physikalische Klasse,* 235–57, 1918.
10. Emmy Noether and Her Influence, Clark Kimberling, in *Emmy Noether, A Tribute to her Life and Work,* eds. James W. Brewer and Martha K. Smith, New York & Basel: Marcel Dekker, 1981.
11. The case for open computer programs, Darrel C. Ince, Leslie Hatton & John Graham-Cumming, *Nature, 482,* 485–88, 2012.

3

Conclusions Based on Principles

Acting on principle can be rewarding but also risky. What if the principle, against all expectations, turns out to be invalid? [a]

On graduating from the Eidgenössische Polytechnikum in the summer of 1900 with a barely passing grade, the 21-year-old Einstein had no prospects of permanent employment. But as he drifted from one temporary position to another, he was at least free to pursue his passion for physics. Less than five months after receiving his diploma, he submitted his first paper to the leading German journal, the *Annalen der Physik*. Like all of his papers published in the subsequent three years, it was grounded in thermodynamics.[6,7,8,9,10]

Thermodynamics is an expression of the conservation of energy as it is converted from one form, such as electrically, mechanically, or chemically stored energy into another. Einstein continued to draw strength from this and other conservation principles for most of the work he would produce. They gave him guidance and assurance as he ventured ever further into uncharted realms.

Rescue of the jobless Einstein came through the father of Marcel Grossmann, Albert's closest friend during their student years at the Polytechnikum. The elder Grossmann was impressed by his son's friend and recommended him to Friedrich Haller, director of the Swiss Patent Office in Bern. When a new position opened there, Einstein at once applied and, in June 1902, won an appointment as patent clerk third-class.[11] He remained immensely grateful to the Grossmanns and, on completing his doctoral work in 1905, dedicated his thesis to his friend.

Einstein took up residence in Bern even before his appointment to the Patent Office became official. He supported himself partly by tutoring students in

---

[a] Readers wishing to view the present chapter in a broader setting of its times, will find the biography of Einstein by Abraham Pais, two autobiographical recollections by Einstein himself, a paper by Helge Kragh and Robert W. Smith, and a historical study by Harry Nussbaumer and Lydia Bieri informative.[1,2,3,4,5]

Figure 3.1. Einstein during his years at the Patent Office in Bern. The photograph was reportedly taken by Lucien Chavan, sometime around 1906–1907. (*Hebrew University of Jerusalem Albert Einstein Archives. Courtesy of the AIP Emilio Segre Visual Archives*).

mathematics and physics. A philosophy student who came to obtain a better grounding in physics was Maurice Solovine, with whom Einstein soon struck a long-lasting friendship. He enjoyed the discussions he, Solovine, and a young mathematician named Konrad Habicht found themselves initiating; from then on he refused payment.

The three friends began to meet regularly to discuss a broad mix of philosophy, literature, and physics.[12] Cheerfully, they called themselves the founders and sole members of the "Olympia Academy." Solovine recorded their reading and discussions in great detail.[13] Sometimes they went paragraph by paragraph or page by page, and for weeks at a time, through the philosophical works of Hume, Mill, Plato, and Spinoza; the literary works of Cervantes, Dickens, Racine, and Sophocles; and, more important for present purposes, the works of Ampère, Dedekind, Mach, Poincaré, and Riemann.

In particular, they read Henri Poincaré's 'La Science et l'Hypothèse,' which had just appeared in 1902. There we may read[14,b]:

---

[b] In 1905 the book was translated into English under the title 'Science and Hypothesis.' The quoted text is from p. 90 of the edition published by Dover in 1952.

1. There is no absolute space, and we only conceive of relative motion; and yet in most cases mechanical forces are enunciated as if there is an absolute space to which they can be referred. 2. There is no absolute time. When we say that two periods are equal, the statement has no meaning, and can only acquire meaning by a convention. 3. Not only have we no direct intuition of the equality of two periods, but we have not even direct intuition of the simultaneity of two events occurring in two different places.... 4. Finally, is not our Euclidean geometry only a kind of convention of language? Mechanical facts might be enunciated with reference to a non-Euclidean space which would be less convenient but quite as legitimate as our ordinary space; the enunciation would become more complicated, but it still would be possible.

Ernst Mach similarly thought about problems of relative motion in his book, 'Die Mechanik in Ihrer Entwicklung – Historisch-Kritisch Dargestellt,' which the three members of the Olympia Academy also read and discussed. They most probably had at hand the 1883 edition of the book.[15] There, Mach extensively quoted Newton on space, time, and motion and their relative measures. He also cited Newton's observation that water in a vessel suspended by a string and set into rotation will have its surface rise at its outer radius once its rotational velocity relative to that of the vessel is damped. Mach asked: How does the water know it is rotating? Does the structure of the Universe, perhaps the distribution of the fixed stars, dictate a preferred rest frame? If so, how did this come about? He emphasized that this was still an open question requiring further study.[16]

From Einstein's later recollections, Mach's Principle, the idea that the Cosmos in some way not yet understood influences physical conditions here and now, made a lasting impression.

The Relativity Principle

As Einstein recalled many decades later, in a series of interviews with the American physicist and historian Robert Sherwood Shankland, he had begun thinking about the propagation of *electromagnetic waves* through the aether by age 16.[17] The recollection is confirmed by a scientific essay Einstein wrote at age 15 or 16 and sent to his uncle Cäsar Koch in Antwerp in 1894 or 1895. This first tentative scientific work was published posthumously in 1971.[18]

Einstein's reading of the works of Maxwell, Hertz, and Lorentz, as well as the books by Poincaré and Mach, evidently were only part of the accumulating perspectives on which his theory of relativity was based.

By the time he was finally ready to write an account he considered fully satisfactory, Einstein, aged 26, was in his third year at the Swiss Patent Office. For the past year he had been brooding about space and time. If there was no favored time, place, or velocity to define the motion of an observer, then surely some principle should exist, which would allow observers in different states of motion to derive identical laws of Nature, regardless of the absolute state of motion in which they found themselves. Yet there appeared to be one measure that seemed absolute – the velocity of light as expressed in Maxwell's electromagnetic theory. Taken at face value, Maxwell's theory required light to always travel at an absolute speed $c \sim 300,000$ kilometers per second. Was there some way in which this speed might appear identical as viewed by different observers, regardless of their motion, and regardless of the motion of the source of light? Initially this idea seemed paradoxical. How could this speed be the same for observers moving at any arbitrary speed? Einstein recalled[19]:

> I spent almost a year in vain trying to modify the idea of Lorentz in the hope of resolving this problem. By chance a friend of mine in Bern (Michele Besso) helped me out. It was a beautiful day when I visited him … We discussed every aspect of this problem. Then suddenly I understood where the key to this problem lay. Next day I came back to him again and said to him, without even saying hello, 'Thank you. I've completely solved the problem.'

As Einstein remarked in his interviews with Shankland, "at last it came to me that time was suspect." What Einstein had concluded was that, if he correctly interpreted how time is read off moving clocks, and how the lengths of similarly moving rods are measured, then the addition of two collinear velocities, v_1 and v_2, should result in a measured velocity v with a value

$$v = \frac{v_1 + v_2}{1 + \frac{(v_1 v_2)}{c^2}} \; . \tag{3.1}$$

Here c is the speed of light in vacuum; its speed through air is very nearly the same.

That this formula kept the observed speed of light constant, no matter what the relative velocities of an observer and a propagating light beam might be, is easy to verify. By letting either v_1, or v_2, or even both these velocities, equal the speed of light c, the equation yields a sum of velocities, v, systematically equal to the speed c.

Once Einstein realized how the relative velocities of moving systems add, the solution to his difficulties became easy. For any observer moving relative to a beam of light, the speed of light will always appear to be c. Only the frequency of the light changes – increasing with velocity if the observer and light source move toward each other, and decreasing if they mutually recede. This change in frequency, termed the *Doppler shift* had already been verified through spectroscopic observations of stars. The shift was already known from Newtonian theory and, at the low relative velocities of nearby stars and Earth, the term $(v_1 v_2)/c^2$

in Einstein's equation (3.1) is so small that the theories of Newton and Einstein predicted practically indistinguishable observational results.

* * *

Einstein's 1905 paper on relativity has two major parts.[20] The first deals with *kinematics*. It demonstrates how the length of solid rods, and time measured by clocks attached to the ends of these rods, change depending on the relative motion of an observer. Einstein found that the lengths of rods aligned perpendicular to the direction of motion remain constant, but that rods aligned along the direction of motion would appear shortened. In addition, clocks in motion relative to the observer systematically appear to run more slowly. Today, these respective effects are called the *Lorentz contraction* and *time dilatation*. The great Dutch theoretical physicist Hendrik A. Lorentz had previously suggested the existence of this peculiar contraction, but without noting that it was but a symptom of a wider set of effects that the kinematics of Einstein's relativity principle was introducing. Henri Poincaré had independently come to conclusions similar to those of Lorentz.

The second part of Einstein's paper applied the kinematics he had just derived to elucidate Maxwell's electromagnetic equations. He inserted his newly obtained relations between space and time into these equations, to show how electric and magnetic fields, electric charges, and electric currents would interact in relative motion; how the mass of an electron would appear changed as forces longitudinal and transverse to its direction of motion were applied; how the energy of radiation and the pressure it applied to a body in relative motion would appear to change as a function of velocity; and how the velocity of light could never be exceeded no matter how strongly a physical object was accelerated. All this, Einstein was able to derive solely on the basis of the *principle of relativity*, the requirement that the laws of physics, including the constancy of the speed of light, be identical in any and all coordinate frames in constant relative motion. All of Einstein's considerations in this paper, including expression (3.1), applied to arbitrary constant velocities, but not to accelerating motions. In this sense they represented a set of special states of motion – hence the name *special relativity*. What the relativity of accelerated motions might entail was a question Einstein could not yet tackle.

* * *

Some months after submitting this paper for publication, Einstein returned to make one more prediction based on his principle of relativity. In a paper only three pages long, he showed that the absorption or emission of radiation with energy E, respectively, increases or decreases the mass m of a body by an amount (m/c^2).[21]

The principle of relativity had led Einstein to a widening understanding of two principles that had always been regarded as separate: the principle of conservation of energy, namely that energy could neither be created nor destroyed and the principle of conservation of mass – that mass could neither be created nor destroyed. Relativity theory now predicted that if one dug rather deeper, one would find a more encompassing relation, still in harmony with available observations but leading to a unifying view of the world requiring fewer arbitrary factors. Mass and energy no longer were independent of each other. Only their sum should be conserved. Emission of energy should result in a compensating loss of mass. Radioactive decay might verify this predicted mass loss.[c]

* * *

Three years later, the gifted mathematician Hermann Minkowski, in a major address at the 80th *Naturforscherversammlung zu Köln* (Congress of Scientists in Cologne) held on September 21, 1908, found a further unification implied by Einstein's theory and the underlying symmetry on which Einstein's principle of relativity was based.

Dramatically, Minkowski declared that "Henceforth space by itself, and time by itself, are doomed to fade away into mere shadows, and only a kind of union of the two will preserve an independent reality."[22,d]

Looking beyond the simplest examples given in his address, Minkowski proposed that similar rotational transformations hold for all other physical quantities as well. Max Planck, the leading thermodynamicist of his era, had recently demonstrated that the principle of relativity also elucidated intricate questions of thermodynamics that had previously remained unresolved.[23]

Minkowski went on to point out that a simple rule of thumb, which at least held consistently in electromagnetic theory, showed how a *ponderomotive force* transforms when viewed from a coordinate frame in constant motion relative to some original coordinate frame. If he confined himself to working with transformations among four-dimensional vectors, then the rule of thumb permitted him to define a particular vector component that always remained invariant under arbitrary transformations. He took this to be the energy representing the work done by the force. By assigning this component time-like coordinates, the three remaining

---

[c] Contrary to popular belief, this paper nowhere contains the equation $E = mc^2$ that brought Einstein fame. In the notation he preferred, it showed instead that radiation of an amount of energy L would reduce the mass of a body by an amount L/V^2, where V designated the speed of light.

[d] Minkowski pointed out that, rather than considering two events separated by a spatial distance $\sqrt{(x^2 + y^2 + z^2)}$ and a time difference t, as seen by an observer O, a single space–time interval $\sqrt{(c^2t^2 - x^2 - y^2 - z^2)}$ should be considered separating the events. A second observer O′, in motion relative to the coordinate system $(t, x, y, z,)$ of observer O, might employ coordinates (t', x', y', z') to describe the same interval and would then find that the interval he measured between events, amounting to $\sqrt{(c^2t'^2 - x'^2 - y'^2 - z'^2)}$, was precisely identical to the interval measured by O. The transformation of the interval measured, respectively by observers O and O′, corresponded simply to a rotation in a four-dimensional Cartesian coordinate system, in which the spatial coordinates were expressed in real numbers, the time coordinate in an imaginary number, and the length of the rotating radius vector remained invariant.

orthogonal space-like components had to represent the ponderomotive force. As Minkowski asserted, he could thus satisfactorily reproduce a dynamical system's laws of motion viewed from different frames in constant relative motion and re-derive Einstein's total energy as consisting of its mass–energy plus a Newtonian kinetic energy – at least to an accuracy of order $1/(c)^2$.

A decade earlier Minkowski, who now was working at the University of Göttingen, had been one of Einstein's mathematics professors at the ETH. His talk now called special attention to Einstein's contributions to the establishment of this new four-dimensional way of viewing the world.

Less than four months after delivering this talk, Hermann Minkowski was felled by appendicitis and died at age 44.

What Makes the Sun Shine?

For astronomy, an immediate consequence of the new relativity principle and the symmetries it entailed was the possibility that mass might be converted into energy with the liberation of enormous amounts of energy.

The heat radiated by the Sun had first been measured with reasonable accuracy in 1838. In France, Claude Servais Mathias Pouillet had developed a special *pyrheliometer* for this purpose.[24] The solar radiation incident on a given area on Earth, which he called the *solar constant*, corresponded to ~1.76 calories per square centimeter per minute.[e] Given the distance of the Sun, it implied a solar energy output of roughly 5×10^{27} calories per minute.

For some years this did not cause undue alarm; but in an essay of 1848, the German physician Julius Robert Mayer, a man of broad scientific interests, calculated that if the Sun were heated by the burning of coal, or simply cooled passively through radiative cooling, it could not have radiated for more than a few thousand years.[25] He concluded that another source of heat must be at work to keep the Sun shining.

Mayer was basing his conclusions on a new principle he had proposed a few years earlier. Soon it would be accepted as the most universally applicable finding of mid-nineteenth century science. It was his law governing the conservation of energy. In an article published in May 1842, 'Bemerkungen über die Kräfte der unbelebten Natur' (Remarks on the Forces of Inanimate Nature), he laid out his hypotheses.[26]

Energy could not be created out of nothing. One could only transform one type of energy into another. His use of the word *Kräfte*, or *forces*, instead of *energy* would today be considered misleading; but scientists of the nineteenth century were groping for new ways to think about the behavior of matter, and a precise terminology eluded them for some time.

---

[e] This was remarkably accurate, being only ~10% lower than the mean value accepted today.

By 1848, the existence of *meteorites* – pieces of rock or iron orbiting the Sun and occasionally falling on Earth – had been well established. Mayer thought that meteorites similarly falling on the Sun at sufficient rates could provide the requisite heat. But the infall rate had to be enormous, of the order of trillions of tons of meteoritic matter per second, comparable to the mass of the Moon falling into the Sun every two years. This large an infalling mass rate would appreciably increase the Sun's gravitational pull and, as Mayer acknowledged, would increase the orbital speed of Earth about the Sun, reducing the length of each successive year by a significant fraction of a second. Because this contradicted experience, the increase in mass would somehow have to be compensated, and Mayer thought that the light radiated by the Sun might carry away that mass.[27]

Ways to avoid an increase in the Sun's mass while maintaining an abundant source of energy were proposed by the Scottish scientist and engineer John James Waterston in 1853 and by Hermann von Helmholz in Germany in 1854. Both proposed a gradual solar contraction that would radiate away the loss of gravitational energy. This process could extend the potential life of the Sun to something like 20 million years before that source of energy would similarly vanish as the Sun first shrank to half its size and later became as dense as Earth. Thereafter, further contraction was considered unlikely.

At the end of the nineteenth century, the solar contraction theory provided the only viable explanation for the source of energy sustaining the Sun's radiation, even though its conclusions defied the protestations of geologists and evolutionary biologists that the age of the Earth must far exceed 20 million years.[28]

The discovery of radioactivity by Henri Becquerel in 1896, and the discovery of radium by Pierre and Marie Curie two years later, suggested a different, almost inexhaustible, source of energy that potentially could sustain solar radiation for far longer periods. By 1903, Sir George Darwin, Plumian professor of astronomy at Cambridge (and son of the naturalist Charles Darwin), was postulating that radioactivity might power the Sun. In a short note published in *Nature*, he pointed to an estimate by the pioneer of experimental nuclear physics, Ernest Rutherford, that one *gram* of radium would emit 10^9 calories of energy, compared to $\sim2.7 \times 10^7$ calories per gram that could be produced by gravitational contraction.[29] Accordingly, if the Sun contained an appreciable amount of radioactive matter, he wrote, "I see no reason for doubting the possibility of augmenting the estimate of solar heat as derived from the theory of gravitation by some such factor as ten or twenty."

Although Darwin's conjecture preceded Einstein's prediction, nobody understood where the energy liberated in radioactivity originated. Nor could anyone fully appreciate the enormous amounts of energy that the expenditure of mass could more generally offer. Once this possibility was accepted, a variety of loosely defined proposals surfaced to explain solar longevity.[30] One was the *annihilation of matter*. With Einstein's enormous conversion factor, predicting an extraction of $c^2 \sim 9 \times 10^{20}$ ergs or 2.15×10^{13} calories per gram of annihilated matter, annihilation could keep the Sun shining for trillions of years.[f]

A Theory of Gravity

There is little doubt, today, that Einstein's greatest contribution to our understanding of the Universe was his extension of the relativity principle to derive a theory of gravitation – his *general theory of relativity*. All modern cosmological theories are founded on this theory. Mathematically it led to Einstein's greatest struggles.

A first step came as early as 1907. Einstein had been invited to write a review article on relativity for the *Jahrbuch der Radioaktivität und Elektronik*.[31] As he was organizing his thoughts, he experienced what he later called "the happiest thought of my life": While musing at his desk at the Swiss Patent Office, it occurred to him that a man in free fall has no sense of gravitational attraction.[32]

Einstein pictured the man gently releasing a number of balls that continued to fall alongside him, regardless of their material composition, just as Galileo's experiments had shown three centuries earlier. More precise experiments, conducted by the Hungarian Baron Roland von Eötvös toward the end of the nineteenth century, had further shown that bodies of widely varying compositions were accelerated by gravity identically to within a possible experimental error no larger than around one part in a hundred million.

To Einstein this imagined vision was an epiphany. If a man and all objects in his immediate vicinity are falling at identical rates, none of them accelerating with respect to each other, there is no evidence that gravity is exerting any effect. The man feels no forces and judges himself at rest. Under these conditions, the relativity principle should hold at least in the man's immediate vicinity. This single thought – Einstein termed it the *equivalence principle*, implying that the principle of relativity equally holds for systems in gravitationally induced free fall – suggested that a theory of gravitation might be constructed along such lines.

Some simple consequences of this might be stated at once. Light emitted by the Sun ought to lose energy as it works its way out against the Sun's gravitational attraction to reach Earth. This loss of energy should express itself as a shift of light toward longer wavelengths, that is, toward the red end of the visible *spectrum* – in short, a *gravitational redshift*. Similarly, the path followed by a beam of light passing close to the Sun ought to be bent by the Sun's gravitational attraction, just as the trajectory of a falling massive object would be diverted by the Sun's attraction.

Newtonian gravity made similar predictions. Extending such ideas into a coherent theory of gravitation incorporating the relativity principle was needed to

---

[f] At the time, the term *annihilation* meant any kind of conversion of mass into energy; it did not, as today, connote annihilation of matter by *antimatter*. The existence of antimatter was not discovered until 1932.

determine whether and to what extent the predictions of Newton's theory of gravitation would need revision.[33]

Einstein sporadically continued working on relativity while also concentrating on quantum theory in the years after 1907. But by 1912, when he was appointed professor of physics at the ETH in Zürich, he realized that he had run into a brick wall. He lacked the mathematical tools to go forward and was at a loss about whether the appropriate tools even existed. In despair, he turned to his friend Marcel Grossmann, by now a professor of mathematics, also at the ETH, pleading "Grossmann, Du musst mir helfen, sonst werd' ich verrückt!" ("Grossmann, you have to help me, I'm going out of my mind!")[34] Grossmann soon identified Riemannian geometry as the tool that would do the job, and the two long-time friends published a joint paper the following year, in which they first applied this form of geometry to the gravity problem.[35] A year later, they published their second and last joint paper on the subject.[36]

In the summer and fall of 1915, Einstein was struck by shortcomings he still perceived in this joint work. He had moved on to Berlin in 1914 to accept the directorship of physics at a new institute established for him at the Kaiser Wilhelm Gesellschaft (the Emperor Wilhelm Society) in Berlin. It gave him the right but no obligation to lecture at the University of Berlin and also made him a member of the Prussian Academy of Sciences. This was an exceptional arrangement for Einstein. He enjoyed giving talks and debating science with colleagues on whatever topics currently interested him. But he found routine lecturing to students an unwelcome interruption to the concentration that he felt work on his theories required.

Earlier that summer of 1915, Einstein had delivered a series of six two-hour lectures on general relativity in Göttingen. These were attended by the two leading German mathematicians, David Hilbert and Felix Klein. Einstein was enthusiastic because he felt he had won the mathematicians over with his Riemannian geometric approach.[37] Perhaps this was also where he became uncomfortable with some of the work on the general theory he had presented. Hilbert's interest in the apparent discrepancies was also aroused. A rapid exchange of correspondence between Einstein and Hilbert in November that year shows that both men were working, perhaps racing, along separate lines to derive the correct defining equations of general relativity.[38]

On November 18, Einstein found that his new variant of the theory allowed him to calculate the *perihelion precession* of Mercury.[g] The excellent agreement he found for this long-observed but equally long unexplained astronomical finding had a profound effect on him. He felt his heart beating wildly! Nature had finally vindicated him and shown that his years-long efforts on the theory were on the right track![39]

---

[g] Mercury's perihelion is the point along its orbit where the planet most closely approaches the Sun. This point of nearest approach slowly circles – *precesses* – about the Sun, requiring roughly three million years to complete a full circuit.

That same day, he also found that his previously calculated deflection of light by the Sun, which had agreed with a value one could derive with Newtonian physics, had been too low by a factor of 2. This provided a second and previously unpredicted test that could verify his theory. Confident of final success with both these of findings, he submitted a short paper to the Academy before day's end.[40]

The following week, on November 25, Einstein completed his theory of general relativity in its final form and again submitted it the same day to the Academy of Sciences.[41]

But five days earlier, Hilbert had submitted his own version of the theory, 'Die Grundlagen der Physik' (The Foundations of Physics), to the Gesellschaft der Wissenschaften in Göttingen.[42] Einstein's completed version of his theory appeared in print, soon thereafter, as 'Die Grundlage der allgemeinen Relativitätstheorie' (The Foundation of the General Theory of Relativity).[43]

As the ambitious title of Hilbert's paper suggests, the great mathematician was seeking a unified field theory combining electromagnetism and gravity in a coherent theory of all physical phenomena. In this he did not succeed. Nor did Einstein, who later had similar ambitions. We still lack such a theory today. But it is clear that Hilbert's approach considerably influenced Einstein in their race to develop the final form of the general relativistic field equations, although Hilbert always acknowledged that it was Einstein's physical insights that had aroused his own interests in finding such a set of equations.

* * *

In 1971, the Princeton theoretical physicist Eugene Wigner wrote the historian of science Jagdish Mehra to ask about Hilbert's independent discovery, which he had never heard any physicist mention.[44]

In a detailed response, Mehra wrote an article that may still be one of the most informed and even-handed analyses of the contributions respectively made to the theory by Einstein and Hilbert.[45] But because Einstein and Hilbert exchanged notes with each other, during their four weeks of intense activity in the fall of 1915, these respective contributions have been somewhat difficult to disentangle, and have led to many further analyses.[46]

As Einstein pointed out in a long paper he published in 1916, both Hilbert and by then also the Dutch theoretical physicist Hendrik Antoon Lorentz had independently derived the fundamental formulation of the general theory, by making use of a single overarching principle. Einstein now proceeded to do the same, basing his approach somewhat differently on the use of the *Hamiltonian Principle*, originally developed by the nineteenth century Irish physicist, mathematician, and astronomer William Rowan Hamilton.[47]

* * *

A most important immediate consequence of Einstein's completed theory was an entirely novel application developed by Karl Schwarzschild, early in 1916, only weeks after Einstein's new theory had appeared in print. Schwarzschild derived the gravitational field in the vicinity of a massive compact body. It became the basis of all subsequent work on what are now called *black holes*. In Chapter 5, I will describe Schwarzschild's work and its appreciating importance to astrophysics in the advancing twentieth century.

The Bending of Light, the Motion of Mercury, and the Gravitational Redshift

General relativity produced three immediate predictions, all of which were verifiable astronomically, albeit with considerable difficulty. The first concerned the aforementioned, small but puzzling anomaly in the motion of the planet Mercury about the Sun – a gradual orbital shift of the point of Mercury's closest approach to the Sun. This anomaly in the *perihelion precession* amounted to ~43 seconds of arc per century, out of a total observed precession appearing to be 5599.74 ± 0.41 *arc seconds* per century, 5557.18 ± 0.85 of which could be attributed to the effects on Mercury's orbit due to the tug of the other planets orbiting the Sun.[48]

Einstein's general theory of relativity not only could explain, but actually demanded this precession. Further, his calculated magnitude of the observed precession was 43 seconds of arc per century, in satisfactory agreement with observations.[49] Explanations of well-known observed effects, however, tend to come cheaply and often turn out to be wrong. So, the explanation Einstein gave

Figure 3.2. Arthur Stanley Eddington, the most influential astrophysicist in Europe in his time. (*Courtesy of the Master and Fellows of Trinity College Cambridge*).

of Mercury's perihelion precession had no more than limited impact. A more convincing validation of a theory is a correct prediction of an effect that has never been previously observed.

Einstein's second prediction involved just such an effect – the deflection of light by the Sun. Newton's theory of gravitation predicted that the gravitational attraction of the Sun should slightly bend light from a distant star as the light passed close to the solar *limb*. This would make it appear that the star's position in the sky had shifted relative to stars at greater angular distances from the Sun. Einstein made a similar prediction, but his general relativity predicted an angular deflection twice as large as that required by Newton's theory, and approximately 1.7 arc seconds in magnitude.[50] The effect was difficult to measure. It meant accurately detecting the position of a distant star, in the intense glare of sunlight as Earth's annual motion about the Sun brought the star's image ever closer to the solar limb. The only appropriate time to do this appeared to be during a total eclipse of the Sun by the Moon.

Eddington and the Deflection of Light by the Sun

In the Netherlands, a nonaligned nation in World War I, Einstein's work had an avid following. In 1916, shortly after the publication of Einstein's 'Die Grundlage der allgemeinen Relativitätstheorie,' the Dutch astronomer Willem de Sitter wrote a lucid summary of the paper and its implications for astronomy, and published it in a series of three articles in the *Monthly Notices of the Royal Astronomical Society*.[51] This brought general relativity to the attention of British astronomers. Arthur Stanley Eddington, a conscientious objector to the war and by then the leading astrophysicist in Cambridge, at once recognized its significance.

Born in 1882 into a Quaker family in Kendal, Westmorland, England, Eddington had been brought up by his mother. His father, Arthur Henry Eddington, headmaster of Friends' School in Kendal, had died during a typhoid epidemic in 1884. An excellent student, young Arthur secured scholarships that helped him through his university studies and allowed him to graduate with an MA from Trinity College, Cambridge in 1905. By then, he had distinguished himself as Senior Wrangler in the Mathematical Tripos of 1904.[52]

Through de Sitter's articles, and papers by Einstein himself, Eddington immersed himself in the new physics. He lectured on relativity at the British Association meeting in 1916, provided a comprehensive report on the subject to the Physical Society in 1918, and raised support for a two-pronged expedition to measure the deflection of light by the Sun during the anticipated solar eclipse on May 29, 1919. One expedition, organized by the Astronomer Royal, Frank Watson Dyson – but not joined by him – set out for the town of Sobral, in the state of Caera in the north of Brazil; the other, led by Eddington, journeyed to Principe Island in West Africa.

Careful calibration of the photographic plates obtained took several months, but on October 30, 1919, Eddington and his two co-authors submitted their results for publication. The mean deflection of light measured at Sobral was 1.98 arc seconds, with a probable error estimated as ±0.12 arc seconds; that at Principe 1.61 ±0.30 arc seconds. They thus straddled Einstein's predicted deflection of 1.75 arc seconds, in satisfactory agreement given the observational uncertainties.[53]

When Dyson and Eddington announced the results of the expedition at a joint meeting of the Royal Society and the Royal Astronomical Society on November 6, 1919, their findings led to an outburst of public enthusiasm for both Einstein and Eddington.[54] Two nations had fought a bitter war for four years. Yet here were two scientists, both avowed pacifists, one who had been working throughout the war at the Kaiser Wilhelm Society in Berlin, the other in Cambridge. Between them they had devised and verified the existence of a totally new way of thinking about light, gravity, and ultimately the Universe.

The Gravitational Redshift

Einstein's third prediction, as we saw earlier, was that light must lose energy on leaving the surface of a compact, massive star; the loss of energy would express itself as a shift of the light toward longer wavelengths, that is, toward the red.

Although the simplest of Einstein's three predictions, this redshift was last to be observed. Ordinary stars proved unreliable because their atmospheres exhibit transmission to different atmospheric depths for different spectral lines. Convective motions in the star's atmosphere can then generate Doppler shifts that confuse the measurements. But a class of highly compact stars that had gradually come to be known might serve the purposes of a test.

In 1914, Walter S. Adams, a gifted astronomer at the Mount Wilson Observatory, wrote a brief note on the stellar binary o_2 Eridani. The light from both its stars is white. But contrary to common expectation at the time, Adams remarked that one of the two stars was much fainter than the other. [55] Two years later, in 1916 Ernst Öpik, a 22-year-old Estonian student just finishing his studies in Moscow, published an article in which he estimated the relative densities of a number of stars gravitationally bound to each other in stellar binaries. Most exhibited quite normal densities, but the fainter star in o_2 Eridani was puzzling. Its density appeared to be 25,000 times higher than that of the primary, though both exhibited similar spectra. Öpik concluded "This impossible result indicates that in this case our assumptions are wrong."[56] Actually, nothing was patently wrong. As other extremely dense white stars were gradually discovered, they came to be known as *white dwarfs*.

Inspired by a paper Eddington had written the previous year, Walter Adams in 1925 provided the first seemingly credible gravitational redshift measurement for a white dwarf, the companion to the star Sirius.[57] But Adams warned that the redshift measurements had been exceedingly difficult, and for some decades the astronomical tests remained quite uncertain.[58]

The predicted gravitational redshift was fully verified only in 1960, five years after Einstein's death. It required an ingenious laboratory experiment conducted by Robert Vivian Pound and his PhD student Glen A. Rebka Jr., at Harvard.[59] Their laboratory measurement was based on the *Mössbauer effect*, pioneered in 1958 by the 29-year-old Rudolf Mössbauer working on his PhD thesis in the Institute for Physics at the Max-Planck Institute for Medical Research in Heidelberg.[60]

A reliable astronomical measurement of the stellar redshift did not become available until 2005, when finally the Hubble Space Telescope provided a breakthrough. Although the results of Adams, eight decades earlier, had appeared to confirm Eddington's predictions, his warning about them was both conscientious and correct: they turned out to be off by a factor of 4.[61]

Two more remarks are worth mentioning: Isaac Newton's theory of gravity is based on a gravitational force of attraction, F, between between two masses M and m. The force, proportional to the product of the masses and the inverse square of the distance r separating them, is $F = MmG/r^2$, where G is the universal constant of gravity. The acceleration, $\ddot{r}$, experienced by the mass m subjected to this force is given by

$$m\ddot{r} = MmG/r^2 \, , \tag{3.2}$$

directed toward M.

Newton had assumed that the mass m on the left side of the equation equals the mass m on the right side. But, as Einstein emphasized, these are two different traits of a body. On the right side of equation (3.2) m is a measure of the gravitational force M exerts on mass m, and one can think of m as the body's *gravitational mass*. On the left side of the equation m represents the inertia, which keeps the acceleration $\ddot{r}$ of m in check; this value of m is called the body's *inertial mass*. As Einstein noted, the equality of these two masses was not obvious, and was firmly established only through the careful experiments of Eötvös.

The equality of inertial and gravitational mass is thus a fortuitous and as yet unexplained property of all matter, and a basic assumption on which general relativity is founded. It led to the notion that a relativity of accelerated motion could be intimately related to gravitational fields defined by the distribution of masses in the Universe. In turn, this led to Einstein's principle of equivalence, the notion that physical processes in the vicinity of bodies freely falling in a gravitational field would be indistinguishable from similar processes in the vicinity of bodies moving with constant velocity in the absence of gravitational forces. This generalization enabled application of the principle of relativity to a wider range of phenomena and provided new insights on the nature of the Universe.

Relativistic Cosmology

A year after its completion, Einstein showed how general relativity could reshape our understanding of the Universe, thus laying the foundation of all modern cosmology. For this, the theory would have to make sense on all scales.

Einstein had already shown that general relativity must be the correct theory on the scale of *planetary systems*; it properly accounted for the perihelion advance of Mercury. Now, he needed to show that the theory also correctly depicted the entire Universe stretching out beyond the realm of the planets.

To do this, he needed to surmount a problem that Newton's theory of gravitation had left unexplained: If all matter attracts, as Newton had found, why did the Universe not collapse?

Einstein considered a number of ways this puzzling instability might be circumvented. None satisfied him until he gave up the intuitively appealing requirement that the geometry of the Universe is Euclidean – meaning that light inexorably travels along straight lines out to infinity, as long as it avoids close passage by stars that might gravitationally bend its trajectory.

He now guessed that a gravitational collapse could be avoided if he postulated that the Universe is a spatially closed sphere. The topology of a sphere, however, required a slight modification of the *field equations* of gravity that he had derived a little more than a year earlier, in late 1915. To his original equations he now added the product of some unknown constant Λ and a quantity defining the curvature of the space.[62] Einstein referred to Λ as the *universal constant*. Nowadays it is universally called the *cosmological constant*.

The value of Λ had a simple relation to the radius of curvature of the Universe, R:

$$\Lambda = \frac{4\pi G\rho}{c^2} = \frac{1}{R^2} \, , \tag{3.3}$$

where G is the gravitational constant; ρ is the mass-density of the Universe – roughly speaking the total mass of all the stars and other matter in the Universe, divided by the cosmic volume; and c again is the speed of light in vacuum.

For a universe with a sufficiently large radius of curvature, R, the cosmological constant Λ could be quite small. The modification of the field equations of general relativity then was so slight as to be essentially unobservable, at least on the scale of the *Solar System*. Einstein's previously derived perihelion precession of Mercury or the bending of light around the Sun would, for all practical purposes, remain unaffected.

The universe Einstein described was static. How could anyone have predicted that it expands? But his application of the general relativistic equations inspired others, mathematicians as well as astronomers, to seek variations of his equations that could describe nonstatic, expanding, contracting, and oscillating models of the universe, thus further enriching the astronomer's toolkit for probing the Cosmos. In 1917, the size of the Universe was unknown; there clearly were relative motions of stars gravitationally attracted to each other; but the thought that some dominant, large-scale motions might exist had not yet occurred to anyone.

The Cosmic Models of de Sitter, Friedmann, and Lemaître

It took the Dutch astronomer Willem de Sitter only nine months after the publication of Einstein's general-relativistic depiction of a static universe to describe an expanding universe that also was compatible with general relativity. Einstein's was only one model of the Universe among three different classes, A, B, and C, which de Sitter considered in the last of his three articles submitted to the *Monthly Notices of the Royal Astronomical Society* in 1917. All three classes were compatible with general relativity.[63]

If de Sitter assumed the spiral nebulae to be distant systems of stars, similar to the Milky Way, and if "observation should confirm the fact that spiral nebulae have systematic positive radial velocities," as the scant data then available suggested to him, he concluded that "this would certainly be an indication to adopt the hypothesis B." Model B was somewhat strange; the Universe was devoid of matter; specifically, the density of matter was zero. But because the density of matter in the Universe was known to be small, this did not seem a crippling defect.[64]

Observational astronomy provided only the most meager evidence to support either an expanding or a static universe. In 1912 Vesto Melvin Slipher of the Lowell Observatory in Flagstaff, Arizona, had obtained the first photograph of the spectrum of a galaxy suitable for determining a velocity from Doppler-shifted spectral lines. Slipher had been observing the nearest large *spiral galaxy*, the Andromeda Nebula, and found that it approaches at a speed of 300 kilometers per second. In addition to the roughly −300 km/sec approach velocity for the Andromeda nebula, the mean *recession velocities* observed by three observers for NGC 1068 was +925 km/sec, and the mean of two independent observations of NGC 4594 was +1185 km/sec. Here a plus sign (+) indicates recession and a minus sign (−) denotes approach. Slipher's observations had been included in determining each of these mean velocities. de Sitter cited these earliest results of Slipher in partial, if tentative, support of his contention that the Universe might be expanding. Slipher was also reported to have observed a total of 15 galaxies, of which only the three listed here had been confirmed. The average recession velocities he had measured were between +300 and +400 km/sec.[65]

Slipher was assuming, of course, that galaxies are independent stellar systems, far out in the Universe and of grandeur comparable to the Milky Way's. However, as late as 1920 and a famous debate between the astronomers Harlow Shapley and Heber Doust Curtis, respectively at the Mt. Wilson and Lick Observatories, this was far from certain. One could still question whether the spiral galaxies were members of the Milky Way or distant rivals.[66]

In retrospect, one could argue that, after 1922, no further doubt could reasonably have existed. That year, Ernst Öpik derived a distance to the Andromeda

Nebula that took into account not only astronomical data but also physical reasoning. He applied the *virial theorem* to a model that pictured a galaxy as a rotating aggregate of gravitationally bound stars and derived a distance ~450 kpc, in rough accord with a number of less certain estimates that had earlier emerged.[67,h] But Öpik's paper seems to have convinced neither Shapley nor Curtis. The astrophysicist and historian Virginia Trimble recounts that both became believers only after Edwin Hubble's discovery, late in 1923, that the Andromeda Nebula contained Cepheid variables. The apparent magnitude of these stars showed the nebula to be a distant stellar system equivalent to the Milky Way, and indicated that other spiral nebulae should be viewed as independent galaxies as well.[69]

* * *

Starting in 1922, Alexander Friedmann, a gifted mathematician and meteorologist in Russia, found whole families of cosmological models, all compatible with the framework of general relativity, and with sensible values for the mass-density of the Universe.

George Gamow, who later became one of the most imaginative astrophysicists of the twentieth century, was a student of Friedmann's in Leningrad in the 1920s. He recalled Friedmann's interests in a wide variety of mathematical applications, including meteorology.[70] Among astrophysicists and cosmologists, Friedmann is now best known for showing that various models of the Universe could be constructed, some of which expanded whereas others contracted or oscillated – and that all these different possibilities were consistent with Einstein's theory of general relativity.[i]

Friedmann's work was not just idle speculation. He wrote about his findings to Einstein but did not get an answer. So when a theoretical physicist and colleague at the University of Leningrad, Yuri Krutkov, obtained permission to travel to the West, Friedmann asked him to try to see Einstein and talk to him about it. As a result, Friedmann received what Gamow described as a rather grumpy letter from Einstein agreeing with Friedmann's argument. This led Friedmann to publish the work in the German *Zeitschrift für Physik*.[71,72]

* * *

By 1925 Slipher had compiled the radial velocities of 41 galaxies, a list of which had been analyzed and published by Gustav Strömberg, born in Sweden but by then working at the Mount Wilson Observatory in the United States.[73] These line-of-sight velocities ranged from an approach velocity of 300 km/sec for the Andromeda galaxy to a recession of 1800 km/sec for the nebula NGC 584; and most

---

[h] Given the many uncertainties, at the time, about interstellar absorption and the structure of our own galaxy, Öpik's estimate appears in surprisingly good agreement with a value of 774 ± 44 kpc, published more than eight decades later, in 2005.[68]

[i] Alexander Friedmann's name can be encountered in various spellings. In his native Russia, he was referred to as Aleksandr Aleksandrovich Fridman.

Figure 3.3. Edwin Powell Hubble, the leading astronomer charting the extragalactic universe in the 1920s. (*AIP Emilio Segre Visual Archives, Brittle Books Collection*).

of the galaxies were receding. The recession velocities of the galaxies also were several times higher than the more random appearing velocities of the Galaxy's *globular clusters*.

The following year, Edwin Powell Hubble at Mount Wilson published a cosmic distance scale based on the visual *magnitudes* of about 400 galaxies, for which J. Holetschek in Vienna had obtained visual magnitudes by 1907.[74] Hubble found an approximate relation between the angular diameters and visual magnitudes for these galaxies. Noting that, by and large, galaxies of a given type constituted a homogeneous group, and making use of known distances to nearby galaxies, he obtained a magnitude-distance relation for galaxies, from which he derived a number density and also a mass density of galaxies in space.[75]

Meanwhile the Belgian astrophysicist, the Abbé Georges Lemaître, had been pondering the differences between de Sitter's and Einstein's cosmologies. Both models were governed by a cosmological constant Λ, and both were homogeneous and isotropic. *Homogeneity*, as applied in cosmology, means that any part of the Universe appears to be essentially identical to any other, at least when we compare sizable regions within which differences on small scales mutually compensate. *Isotropy* implies that the Universe exhibits no favored directions: The speed of light, or the strength of the gravitational field a massive body exerts on its surroundings, is identical along all directions. Whereas de Sitter's model was in general agreement with a cosmic expansion, apparently reflected in Slipher's

Figure 3.4. Georges Lemaître in the early 1930s, soon after he had proposed a physical model to explain an expanding universe. (*Courtesy of the Catholic University of Louvain Archives Picture Gallery*).

observations and Strömberg's list, it lacked the symmetry between time and space that general relativity demanded. In contrast, Einstein's cosmology preserved that symmetry, but was static. By 1927, Lemaître had resolved these differences and devised a model of an expanding universe that did preserve Einstein's required symmetries. With this in hand, he fitted the velocities compiled by Strömberg and the mass densities obtained by Hubble during the preceding two years. Between them they yielded the insight that galaxies recede from us at a rate not only proportional to their distance, but also consistent with the cosmic density general relativity demanded. Lemaître thus proposed a self-consistent general relativistic model of the Universe that accounted for the cosmic expansion in terms of observed mass densities.[76]

It was an impressive paper!

The historian of science Helge Kragh has pointed to two of Lemaître's significant innovations.[77] Lemaître's article was the first to apply thermodynamics to relativistic cosmology. He derived the expression for energy conservation

$$\dot{\rho} + \frac{3\dot{R}}{R}\left(\rho + \frac{P}{c^2}\right) = 0 \,, \tag{3.4}$$

in which the pressure, P; the mass density, ρ; as well as the rate at which density changes with time, $\dot{\rho}$, are related to the cosmic expansion rate $\dot{R}/R$. This is a particularly powerful equation because it holds true for any homogeneous, isotropic universe, independent of whether its expansion is governed by a cosmological constant Λ. Strömberg's compilation of data hinted that increasingly distant galaxies were progressively more redshifted. Lemaître interpreted this as a sign of cosmic expansion which permitted him to estimate the cosmic rate of expansion, $\dot{R}/R \sim 625$ km sec^{-1} Mpc^{-1}.

Kragh minces no words: "The famous Hubble law is clearly in Lemaître's paper. It could as well have been named Lemaître's law."

Regrettably, as Harry Nussbaumer and Lydia Bieri have documented in their insightful book, 'Discovering the Expanding Universe,' nobody seemed to have taken any notice.[78] Lemaître sent copies of his paper to Eddington and de Sitter, neither of whom apparently read it. Buried in the *Annales de la Société scientifique de Bruxelles*, where Lemaître had chosen to publish, few others would have come across it either. It was not until 1931, when Lemaître brought the paper to Eddington's attention again, that Eddington recognized its importance.

William Marshall Smart, editor of the *Monthly Notices of the Royal Astronomical Society*, then wrote Lemaître to ask whether he would give permission for the article to be reprinted in the *Monthly Notices of the Royal Astronomical Society*, preferably in English and with any annotated modifications Lemaître might wish to make. Lemaître accepted this invitation with pleasure, translated the paper himself, but replaced his own quantitative calculations from 1927, in which he had used astronomical data available at the time to derive the cosmic expansion rate. He wrote Smart "I did not find advisable to reprint the provisional discussion of radial velocities which is clearly of no actual interest..."[j] Instead, Lemaître substituted data that Hubble had later published in 1929. The astronomer Mario Livio, who unearthed Lemaître's letter to Smart, rightly finds Lemaître's attitude admirable.[79] But, as a result of Lemaître's sensible approach, Hubble is now usually credited with the discovery.

In 1929, using newer, improved observations, Hubble had independently discovered the same relationship between the redshifts and the distances of extragalactic nebulae – as he preferred to call the galaxies – that Lemaître had previously established. He explicitly took these redshifts to signify their "apparent velocities." But, as Kragh and Robert W. Smith have pointed out, Hubble to the end of his life shied away from explicitly concluding that these velocities also implied cosmic expansion.[80,81] Lemaître now evidently felt that Hubble's newer data were more reliable than those he himself had gathered from the available literature in 1927. The English version of Lemaître's article regrettably suggests, contrary to fact, that Hubble had proposed the cosmic expansion before Lemaître.[82]

Credit for a discovery often goes to those whose papers are most accessible, and Hubble's presentation was both accessible and convincing. This is why one speaks of *Hubble's law of expansion* instead of *Lemaître's law*. More accurately, as Kragh proposed, one should probably assign Lemaître credit for the discovery.

[j] As many French speakers have noted, Lemaître undoubtedly meant "current" rather than "actual." The French word "actuel" translates as "current," or "present."

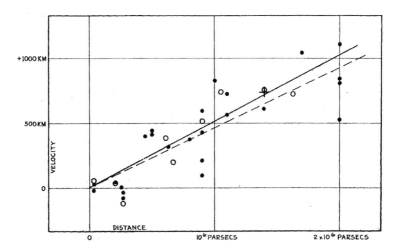

Figure 3.5. The Velocity–Distance Relation among Extra-Galactic Nebulae. Edwin Hubble's diagram showing the distances of galaxies and their apparent velocities judged by their redshifts. The black dots and solid line are a best linear fit to individual galaxies. The circles and dashed line represent a fit obtained by combining galaxies into groups. The cross represents the mean velocity corresponding to 22 galaxies whose distances could not be estimated individually. (*Reproduced from the Proceedings of the National Academy of Sciences of the USA*, 15, 172, 1929).

Lemaître apparently was out of luck. But so was Friedmann, whose papers Lemaître had never read either. And Alexander Friedmann could not even savor some private satisfaction. He did not live to see the success of his predictions. Friedmann's many interests extended to meteorological balloon flights. On one of his ascents, as Gamow recalls, "Friedmann received a severe chill which resulted in pneumonia." A month later, in September 1925, Friedmann died of a typhus infection.[83,84]

After the appearance of the English version of Lemaître's paper, leading theorists quickly accepted the expansion of the Universe and its general relativistic interpretation. By 1931, even Einstein, who had initially resisted Friedmann's and Lemaître's cosmological models as somehow unphysical, was persuaded and wrote a brief article adding his approval.[85]

Perhaps most important to Einstein, in this article, was the need to point out that the cosmological constant was no longer required. He had inserted this into his equations in 1917 only because he thought the spatial structure and density of the Universe should remain constant over time. Once this criterion was dropped, as Hubble's findings now showed necessary, the cosmological constant served no further purpose and, in Einstein's view, had always been unsatisfactory (*unbefriedigend*).

Today, most astrophysicists would disagree. The observed current expansion of the Universe seems to require inclusion of something that at least closely resembles Einstein's cosmological constant.

Mach's Principle and the Principle of Relativity

When Einstein first set out on his theoretical depictions of gravitation and the Cosmos, he expected that Mach's perceptions about the unity of the Universe would emerge. Just what form this would take was not clear. Mach had assumed that at least the coordinate system that told us whether a bucket of water was rotating or at rest would be determined by the large-scale structure of the Universe. Surely the rising water level at the circumference of a rotating bucket must be due to some universal distinction between rotating and nonrotating coordinate frames. Could this not also mean that the structure and mass of atomic matter, the charge of the electron, and the speed of light might all be explained in terms of some grand cosmic theory?

But, even as Einstein's efforts continued to bear fruit, none of Mach's predictions seemed to be playing a role. In the end, Einstein's expectations declined. Mach's principle appeared inconsequential.

Why?

Even today we have no answer.

Einstein had followed the consequences of the principle of relativity to its logical conclusions. The physical laws governing the conduct and evolution of the Universe and its contents had to be invariant, agreed to, and understood by any and all observers, independent of their location or motion. Were this not so, there could be no agreed on basis for science.

In the days of the Olympia Academy, around 1903, when its three members, Konrad Habicht, Maurice Solovine, and Albert Einstein were studying the works of Ernst Mach and Henri Poincaré, both Mach's principle and the principle of relativity had beckoned persuasively. But only the principle of relativity led forward; Mach's principle seemingly led nowhere. Someday, perhaps it still might. But, by now, we know that, over the eons, the expansion of the Universe has had no notable effect on the electron's charge e, Planck's constant h, the gravitational constant G, the speed of light c, or on any law of physics. The structure of the Universe might still determine which coordinate systems are rotating and which are not. But even this is not clear.

Einstein evidently was lucky or else profoundly perceptive in choosing first to follow where the principle of relativity would lead. Had he taken the road suggested by Mach, he would not have learned much.

In science, guiding principles can be invaluable.

They can also misguide!

*　*　*

Earlier, I gave a somewhat detailed introduction to Einstein and his way of puzzling about matters. His thirst for learning and his need to "brood"[k] over difficult problems for years at a time appear to have been central to his work.

---

[k] Einstein used the German expression "grübeln."

Perhaps the simplest way to understand Einstein's style is to recognize him as primarily interested in understanding the *structure* of science – the grander proportions that relate its different parts and make the world understandable.

Einstein's biographer, the theoretical physicist Abraham Pais, and the relativist, Kip Thorne, both have remarked on how easily Einstein's early work seemed to flow from his mind and pen. And how belabored the later work on general relativity became. Continual small augmentations, retractions, confidence that another step would work, only to have that hope dashed once again. His last four papers on general relativity, written in late 1915, were submitted only days apart to the *Sitzungsberichte*. In his race against Hilbert to obtain the general relativistic field equations one sees him almost stumbling across the finish line.

Pais reflects on Einstein's early papers, citing them as having the qualities of the music of Mozart, flowing with apparent supreme ease from the composer's consciousness, while the later work on general relativity suggests to him the more labored compositions of a Beethoven – monumental works but also the product of a tortured mind.

The one exception to Pais's reflection might be the triumphant short paper of 1915, at the end of his race against Hilbert, in which he derived the perihelion precession of Mercury and the bending of light past the Sun.[86] Here one palpably feels Einstein's exhaustion but also his reemerging enthusiasm and pride of work. He has found and proposed two new tests of a novel theory that he is confident will be found correct.[87] One of them, the perihelion precession, he knows must be right. The other will seal the theory's fate and distinguish it clearly from Newton's work. He seems like a man immensely relieved.

But his long and arduous battle to devise the general theory of relativity had also taught Einstein a lesson about himself, which he now recognized and expressed with deepest regret. It was a lesson that others might not have noted.

At a meeting held in Trieste in 1979, in honor of the centenary of Albert Einstein's birth, the great mid-twentieth-century theoretical physicist Chen-Ning Yang remarked on Einstein's influence on current thought. Einstein, in Yang's view, emphasized for theorists "that symmetry dictates interactions, that geometry is at the heart of physics, and that formal beauty plays a role in describing the world."[88]

Einstein probably would have been pleased with Yang's assessment; but late in life he saw matters differently, regretfully ruminating that he had not sufficiently early understood the supreme importance of mathematics to his work. Where Yang praised Einstein's appreciation of symmetries, Einstein recognized that his appreciation of them had been too limited, and had perhaps even failed him where he might have achieved more had his knowledge of, and sense for, mathematical forms been better developed.

This was not false modesty. Einstein simply realized that he did not have the wide-ranging familiarity with and intuition for mathematics that he had gained in physics. He may have noted this when Hermann Minkowski showed how the special theory of relativity could be viewed most simply in terms of a rotational symmetry between space and time.[89] As Einstein later recalled, he had initially dismissed Minkowski's approach as superfluous erudition.[l] Not until years later, when he began to work on general relativity, did Einstein fathom its importance.[90]

Having to request the help of his friend Marcel Grossmann, in 1912, to find a mathematical structure with which to begin work on a general theory of relativity, would probably have reinforced the realization. But for another four decades, Einstein remained silent on the help he had received from his friend, to whom fate had been unkind. Around 1920, Grossmann began exhibiting symptoms of multiple sclerosis. Speech became progressively more difficult. In 1927 he resigned his position at the ETH and ultimately succumbed to the disease in 1936.[91] Even then, Einstein's debt to him remained unpaid.

Not until 1955, when he wrote a brief autobiographical essay, shortly before the end of his life, did Einstein acknowledge his great indebtedness to Grossmann. "The courage to write this somewhat motley autobiographical sketch came from my need to express, at least once in my life, my gratitude to Marcel Grossmann."[92:m] Ruefully, Einstein acknowledged Bernhard Riemann's great contribution, "From this one could determine the form that the field equations of general relativity would have to assume if invariance under the group of all continuous coordinate transformations was to be required. That this requirement was justified, however, was not easy to see, particularly since I thought I had found countervailing arguments. These, admittedly erroneous, considerations brought about that the theory only appeared in its final form in 1916."[n]

Finally, Einstein's weakness in finding transparent mathematical forms must have become painfully clear to him as he was racing the great twentieth century mathematician David Hilbert to the final formulation of general relativity.

Writing late in life, Einstein recalled his excellent mathematics professors when he was a student at the Eidgenössische Polytechnikum in Zürich, Adolf Hurwitz and Hermann Minkowski. Nevertheless, he had spent most of his time familiarizing himself with physics. He recalled[93]:

> The fact that I neglected mathematics to a certain extent had its cause
> not merely in my stronger interest in the natural sciences than in math-
> ematics, but also in … that mathematics was split up into numerous

---

[l] Überflüssige Gelehrsamkeit

[m] Den Mut, diese etwas bunte autobiographische Skizze zu schreiben, gab mir das Bedürfnis, wenigstens einmal im Leben meiner Dankbarkeit für Marcel Grossmann Ausdruck zu geben.

[n] Daraus war zu ersehen, wie die Feldgleichungen der Gravitation lauten müssen – falls Invarianz gegenüber der Gruppe aller kontinuierlicher Koordinaten-Transformationen gefordert wird. Dass diese Forderung gerechtfertigt sei, war aber nicht so leicht einzusehen, zumal ich Gründe dagegen gefunden zu haben glaubte. Diese, allerdings irrtümlichen, Bedenken brachten mit sich, dass die Theorie erst 1916 in ihrer endgültigen Form erschien.

specialties. . . . [M]y intuition was not strong enough in the field of mathematics to differentiate clearly the fundamentally important, . . . and it was not clear to me as a student that the approach to a more profound knowledge of the basic principles of physics is tied up with the most intricate mathematical methods. This dawned upon me only after years of independent scientific work.

Realization of those missed mathematical opportunities evidently troubled Einstein deeply. He recognized the great achievements that had been his, but wished they might have been even greater.

Notes

1. *Subtle is the Lord – The Science and the Life of Albert Einstein*, Abraham Pais. Oxford University Press, 1982.
2. Autobiographical note in *Helle Zeit, dunkle Zeit – in memoriam Albert Einstein*, A. Einstein (Carl Seelig, editor). Zürich: Europa Verlag, 1956, p. 16.
3. Autobiographical Notes, Albert Einstein, in *Albert Einstein: Philosopher Scientist*, Paul A. Schilpp ed., 2nd edition. New York: Tudor Pub., 1951, pp. 2–95.
4. Who Discovered the Expanding Universe? Helge Kragh & Robert W. Smith, *History of Science*, 42, 141–62, 2003.
5. *Discovering the Expanding Universe*, Harry Nussbaumer & Lydia Bieri, Cambridge University Press, 2009.
6. Folgerungen aus den Capillaritätserscheinungen, Albert Einstein, *Annalen der Physik*, 4, 513–23, 1901.
7. Thermodynamische Theorie der Potentialdifferenz zwischen Metallen und vollständig dissoziierten Lösungen ihrer Salze, und eine elektrische Methode zur Erforschung der Molekularkräfte, A. Einstein, *Annalen der Physik*, 8, 798–814, 1902.
8. Kinetische Theorie des Wärmegleichgewichtes und des zweiten Hauptsatzes der Thermodynamik, A. Einstein, *Annalen der Physik*, 9, 417–33, 1902.
9. Theorie der Grundlagen der Thermodynamik, A. Einstein, *Annalen der Physik*, 11, 170–87, 1903.
10. Allgemeine molekulare Theorie der Wärme, A. Einstein, *Annalen der Physik*, 14, 354–62, 1904.
11. Ibid., *Subtle is the Lord . . .*, Pais, p. 46.
12. Ibid., *Subtle is the Lord . . .*, Pais, pp. 46–7.
13. *Albert Einstein, Lettres a Maurice Solovine*, ed. Maurice Solovine, Paris: Gauthier-Villars, 1956.
14. *La Science et l'Hypothèse*, H. Poincaré, 1902; translated into *Science and Hypothesis*, H. Poincaré, 1905; Mineola NY: Dover Publications, 1952.
15. *Die Mechanik in Ihrer Entwicklung – Historisch-Kritisch Dargestellt*, Ernst Mach, Brockhaus, Leipzig, 1883, reissued 1897, pp. 216–23.
16. Ibid., *Die Mechanik . . .*, Mach, p. 232.
17. Conversations with Albert Einstein, R. S. Shankland, *American Journal of Physics*, 31, 47–57, see p. 48, 1963.
18. Albert Einstein's erste wissenschaftliche Arbeit, Jagdish Mehra, *Physikalische Blätter*, 27, 385–91, 1971.
19. How I created the theory of relativity, Albert Einstein, Translated from the German into Japanese by the physicist, J. Ishiwara, and published in the Japanese periodical *Kaizo* in 1923;

then translated from the Japanese into English by Yoshimasa A. Ono and published in *Physics Today*, 35, 45–47, 1982.

20. Zur Elektrodynamik Bewegter Körper, A. Einstein, *Annalen der Physik*, 17, 891–921, 1905, translated as On the Electrodynamics of Moving Bodies in *The Principle of Relativity*, edited by A. Sommerfeld, translated by W. Perrett and G. B. Jeffery, 1923, Dover Publications.

21. Ist Die Trägheit eines Körpers von seinem Energieinhalt abhängig? A. Einstein, *Annalen der Physik*, 18, 639–41, 1906, translated as Does the Inertia of a Body Depend upon its Energy Content? Ibid., Sommerfeld, 1923, Dover.

22. Raum und Zeit, address by Hermann Minkowski reprinted in *Physikalische Zeitschrift*, 10, 104–11, 1909, translated as Space and Time, Ibid., Sommerfeld, 1923, Dover.

23. Zur Dynamik bewegter Systeme, M. Planck, *Annalen der Physik*, 26, 1–35, 1908.

24. Mémoire sur la chaleur solaire, sur les pouvoirs rayonnants et absorbants de l'air atmosphérique, et sur la température de l'espace, Pouillet, *Comptes Rendus des Séances de L'Academie des Sciences*, 7, 24–65, 1838; see also *A Popular History of Astronomy during the Nineteenth Century*, Agnes M. Clerke, London: Adam and Charles Black, 1908, Republished by Scholarly Press, St. Clair Shores, Michigan, 1977, p. 216.

25. Beiträge zur Dynamik des Himmels in populärer Darstellung, Robert Mayer, Heilbronn, Verlag von Johann Ulrich Landherr, 1848; reissued in the series *Ostwald's Klassiker der Exakten Wissenschaften* Nr. 223 as *Beiträge zur Dynamik des Himmels und andere Aufsätze*, Robert Mayer; published by Bernhard Hell. Leipzig: Akademische Verlagsgesellschaft m.b.h., 1927, pp. 1–59.

26. *Bemerkungen über die Kräfte der unbelebten Natur*, J. R. Mayer, (Remarks on the Forces of Inanimate Nature), *Liebigs Annalen der Chemie*, 1842, p. 233 ff.

27. See also *A Popular History of Astronomy during the Nineteenth Century*, Agnes M. Clerke. London: Adam and Charles Black, 1908. Republished by Scholarly Press, St. Clair Shores, Michigan, 1977, p. 310 ff.

28. Astronomers take up the stellar energy problem 1917–1920, Karl Hufbauer *Historical Studies in the Physical Sciences*, 11, part 2, 273–303,1981.

29. Radio-activity and the Age of the Sun, G. H. Darwin, *Nature*, 68, 222, 1903.

30. The state of affairs in this debate at the start of the 20th century has been recorded in a number of detailed publications cited and summarized by Karl Hufbauer: See Ibid. Astronomers take up the stellar..., Hufbauer, pp. 277–303.

31. Über das Relativitätsprinzip und die aus demselben gezogenen Folgerungen (The Relativity Principle and its Consequences), A. Einstein, *Jahrbuch der Radioaktivität und Elektronik*, 4, 411–462, 1907; *Berechtigungen* (errata), 98–99, 1908.

32. Ibid. How I created..., Einstein, pp. 45–47, 1982.

33. Über den Einfluss der Schwerkraft auf die Ausbreitung des Lichtes, A. Einstein, *Annalen der Physik*, 35, 898–908, 1911, translated as On the Influence of Gravitation on the Propagation of Light, Ibid., Sommerfeld, 1923, Dover.

34. Erinnerungen eines Kommilitonen, Louis Kollros, in *Helle Zeit – Dunkle Zeit, in Memoriam Albert Einstein*, editor Carl Seelig. Zürich: Europa Verlag, 1956, p. 27.

35. Entwurf einer verallgemeinerten Relativitätstheorie und eine Theorie der Gravitation I. Physikalischer Teil von A. Einstein; II. Mathematischer Teil von M. Grossmann, *Zeitschrift für Mathematische Physik*, 62, 225–61, 1913.

36. Kovarianzeigenschaften der Feldgleichungen der auf die verallgemeinerte Relativitätstheorie gegründeten Gravitationstheorie, A. Einstein & M. Grossmann, *Zeitschrift für Mathematische Physik*, 63, 215–25, 1914.

37. Ibid., *Subtle is the Lord*, Pais p. 259.

38. Ibid., *Subtle is the Lord*, Pais pp. 259–60.

39. Ibid., *Subtle is the Lord*, Pais p. 253, quoting A. D. Fokker's article in *Nederlands Tijdschrift voor Natuurkunde*, *21*, see p. 126, 1955.

40. Erklärung der Perihelbewegung des Merkur aus der allgemeinen Relativitätstheorie, A. Einstein, *Sitzungsberiche der Königlich-Preussischen Akademie der Wissenschaften zu Berlin*, *44*, 831–39, 1915.

41. Die Feldgleichungen der Gravitation, A. Einstein, *Sitzungsberichte der Königlich-Preussischen Akademie der Wissenschaften zu Berlin*, *44*, 844–47, 1915 (see also Ibid., *Subtle is the Lord*, Pais p. 257.)

42. Die Grundlagen der Physik (The Foundations of Physics), David Hilbert, *Nachrichten der Königlichen Gesellschaft der Wissenschaften zu Göttingen, Mathematisch-physikalische Klasse*, 395–407, 1915.

43. Die Grundlage der allgemeinen Relativitätstheorie, A. Einstein, *Annalen der Physik*, *49*, 769–822 ,1916; translated as The Foundation of the General Theory of Relativity, Ibid., Sommerfeld, 1923, Dover.

44. Letter from E. Wigner to J. Mehra, November 29, 1971, reproduced in Einstein, Hilbert, and the Theory of Gravity, Jagdish Mehra, in *The Physicist's Conception of Nature*, ed. Jagdish Mehra, Dordrecht-Holland: D. Reidel Publishing Company, 1973, 92–178, see especially pp. 174–78.

45. Ibid., Einstein, Hilbert and the Theory of Gravity, Mehra.

46. A set of references to the debate is included in an article by Ivan T. Todorov, Einstein and Hilbert: The Creation of General Relativity, Ivan T. Todorov, http://lanl.arxiv.org/pdf/physics/0504179v1.pdf

47. Hamiltonsches Princip und allgemeine Relativitätstheorie, A. Einstein, *Sitzungsberichte der Königlich-Preussischen Akademie der Wissenschaften zu Berlin*, 1916; translated by W. Perrett & G. B. Jeffery, as Hamilton's Principle and the General Theory of Relativity, Ibid., Sommerfeld, 1923, Dover.

48. The Relativity Effect in Planetary Motions, G. M. Clemence, *Reviews of Modern Physics*, *19*, 361–63, 1947. Although the observed precession values cited here correspond to a re-evaluation in 1947, they do not appreciably differ from those available to Einstein in 1915.

49. Ibid., Die Grundlage . . . , Einstein.

50. Ibid., Erklärung . . . , Einstein.

51. On Einstein's Theory of Gravitation, and its Astronomical Consequences, W. de Sitter, *Monthly Notices of the Royal Astronomical Society*, *76*, 699–728, 1916; *77*, 155–84, 1916; *78*, 3–28, 1917.

52. Sir Arthur Stanley Eddington, O.M., F.R.S., Obituaries written by several astronomers in *The Observatory*, *66*, 1–12, February 1945.

53. A Determination of the Deflection of Light by the Sun's Gravitational Field, from Observations at the Total Eclipse of May 29, 1919, F. W. Dyson, A. S. Eddington, & C. Davidson, *Philosophical Transactions of the Royal Society of London*, *220*, 291–333, 1920.

54. Joint Eclipse Meeting of the Royal Society and the Royal Astronomical Society, *The Observatory*, *42*, 388–98, 1919.

55. An A-Type Star of Very Low Luminosity, Walter S. Adams, *Publications of the Astronomical Society of the Pacific*, *26*, 198, 1914.

56. The Densities of Visual Binary Stars, E. Öpik, *Astrophysical Journal*, *44*, 292–302, 1916.

57. On the Relation between the Masses and Luminosities of the Stars, A. S. Eddington, *Monthly Notices of the Royal Astronomical Society*, *84*, 308–32, 1924.

58. The Relativity Displacement of the Spectral Lines in the Companion of Sirius, Walter S. Adams, *Proceedings of the National Academy of Sciences of the USA*, *111*, 382–89, 1925.

59. Apparent Weight of Photons, R. V. Pound & G. A. Rebka, *Physical Review Letters*, 4, 337–41, 1960.

60. Kernresonanzfluoreszenz von Gammastrahlung in Ir^{191}, R. L. Mössbauer, *Zeitschrift für Physik*, *151*, 124–43, 1958.

61. Hubble Space Telescope spectroscopy of the Balmer lines in Sirius B, M. A. Barstow, et al., *Monthly Notices of the Royal Astronomical Society*, *362*, 1134–42, 2005.

62. Kosmologische Betrachtungen zur allgemeinen Relativitätstheorie, (Cosmological Considerations on the General Theory of Relativity) A. Einstein, *Sitzungsberichte der Königlich-Preussischen Akademie der Wissenschaften zu Berlin*, 142–52 (1917) translated from the German in *The Principle of Relativity*, A. Sommerfeld (ed.), Mineola NY: Dover Publications.

63. On Einstein's Theory of Gravitation and its Astronomical Consequences, W. de Sitter, *Monthly Notices of the Royal Astronomical Society*, *78*, 3–28, 1917.

64. Ibid., On Einstein's Theory, de Sitter.

65. Council note on The Motions of Spiral Nebulae, Arthur Stanley Eddington, *Monthly Notices of the Royal Astronomical Society*, *77*, 375–77, 1917.

66. The 1920 Shapley-Curtis Discussion: Background, Issues, and Aftermath, Virginia Trimble, *Publications of the Astronomical Society of the Pacific*, 1133–44, 1995.

67. An Estimate of the Distance of the Andromeda Nebula, E. Oepik, *Astrophysical Journal*, *55*, 406–10, 1922.

68. First Determination of the Distance and Fundamental Properties of an Eclipsing Binary in the Andromeda Galaxy, Ignasi Ribas, et al., *Astrophysical Journal*, *635*, L37–L40, 2005

69. Ibid. The 1920 Shapley-Curtis Discussion, Trimble, see p. 1142.

70. *My World Line*, George Gamow. New York: Viking Press, 1970, pp. 42–45.

71. Über die Krümmung des Raumes, A. Friedman, *Zeitschrift für Physik*, *10*, 377–86, 1922.

72. Über die Möglichkeit einer Welt mit konstanter negativer Krümmung des Raumes, A. Friedmann, *Zeitschrift für Physik*, *21*, 326–32, 1924.

73. Analysis of Radial Velocities of Globular Clusters and Non-Galactic Nebulae, Gustaf Strömberg, *Astrophysical Journal*, *61*, 353–62, 1925.

74. Beobachtungen über den Helligkeitseindruck von Nebelflecken und Sternhaufen in den Jahren 1886 bis 1906, J. Holetschek, *Annalen der K. & K. Univeristäts-Sternwarte Wien (Wiener Sternwarte)*, *20*, 1–8, 1907, digitally available through the Hathi Trust Digitial Library.

75. Extra-Galactic Nebulae, Edwin Hubble, *Astrophysical Journal*, *64*, 321–69 and plates P12–P14, 1926.

76. Un univers homogène de masse constante et de rayon croissant, rendant compte de la vitesse radiale des n'ebuleuses extra-Galactiques, Abbé G. Lemaître, *Annales de la Société scientifique de Bruxelles, Série A*, *47*, 49–59, 1927

77. *Cosmology and Controversy – The Historical Development of two Theories of the Universe*, Helge Kragh. Princeton University Press, 1996, pp. 29–30.

78. *Discovering the Expanding Universe*, Harry Nussbaumer & Lydia Bieri. Cambridge University Press, 2009.

79. Mystery of the missing text solved, Mario Livio, *Nature*, *479*, 171–73, 2011.

80. Who Discovered the Expanding Universe?, Helge Kragh & Robert W. Smith, *History of Science*, *41*, 141–62, 2003.

81. A Relation between Distance and Radial Velocity among Extra-Galactic Nebulae, Edwin Hubble, *Proceedings of the National Academy of Sciences of the USA*, *15*, 168–73, 1929.

82. A Homogeneous Universe of Constant Mass and Increasing Radius accounting for the Radial Velocity of Extra-galactic Nebulae, Abbé G. Lemaître, *Monthly Notices of the Royal Astronomical Society*, *91*, 483–90, 1931.

83. Ibid., *My World Line*, Gamow.
84. Alexander Friedmann and the origins of modern cosmology, Ari Belenkiy, *Physics Today*, *65*, 38–43, October 2012.
85. Zum kosmologischen Problem der allgemeinen Relativitätstheorie, A. Einstein, *Sitzungsberichte der Deutschen Akademie der Wissenschaften*, 235–37, 1931.
86. Erklärung der Perihelbewegung des Merkur aus der allgemeinen Relativitätstheorie, A. Einstein, *Sitzungsberichte der Königlich-Preussischen Akademie der Wissenschaften zu Berlin*, *44*, 831–39, 1915.
87. Albert Einstein 14 Maart 1878 - 18 April 1955, A. D. Fokker, *Nederlands Tijdschrift voor Natuurkunde*, *21*, 125–29, 1955.
88. Einstein's Impact on Theoretical Physics Chen-Ning. N. Yang, *Physics Today*, *33*, 42–49, June 1980.
89. Raum und Zeit, H. Minkowski, *Physikalische Zeitschrift*, *10*, 104–11, 1909.
90. Ibid., *Subtle is the Lord*, Pais p. 152.
91. Ibid., *Subtle is the Lord*, Pais p. 224.
92. Autobiographische Skizze in *Helle Zeit, Dunkle Zeit - in memoriam Albert Einstein*, A. Einstein (Carl Seelig, editor). Zürich: Europa Verlag, 1956, p.16.
93. Autobiographical Notes, Albert Einstein, Paul A. Schilpp ed., 1951, 2nd edition. New York: Tudor Pub., 1951, pp. 14–17.

4

Conclusions Based on a Premise

Scientific work is generally based on a premise, a hunch that certain assumptions can safely lead to useful advances.[a] As new observations, experimental evidence, or calculations accumulate, the premise may then lead to a coherent view that includes predictions concerning future observations. When those observations only partially vindicate the evolving world view, corrections or modifications may be necessary, and so an increasingly complex world view emerges, whose intricate details need to be understood in terms of an overarching theory. Where the theory appears enigmatic, new physical processes may be hypothesized to make the theory more understandable. A search is set in motion to identify those processes, and so the work continues.

Two Legacies from the Nineteenth Century
The Chemical Elements in the Sun

In 1859, the Heidelberg physicist Gustav Robert Kirchhoff and his colleague, the chemist Robert Wilhelm Bunsen, discovered that a minute trace of strontium chloride injected into a flame gave rise to a spectrum that clearly identified the presence of the element strontium. Sodium was also readily identified by a spectrum exhibiting two bright, closely spaced features, whose wavelengths corresponded precisely to two dark features in the spectrum of the Sun.

Later that year, Kirchhoff passed a beam of sunlight through a flame containing vaporized sodium. To his surprise, instead of filling the dark solar features with compensating light emitted by the flame, the solar features actually darkened. The faint light emanating from the Sun appeared to be absorbed by the flame.

---

[a] Readers seeking the broader setting for events recounted in this chapter will find informative accounts in the articles of David H. DeVorkin and Ralph Kenat; DeVorkin's biography of Henry Norris Russell; and Karl Hufbauer's review of stellar structure and evolution early in the twentieth century.[1,2,3,4]

The conclusion was inescapable, particularly for some of the elements pursued with greatest vigor in the laboratory. On behalf of Bunsen and himself, Kirchhoff wrote, "The dark D lines in the solar spectrum lead to the conclusion that the solar atmosphere contains sodium."[b] The same conclusion also held for potassium.[5]

Kirchhoff and Bunsen later were able to show that other metals, including iron, magnesium, calcium, chromium, copper, zinc, barium, and nickel were also present in the Sun. They divined that no longer would the constitution of the stars be unfathomable; step by step we'd finally know the chemical constitution of the stars!

One can only imagine their excitement and awe. Nature's silence had finally been broken. Enthusiastically they announced that we were now poised to determine the chemical composition of the world "far beyond the limits of the earth, or even our solar system."[6]

The Temperature of Stars

Although astronomers of the late nineteenth century had begun to estimate stellar temperatures by matching the color of a star to that of a thermally glowing solid of known temperature, the process had been largely empirical.

A second empirical approach was that of the Viennese scientist Josef Stefan, whose remarkably accurate estimate of the Sun's surface temperature ~6000 degrees Kelvin was based on his determination that the power radiated by a heated body was proportional to the fourth power of the temperature, σT^4, where σ is a universal constant that holds for any perfectly black surface.[7] But few astronomers of the nineteenth century had much confidence in Stefan's work, and so competing estimates of the Sun's surface temperature could range over many tens of thousands of degrees Kelvin.

On December 14, 1900, Max Planck, professor of theoretical physics at the University of Berlin, gave a talk to the German Physical Society in Berlin, in which he provided a theoretical derivation of the experimentally observed spectrum of *blackbody radiation*, the radiation emitted by a heated, perfect black body in thermal equilibrium with its surroundings.[8] Planck referred to this spectrum as the *Normalspectrum*, but the term *blackbody spectrum* soon became universally accepted. In this context, the depiction "black" indicated the ability to absorb completely all incident radiation. Planck's derivation represented a culmination of work on thermal radiation that had been gradually emerging in the late nineteenth century from concerted efforts in thermodynamics, *statistical mechanics*, the *kinetic theory of gases*, and the electromagnetic theory of James Clerk Maxwell.[9]

Planck's work, and particularly his demonstration that Stefan's law could be derived on the basis of the electromagnetic theory of light, gave strong support to the use of that law for deriving temperatures. Nevertheless, astronomers felt

---

[b] [D]ie dunkeln Linien D im Sonnenspektrum lassen daher schliessen, dass in der Sonenatmosphäre Natrium sich befindet.

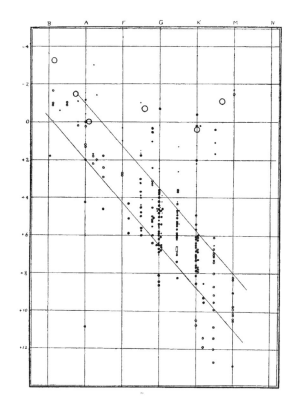

Figure 4.1. The original *Hertzsprung–Russell diagram* for stars with known *parallaxes*, presented by Henry Norris Russell at a joint meeting of the Astronomical and Astrophysical Society of America and Section A of the American Association for the Advancement of Science at Atlanta, Georgia on December 30, 1913. Plotted from left to right are stars whose colors range from blue to red, designated by *spectral types* B, A, F, … M, N. Plotted from top to bottom are the *absolute magnitudes* of the plotted stars, ranging from the most luminous, -4, to the faintest, roughly +9. Different symbols refer to stars located in different stellar groupings on the celestial sphere. (*Reprinted by permission of Macmillan Publishers Ltd from: 'Relations Between the Spectra and Other Characteristics of the Stars,' Henry Norris Russell, Nature 93, 252, 1914 ©1914 Nature Publishing Group*).

uncertain about stellar temperatures partly because no star's spectral energy distribution precisely matches that of a blackbody. As a compromise, they agreed that it was possible to obtain an *effective temperature*, T_{eff}, from the total radiation escaping from the star's surface, as long as this temperature was defined as $(F/\sigma)^{1/4}$, where F is the total power – the radiant flux – escaping unit stellar surface area per second, that is, $F = \sigma T_{eff}^4$.

Giant Stars and Dwarfs

In 1909 Ejnar Hertzsprung in Copenhagen called attention to a new insight. Stars appeared to be grouped in two distinct classes, *giants* and *dwarfs*.

A few stars that appeared to share the same warm colors, and therefore the same temperature and *surface brightness* of others, had very low parallaxes and must therefore be more distant; yet they had apparent magnitudes similar to those of nearer stars with high parallaxes. These rare distant stars had to be far larger than those nearby. In Hertzprung's depiction, they were the whales among the fish.[10]

Hertzsprung's finding took some time to sink in. But four to five years later, in 1913-14, Henry Norris Russell in Princeton came to the same conclusion.[11] Using additional information on stellar masses deduced from the orbits of stellar binaries, and temperatures based on color, he found that (i) the masses of giants and dwarfs of similar color do not differ greatly, (ii) the surface brightness of stars diminishes rapidly with increasing redness, and (iii) *red giants* can have diameters a hundred times larger and densities a million times smaller than those of red dwarfs. Hertzsprung's analogy to whales and fish was not far amiss. Most notably, in an address at a joint meeting of the Astronomical and Astrophysical Society of America and Section A of the American Association for the Advancement of Science at Atlanta, Georgia on December 30, 1913, Russell exhibited these characteristics in a chart, Figure 4.1, now known as the Hertzsprung–Russell diagram, one of the most informative astronomical drawings produced in the course of the twentieth century. His talk, "with a few additions," was reported in three separate issues of *Nature* starting with the April 30, 1914 issue.[12]

The Sun, a spectral type G star, is one of the dwarfs, a class of stars that line the lower, redder portion of the *main sequence* of stars running from the upper left to the lower right in Figure 4.1. The group comprising the giants stretches from stars somewhat more luminous than the Sun and bluer, to luminous stars that are progressively redder and lie on a *red giant branch*, which starts at the main sequence and runs upward and to the right. The Hertzsprung–Russell diagram became one of the most enduring legacies of early twentieth century astrophysics. But, for a number of years, its full significance and implications remained unclear. The prevalence of particular spectral lines in stars of different luminosities also remained unexplained.

Niels Bohr and the Theory of Atomic Spectra

A critical breakthrough of 1913 led to entirely new ways of studying the temperature and chemical abundances of stellar atmospheres. Throughout the fall of 1912, the 27-year-old Danish physicist Niels Bohr was searching for a way to understand the structure of atoms and the radiation they emitted. He had spent the previous year at Cambridge and Manchester, to work, respectively with J. J. Thomson and Ernest Rutherford, whose work had produced two quite different models of atomic structure. Bohr was searching for ways to combine the two approaches, each based on its own set of experimental observations. Thomson's model of 1898 pictured the atom as constituted of a smoothly distributed positive charge in which the electrons were somehow embedded like raisins in an otherwise homogeneous cake. This model seemed unlikely to Bohr, given that Rutherford's experiments on the scattering of beams of *alpha particles* had shown that atoms have a point-like, massive, positively charged nucleus.

Bohr's contribution was based on a physical insight into a relation first discovered empirically in 1885 by the Swiss physicist and teacher in a girl's secondary school, Johann Jakob Balmer. It related the wavelengths λ at which hydrogen emitted spectral lines by a formula that could be rewritten as

$$\frac{c}{\lambda} = R\left(\frac{1}{b^2} - \frac{1}{a^2}\right), \tag{4.1}$$

where c is the speed of light, making c/λ the frequency of light. R is a constant named after the Swedish spectroscopist Johannes Robert Rydberg; b is an integer that can take on values $1, 2, 3, \ldots$; and a is an integer greater than b.

Bohr's great achievement was to provide a physical model of hydrogen that led to an evaluation of R in terms of fundamental constants of nature

$$R = \frac{2\pi^2 me^4}{h^3}, \tag{4.2}$$

where m and e are the mass and charge of the electron, and h is Planck's constant.

Bohr's formula was based on a model of hydrogen in which an electron can stably orbit a much more massive nucleus in any of a number of discrete circular orbits. The radius of each orbit is determined by the electron's energy; the electrostatic attraction of the electron's negative charge to the positively charged nucleus; and the requirement that the angular momentum of the electron in any of its stable orbits be an integer multiple of $\hbar \equiv h/2\pi$.

In jumping from an orbit where its energy is E_a to another, where its energy has a lower value E_b, the electron radiates at a wavelength λ_{ab}. To obtain Balmer's results one needed to set $b = 2$; a had to be an integer greater than 2; and the radiation frequency had to be related to the difference in the orbital energies of the electrons in their respective orbits, $\nu_{ab} \equiv c/\lambda_{ab} = (E_a - E_b)/h$.

This made good sense, as Einstein, in a 1905 paper in which he had explained the emission of electrons from a metallic surface illuminated by ultraviolet radiation and a number of other interactions of light with matter, had found the energy E of an absorbed quantum of light to equal the Planck constant h multiplied by the frequency of the absorbed light ν[13]:

$$E = h\nu. \tag{4.3}$$

There was further evidence as well. In 1896 Charles Pickering at Harvard had found a series of lines in the spectrum of the star $\zeta Puppis$ that did not quite fit Balmer's formula but were thought to be due to hydrogen. Bohr pointed out that these could be attributed to singly ionized helium, for which the nuclear charge

Z would be twice as high as for hydrogen and his formula for R would become

$$R_Z = \frac{2\pi^2 m Z^2 e^4}{h^3} . \tag{4.4}$$

For the nuclear charge of hydrogen, Bohr took $Z = 1$, and $R_{Z=1} = 3.1 \times 10^{15}$ cycles per second, which agreed with observations to within experimental uncertainties for the inserted values of m, e, and h. This was a strong indication that he was on the right track. For singly ionized helium the nuclear charge would be $Z = 2$, and $R_{Z=2} = 4R_{Z=1}$.

Bohr's work appeared in three articles, all published in 1913. In these he expanded his findings on hydrogen to also cover other atomic and molecular topics.[14,15,16]

Responding in the journal *Nature* to the first of Bohr's articles, the British spectroscopist Alfred Fowler asserted that Bohr's conjecture associating some of the observed solar emissions with spectra of helium could not be correct. The laboratory spectra for helium quite clearly showed that the ratio $R_{Z=2}/R_{Z=1}$ was 4.0016, not 4.000 as Bohr had claimed.[17]

In a letter to *Nature* published only four weeks later, Bohr replied that the value of $R_{Z=2}$ his paper had cited was based on the simplifying assumption that the nuclear mass M was infinitely more massive than the mass of the electron. If the actual value for the nuclear masses M for hydrogen and helium were inserted in his theory, and the electron's mass m in his formula was replaced by its *reduced mass*, $m_r = mM/(m+M)$, a quantity familiar to workers on the orbital theory of planetary systems, then the ratio $R_{Z=2}/R_{Z=1} = 4$ would be replaced by 4.00163.[18]

Fowler responded in the same issue of *Nature*, generously writing, "I am glad to have elicited this interesting communication from Dr. Bohr, and I readily admit that the more exact form of his equation given above is in close accordance with the observations of the lines in question."[19]

Bohr's two main results, the derivation of R_Z and the undoubted identification of the helium spectra through the more detailed calculation of the ratio $R_{Z=2}/R_{Z=1}$, could not be ignored. Einstein had been skeptical of the original derivation of R_Z. But on hearing about the helium results from the Hungarian-born chemist, Georg von Hevesy, he reportedly replied, "This is an enormous achievement. The theory of Bohr must then be right."[20]

Bohr's theory, consistent with both Planck's and Einstein's earlier quantum considerations, in some sense *explained*, albeit in a totally unfamiliar logic at the time, why atomic matter absorbed and emitted radiation solely at discrete wavelengths. It also made clear that high-lying energy levels could be populated only by infusing atoms with considerable amounts of energy, meaning that these levels were likely to be populated solely in stellar atmospheres at high temperatures.

Henry Norris Russell and the Chemical Elements of the Sun, Earth, and Meteorites

By the start of the twentieth century, it had become clear that the material that made up the Sun bore considerable resemblance to that on Earth. But the solar spectrum also exhibited puzzling features not duplicated in the laboratory.

Not to be dissuaded by such anomalies, astronomers assumed that the relative abundances of chemical elements in the Sun were similar to those on Earth. The means were not at hand to test this, but Henry Norris Russell at Princeton certainly thought so. It became his premise, the bedrock of his research, for the 15 years from 1914 to 1929.

* * *

Russell, born in 1877 into a Presbyterian minister's family, had received both his undergraduate and graduate education at Princeton, where he completed his

Figure 4.2. Henry Norris Russell, his wife Lucy May, and their four children, about the time he was devising the Hertzsprung–Russell diagram and assembling his initial table of elemental abundances. (*Courtesy of the American Institute of Physics Emilio Segre Visual Archives, Margaret Russell Edmonson Collection*).

doctoral degree in 1899. Returning to Princeton in 1905, from postdoctoral work in England, he soon became the leading figure in American astrophysics.

In 1914, the 37-year-old Russell, by then Director of the Princeton University Observatory, published a comparison of elements from three sources: a compilation of elements that had by then been discovered in the Sun; a listing of elements present in the crust of Earth; and a census of elements found in meteorites – planetary system material that had fallen on Earth.[21] Russell noted the striking resemblance particularly between the first two in the table reproduced in Table 4.1: "[F]ifteen of the sixteen leading metallic elements are common to the two lists." Referring to the eminent Johns Hopkins physicist Henry Augustus Rowland, who in 1899 had become the American Physical Society's first president and had many years earlier compiled the first list of elements found in the Sun, Russell concluded, "In spite of … exceptions, the agreement of the solar and terrestrial lists is such as to confirm very strongly Rowland's opinion that, if the earth's crust should be raised to the temperature of the sun's atmosphere, it would give a very similar absorption spectrum. …"

Eddington and the Internal Constitution of the Stars

By 1916 the Cambridge astronomer Arthur Stanley Eddington had found a tentative way to explain the difference between dwarf and giant stars.[22] He argued that for the low-density giants the pressure exerted by radiation should be significant compared to pressures exerted by their tenuous gas. He further assumed that the transport of heat is the same in all stars and occurs through a flow of radiation from an intensely hot interior to cooler surface layers. The transport of heat through conduction and convection he took to be negligible.

But radiation does not escape as though the star were transparent, and Eddington assumed the existence of an absorption coefficient k per unit mass and unit area, that is, per unit *column density*. He thought that there must be some supply of energy, in the interior of the star, that might be due to radioactivity or simply the generation of heat through stellar contraction, and designated by ϵ the energy generated by unit mass. He admitted not knowing anything about either k or ϵ, but assumed the product $k\epsilon$ to be constant throughout the inner portions of the star.

Eddington further assumed hydrostatic equilibrium, that is, that the pressure at each point within the star is sufficiently high to withstand further compression due to the weight bearing down on it from matter at greater radial distances from the center. This enabled him to make use of a set of equations for the radial distribution of temperature and density in stars compiled in the 1907 treatise 'Gaskugeln' by the Swiss astrophysicist Robert Emden.[23] Eddington's only other assumption was that the star's temperature was much lower in its surface layers

Table 4.1. *Table of Elements published by Henry Norris Russell in 1914*

Solar Spectrum, Dark Lines (Howland)		Chromo-sphere, Bright Lines (Mitchell)	Earth's Crust, Outer 10 Miles (Clarke)		Stony Meteorites (Merrill)	
1	Ca	Fe	O	49.85%	O	35.75%
2	Fe	Ti	Si	26.03	Fe	24.52
3	H	H	AL	7.28	Si	18.20
4	Na	Cr	Fe	4.12	Mg	13.80
5	Ni	Ca	Ca	3.18	S	1.85
6	Mg	V	Na	2.33	Al	1.45
7	Co	Sc	K	2.33	Ca	1.25
8	Si	Zr	Mg	2.11	Ni+Co	1.32
9	Al	C	H	0.97	Na	0.70
10	Ti	Mn	Ti	0.41	Cr	0.34
11	Cr	Mg	Cl	0.20	K	0.27
12	Sr	Ni	C	0.19	P	0.11
13	Mn	Ce	P	0.10		
14	V	Nd	S	0.10		
15	Ba	He	F	0.10		
16	C	Co	Ba	0.09		
17	Sc	Y	Mn	0.08		
18	Y	Sr	Sr	0.03		
19	Zr	Ba	Cr	0.025		
20	Mo	La	Ni+Co	0.018		
21	La	Sa	V	0.015		
22	Nb	Al	Zr	0.013		
23	Pd	Er	Cu	0.010		
24	Nd	Gd	Zn	0.004		
25	Cu	Na	Li	0.004		
26	Zn	Si	Pb	0.002		
27	Cd	Eu	Br	0.0006		
28	Ce	Zn	As	0.0004		
29	Gl	Dy	Cd	0.00002		
30	Ge	Cu				
31	Rh	Pr	Allowance for all other elements 0.38			
32	Ag	Nh				
33	Sn					
34	Pb					
35	Er					
36	K					

Note: The perceived similarities in the relative abundances of the elements he listed convinced Russell that the composition of the Sun must be similar to that of Earth.
Source: Courtesy of the American Association for the Advancement of Science.

than deep in its interior, where typical temperatures ranged into the millions of degrees Kelvin.

In a second paper published the same year, 1916, Eddington corrected an important error he had earlier failed to notice. He realized now that atoms deep in the interior of a star must be fully ionized.[24] The use of Bohr's model, published three years earlier, told him that the gas pressure at the high internal temperatures could not be proportional just to the density of atoms. Instead, in totally ionized matter, it would be proportional to the number density of particles. For a typical atom this increased the number density by a factor $n_e + 1$, where n_e was the number of electrons per atom, and one additional particle accounted for the atomic nucleus.

This raised the gas pressure by more than an order of magnitude because, like Russell in 1914, Eddington assumed stars to consist of heavy elements similar to those on Earth. For fully ionized iron, which has 26 electrons, the pressure would increase 27-fold.

* * *

Two years later, in 1918, Eddington returned to the same subject. By now he had streamlined his presentation and was able to simplify the subject considerably.[25] He still disregarded the possibility of convection and assumed that energy is radiatively transported from the interior of a star to its surface, a condition that we now know holds only in certain portions of some stars. Other stars can be convective throughout, so that his assumptions should not apply. He was aware of this limitation, but his contribution nevertheless provided a new and far-reaching approach.

Eddington's key assumption was that only elements heavier than hydrogen should be present. With this assumption, he could at once determine the mean mass of all the atomic constituents in the star. All elements more massive than hydrogen have a nuclear mass close to $2Am_p$, where A is the number of electrons surrounding the nucleus, and m_p is the nuclear mass of hydrogen – the mass of a proton. The theory held together nicely, independent of the precise chemical composition of stars, as long as they consisted primarily of heavy elements, that is, as long as hydrogen was excluded.

* * *

Eddington now showed that the *luminosity* of a star L could be related to its mass M by an equation

$$L = \frac{4\pi cGM(1-\beta)}{k}, \quad \text{where } (1-\beta) \propto M^2 m^4 \beta^4 . \tag{4.5}$$

Here k, as earlier, is the absorption coefficient for unit column density; c is the speed of light; G is the gravitational constant; M is the star's mass; m is the mean mass of the particles; and β is the fraction of the pressure contributed by gas, the remaining fraction, $(1-\beta)$, being the fraction contributed by radiation.

In Eddington's *mass–luminosity relation* (4.5), the assumption that stars consist mainly of heavy elements meant that the mean mass of particles was approximately $m \sim 2$ *atomic mass units*, or roughly two protonic masses $2m_p$. If the gas was not fully ionized, Eddington estimated m to be somewhat higher, $m = 2.8m_p$. Finally, he also assumed the absorption coefficient k, solely a property of the ionized matter, to be constant throughout the star.

Because equation (4.5) is independent of the star's radius R, Eddington concluded that β could depend only on the star's mass M. In turn, this meant that the product Lk had to be identical for stars of identical mass. This was significant. By comparing the luminosities of the Sun and that of a giant star with a comparable color, or surface temperature, Eddington could obtain two equations enabling him to determine separately the values of both β and k.

For the Sun and a giant of similar color, Eddington's respective estimated masses were 1 and 1.5 $M_\odot$. The respective values of $(1-\beta)$ were 0.106 and 0.174. In his estimate for k Eddington concluded that an absorbing layer containing as little as 1/23 gram per square centimeter would reduce the transmitted radiation to "about a third of its original intensity."[26]

Eddington's theory of 1918 appeared to be a tour de force!

It explained the observed relations between a star's luminosity and its mass, and the differences between giants and dwarf stars. At least the agreement with observations indicated as much.

Regrettably, the details of the theory depended on the assumption that stars are composed largely of terrestrial materials; that they generate energy throughout the star rather than primarily in its central regions; and that the outward transport of heat is dominated by radiative transfer, rather than partly or primarily by convection.

An unfortunate consequence of the success of Eddington's theory was that it confirmed Russell's conviction that stars must have a terrestrial chemical composition. In turn, Russell's convictions about that composition, as we shall see, convinced Eddington that his theory of the structure of stars and their transfer of radiation must be correct as well. The two titans of astrophysics, one in England and the other in the United States, often stood shoulder to shoulder in guiding the thrust of astrophysics – and sometimes misguiding it.

Saha and the Spectra of Ionized Gases

Even as Eddington was trying to understand stellar interiors, Bohr's work and an increasing body of laboratory work began to provide also clues to the chemical nature of stellar atmospheres. Most important to this end was a striking new insight by a young student in India.

Meghnad Saha was born in 1893 as the fifth child in a shopkeeper's family in a village near Dhaka in Eastern Bengal, today's Bangladesh. The young Saha struggled to receive an education.[27] Despite losing a scholarship and getting suspended from school for taking part in anti-British protests, Saha eventually obtained a DSc in applied mathematics at the Calcutta Presidency College in 1919, while teaching thermodynamics and spectroscopy at the newly established University College of Science in Calcutta.

Working in virtual isolation on his doctorate, he had by 1919 established a kinetic theory of ionized gases that provided the first clear insight on physical conditions in the atmospheres of stars. In ground-breaking papers on the theory of stellar spectra, the first of which he published in early 1920, he explained the conundrum he had resolved[28]:

> Up to this time thermodynamics has been confined to the treatment of physical processes like liquefaction and vaporization, or chemical processes like decomposition or dissociation of molecules into atoms. It has thus carried us up to the stage where all substances are broken up into constituent atoms. But what takes place if a gaseous mass consisting of atoms only be continued to be heated?...[T]he first effect of increasing temperature will be to tear off the outermost electron from the atomic system...The next point in question is the identification of the ionised elements in the physical systems in which they may occur. Spectroscopically it is quite an easy matter, for the ionised elements show a system of lines which is quite different from the lines of the neutral atom."

He then proceeded to portray what happens as atoms are gradually excited by continuous heating; the excited energy levels from which they begin to absorb or re-emit radiation; the ionization they eventually undergo as the heating continues, permitting them to begin radiating or absorbing radiation from a series of totally different energy levels belonging to the ionized atom; and how the equilibrium between these various levels can be calculated as a function of temperature and pressure.

In this calculation he showed that the pressure of the gas plays a critical role because electrons released through ionization recombine more readily with the ions at higher pressures. As a result of such calculations, Saha explained, "The physical meaning of the appearance and disappearance of (spectral) lines...now becomes apparent." He tied his theory to observed stellar spectra and concluded, "We have practically no laboratory data to guide us, but the stellar spectra may be regarded as unfolding to us, in an unbroken sequence, the physical processes succeeding each other as the temperature is continually varied from 3000 K to 40,000 K."

Commenting on Saha's approach a few months after its publication, an enthusiastic Russell in 1921 referred to "the immense possibilities of the new field of investigation which opens before us. A vast deal of work must be done before it is even prospected – much less worked out, and the astronomer, the physicist, and the chemist must combine in the attack, bringing all their resources to bear upon this great problem which is of equal importance to us all."[29]

Russell perceived more clearly than most of his contemporaries that astronomy could advance quickly only if it could import the tools and invoke the efforts of the much larger population of scientists working in related fields. Two like-minded others were the Cambridge mathematician and theoretical physicist, Ralph Howard Fowler, a leading researcher on thermodynamics and statistical

Figure 4.3. Cecilia Payne at Harvard in 1924. (*Courtesy Collection of Katherine Haramundanis*).

mechanics, and his colleague, the theoretical astrophysicist Edward Arthur Milne. Together, they refined Saha's theory, finding by 1923 that gas kinetic pressures in stellar atmospheres were far lower than Saha and most astronomers had assumed.

A low-pressure atmosphere was required by the wealth of spectral lines that could be observed at different depths in the atmosphere, corresponding to higher temperatures and thus higher states of ionization at increasing depths. Penetration of radiation to such depths was possible only in highly tenuous atmospheres, in which gas pressures were about ten thousand times lower than on Earth. This meant that radiation pressures, and not just gas kinetic pressures, would need to be considered in modeling stellar atmospheres.[30]

Cecilia Payne's Astonishing Insight on the Chemical Composition of the Stars

Cecilia H. Payne, a graduate student from England, arrived at Harvard in 1923 after undergraduate studies at Cambridge with Eddington. In Cambridge, she had become familiar with the work of Fowler and Milne, and for her PhD thesis she undertook to explore fully their theory and its consequences. She convinced herself that their improvements to Saha's work provided a new way forward. A first task would be to derive the temperatures of stars of different spectral types in the Harvard spectral classification system of stars.

With the large collection of stellar spectra available to her in the Harvard Observatory collection, she was ideally positioned to do this. Soon, she succeeded in establishing a temperature scale for stars of different spectral types.

This in itself was an impressive achievement. But with her work on stellar temperatures completed, Payne set out to apply the Fowler–Milne theory further to estimate the relative abundances of the major chemical constituents in the Sun.[31]

The strengths of spectral features vary with temperature as the relative populations of atomic and ionic levels evolve. Fowler and Milne had shown that the temperature at which a spectral line is barely detectable provides a good measure of the relative abundance of the element it represents; the amount of self-absorption in the lines is then negligible.[32] They also calculated temperatures and pressures at which maximum absorption in a spectral line of a particular state of ionization could be expected.[33]

Using the Harvard spectra, Payne determined the change in the strengths of different spectral lines as a function of the atmospheric temperatures she had just established. She knew that the methods she was using were rough and preliminary, but estimated that hydrogen was at least a hundred thousand times more abundant than elements such as silicon, calcium, or magnesium in their more probable states of ionization, and that singly ionized helium appeared to be more than a million times more abundant. We now know that her estimate for hydrogen was roughly a factor of ten too high, and that for helium a factor of about a hundred; but these strikingly high estimates were still far more suggestive than those of her contemporaries.

Harlow Shapley, Director of the Harvard College Observatory and Cecilia Payne's thesis advisor, forwarded a draft of her doctoral thesis to Russell, who had been Shapley's own thesis advisor at Princeton. Russell emphatically objected to the high hydrogen abundances Payne was advocating. He wrote, "…there is something seriously wrong with the present theory. It is clearly impossible that hydrogen should be a million times more abundant than the metals," and then referred to a recent paper he had published with his Princeton colleague, the physicist Karl Taylor Compton.[34,35]

In that 1924 letter to *Nature*, Russell and Compton maintained that a comparison of the absorption in the spectra of hydrogen and magnesium in the Sun, on the basis of adopted spectral theory, "would demand an absurdly great abundance of hydrogen relative to magnesium (itself an abundant element)." To solve this apparent paradox, they suggested that the *weight*, that is, the propensity of the state giving rise to Balmer absorption in hydrogen "is increased, in some special way by a very large factor."

The letter provided little beyond an emphatic opinion. We saw earlier how Russell had convinced himself in 1914 that the chemical abundances in the atmospheres of stars and the surface layers of Earth were rather similar.[36] He still held this opinion in 1922 when he wrote Saha "…we cannot be sure that the relative proportion of the elements on Earth is similar to that in the star, but I think it is, nevertheless, the best guide that we have."[37] In retrospect, Russell's and Compton's letter to *Nature* appears to have been a last desperate attempt to sustain the assumption that the chemical constitution of the Sun is roughly identical to that of Earth.

* * *

In her impressive PhD thesis, 'Stellar Atmospheres: A contribution to the observational study of matter at high temperatures,' published also as the first of the *Harvard Observatory Monographs* edited by Harlow Shapley, Payne compared the relative abundances of the elements in the stars to terrestrial abundances. For atoms more massive than hydrogen and helium she had found rough correspondence. Hydrogen and helium abundances, however, were orders of magnitude higher, as cited in her Table XXVIII.[38]

In response to Russell's outspoken criticism, the young doctoral student wrote, "Although hydrogen and helium are manifestly very abundant in stellar atmospheres, the actual values derived from the estimates of marginal appearance are regarded as spurious."[39] Two pages later she added, "The enormous abundance derived for these elements in the stellar atmosphere is almost certainly not real."[40] As if to explain these protestations, she referred to Russell and Compton's letter in *Nature*, and then quoted Russell writing her that "there seems to be a real tendency for lines, for which both the ionization and excitation potentials are large, to be much stronger than the elementary theory would indicate."[41]

On the one hand, she sounded unsure of the validity of her results. On the other, she attributed the distrust to Russell. It wasn't quite clear where she stood. In retrospect, this may have been the correct attitude. A radical new result calls for thorough, independent verification, and this was nowhere in sight.

Like Russell, Eddington was also skeptical: In her autobiography, Cecilia Payne recalls visiting Eddington in Cambridge after having completed her book-length thesis. She writes "In a burst of youthful enthusiasm, I told him that I believed that there was far more hydrogen in the stars than any other atom. 'You don't mean *in* the stars, you mean *on* the stars' was his comment.' In this case, indeed, I was right, and in later years he was to recognize it too."[42]

* * *

Payne's contributions were based on her amalgamation of observations and theoretical work. Her successes were based on her recognition of two available resources, which she ably converted into applicable tools. The first of these was the unparalleled collection of stellar spectra archived at the Harvard College Observatory. The second was her appreciation of the novel radiative transfer techniques that Ralph Fowler and E. A. Milne had recently developed in Cambridge.

Together, these resources allowed her to determine the atomic and ionic abundances of different elements in the atmospheres of stars. To her it was all quite straightforward: Hydrogen was by far the most abundant element in the atmospheres of stars, and presumably within them. Her finding and her care in applying her methods should at least have impressed astronomers as experienced as Henry Norris Russell and Arthur Stanley Eddington. And perhaps they did stimulate Russell to dig more deeply, even as they challenged his deepest convictions.

Where Eddington had erred in his theory of stellar structure was his failure to acknowledge that a variant of the theory could work equally well for stars consisting solely of hydrogen, rather than consisting solely of heavy elements as he had assumed. This evidently had not appeared likely to him because, like Russell, he instinctively was sure that stars could not consist largely of hydrogen. Stellar spectra clearly showed the presence of many of the heavy elements also found on Earth. Like Russell, Eddington assumed that these pronounced spectral features corresponded to high abundances of heavy chemical elements, and this made his simplifying assumption $m \sim 2m_p$ so attractive.

The contention that hydrogen vastly outnumbered other elements would have fundamentally changed Eddington's mass–luminosity relation, as this would mean that the mean mass of a star's constituent particles would become half the protonic mass, $m \sim m_p/2$. Because both the radiation pressure and the luminosity in his theory were proportional to the fourth power of mass m, that is, m^4, this would throw both values off by a factor as high as 256. Little wonder then that Eddington did not enthusiastically embrace Cecilia Payne's high abundance of hydrogen in stars. It threatened the beautiful edifice he had erected, as well as the classic book, 'The internal constitution of the stars,' he may have been readying for publication around the time of her visit.[43]

Eddington's formulation of the theory of stellar structure had been influential as long as little was known about chemical abundances and the actual densities, by mass and number, of prevalent ions in stars. His general approach continued to be central when more careful attention to elemental abundances, stellar opacities, and thermal convection were adopted; but in the form he had devised the theory, with the simplifications he had assumed, it was soon outdated.

Henry Norris Russell's Estimates of the Chemical Composition

It took Russell another four years to come to terms with the notion that hydrogen might be the most abundant element in the atmospheres of stars. He appeared indefatigable, studying how the widths of spectral lines might provide clues to abundances, and attacking the problem from many other vantage points. By 1927 his own former PhD student Donald Menzel, his assistant Charlotte E. Moore, Russell's younger Princeton colleague John Quincy Stewart, and the young German theorist Albrecht Unsöld who had just finished his PhD thesis under Arnold Sommerfeld in Munich, had all expressed a need for recognizing a high abundance of hydrogen in stellar atmospheres.[44] But Russell still had to justify a high abundance in his own way, perhaps for his own peace of mind.

Not until 1929 did Russell finally conclude (concede?) that the high hydrogen abundance was real. He summarized the justification for his conversion in a lengthy article in the *Astrophysical Journal*, in which he also provided his estimates for the relative abundances of 56 elements and six compounds.[45]

In table 16 of this long paper, Russell compared his derived abundances of the elements with those of "the most important previous determination of the abundances of the elements by astrophysical means … by Miss Payne" four years earlier. The "very gratifying agreement," as Russell termed it, between the two was surprisingly good, particularly because they had been derived, as he emphasized, by quite different means. Yet, as one notes with the hindsight of eight decades, both their sets of estimates often differ by orders of magnitude from modern values despite their close mutual agreement.

Somewhat dismayingly, perhaps, given Russell's emphatic rejection of a high hydrogen abundance only four years earlier, one finds no explicit acknowledgment from him that Payne had first realized the dominant abundance of hydrogen in stars.

Cecilia Payne may have resented the brush-off her work was given by the two leading astrophysicists of her time. Eddington, at least, had been a wonderful mentor to her when she was an undergraduate at Cambridge, at a time when women were generally discouraged from entering science. Even late in life she considered Eddington to be "the greatest man I have been privileged to know."[46] For Russell, her autobiography has no such kind words.

With her thesis completed, Payne became the first woman to receive her doctorate for work done at the Harvard College Observatory. She stayed on at Harvard which, despite her brilliance, denied her a decent position for another three decades. Only after Shapley and a sequence of university presidents had retired was she finally awarded a position her stature deserved. She became the first woman to be promoted to full professor on Harvard's Faculty of Arts and Sciences, albeit not until age 56, in 1956, after which she also became the first woman to head an academic department at Harvard. The role of women in academic life was arduous, and she had to surmount hurdle after hurdle where there were no precedents.[47]

* * *

A number of features of this story are important to remember.

First, it centered on the highest caliber, most qualified people astronomy had produced. Russell and Eddington were leaders in astrophysics. Payne was a brilliant young pioneer and role model for other women attracted to the field.

Second, the quick assimilation of early quantum theory, and the tools it provided, showed the importance that astronomers, and particularly Russell, placed on rapidly incorporating the most powerful theoretical tools and laboratory procedures in furthering their work. Conversely, the welcome that the work of Bohr, Saha, and Fowler and Milne received encouraged other physicists to take an interest in applying their talents to solving astronomical problems.

Finally, Russell's conversion to the high abundance of hydrogen shows that even a false premise can ultimately be overcome. It may have involved a tortuous path, and taken a measure of realism and perhaps grudging humility, to reverse himself so completely, but ultimately Russell made it work.

* * *

In many ways, Russell understood, long before his contemporaries, that astronomy would advance more quickly if astronomers joined forces not only among observers and theorists, but also with physicists and potentially chemists, to solve major astronomical problems. He was quick to accept Saha's novel ideas on the structure of ionized atoms and their significance to stellar spectroscopy. He worked closely with physicists, including the Harvard spectroscopist F. A. Saunders, with whom he published an influential paper on the theory of spectra of alkaline-earth atoms – among them calcium, strontium, and barium, all of which have a single pair of valence electrons.[48]

In this vein, he also worked tirelessly with observers at Mt. Wilson Observatory, establishing a joint venture that would permit him access to spectra he would need for his own work. He placed former students in positions where he might influence them and their associates. And he supported younger people working on astronomical problems that interested him.

Two of Russell's PhD students, Donald Howard Menzel, later Director of the Harvard College Observatory, and Lyman Spitzer, who eventually succeeded Russell at Princeton, were among the first American-educated astronomers well grounded in modern physics and thus able to take a deeper approach to astrophysical problems.

Russell's biographer, David DeVorkin, provides this summarizing insight on Russell[49]:

> At the midpoint of his career, he set out to rationalize astrophysics, raising it out of its still largely empirical framework ... of astronomical spectroscopy ... to the analytic, causal, and interpretive frame of modern astrophysics. He worked at the interface between observation and theory, making theory more accessible to observing astronomers ... As a result he helped transform American astrophysics into a wholly physical discipline. Establishing this framework, more than any one discovery or application that bears his name, is Russell's greatest legacy to Astronomy. ...

> All-important for a pragmatic theorist like Russell was knowing how to craft technique and evidence so they would be acceptable to the community. To be acceptable something had to be useful. Utility was as important as validity as long as it promised to lead to new knowledge in an accessible manner. This is where the essence of Russell's contribution lies. ...

There are many paths to power, through merit or political patronage, past track record, powers of persuasion, ability to marshal evidence, successful recruitment of allies, and institutional affiliation. Russell's power was a product of all of these, but it rested most firmly upon merit.... [H]e knew how to forge intellectual alliances, gain trust even of his competitors, and take advantage of ripe opportunities.

In his time, there was no one in America who exerted a greater influence on astronomy than Russell. To look for his counterpart in Europe one has to turn to Eddington. In the next chapter we will again encounter this gifted, complex man.

Notes

1. Quantum Physics and the Stars (I): The Establishment of a Stellar Temperature Scale, David H. DeVorkin and Ralph Kenat, *Journal for the History of Astronomy*, xiv, 102–32, 1983.

2. Quantum Physics and the Stars (II): Henry Norris Russell and the Abundances of the Elements in the Atmospheres of the Sun and Stars, David H. DeVorkin and Ralph Kenat, *Journal for the History of Astronomy*, xiv, 180–222, 1983.

3. *Henry Norris Russell, Dean of American Astronomers.* David H. DeVorkin, Princeton University Press, 2000.

4. Stellar Structure and Evolution, 1924–1939, Karl Hufbauer, *Journal for the History of Astronomy*, 37, 203–27, 2006.

5. Über die Fraunhoferschen Linien, *Monatsberichte der Königlich Preussischen Akademie der Wissenschaften zu Berlin*, 662–65, 1859.

6. Chemical Analysis by Spectrum-observations, Gustav Kirchhoff & Robert Bunsen, *Philosophical Magazine* 4th Series 20, 1860, pp. 89–109. (See also *A Popular History of the Nineteenth Century*, Agnes M. Clerke, Adam & Charles Black,1908, pp. 132–35, republished by Scholarly Press, Inc. St. Claire Shores, Michigan, 1977).

7. Über die Beziehung zwischen Wärmestrahlung und der Temperatur, Josef Stefan, *Sitzungsberichte der Akademie der Wissenschaften in Wien*, lxxix, part 2, 391–428, 1879.

8. Über das Gesetz der Energieverteilung im Normalspectrum, Max Planck, *Annalen der Physik*, 4th Series, 4, 553–63, 1901.

9. *The Historical Development of Quantum Theory*, Jagdish Mehra and Helmut Rechenberg, volume 1, part 1, chapter 1. New York: Springer Verlag, 1982.

10. Über die Sterne der Unterabteilungen c und ac nach der Spektralklassifikation von Antonia C. Maury, Ejnar Hertzsprung, *Astronomische Nachrichten*, volume 179, Nr. 4296, columns 373–80, 1909.

11. "Giant" and "Dwarf" Stars, Henry Norris Russell, *The Observatory*, 36, 324–29, 1913.

12. Relations Between the Spectra and other Characteristics of the Stars, Henry Norris Russell, *Nature*, 93, 252–58, 1914; essentially the same material was also published in *Popular Astronomy*, in two articles, the first of which appeared in the May 1914 issue *Popular Astronomy* 22, 275–94, 1914.

13. Über einen die Erzeugung und Verwandlung des Lichtes betreffenden heuristischen Gesichtspunkt, A. Einstein, *Annalen der Physik*, 17, 132–48, 1905.

14. I. On the Constitution of Atoms and Molecules, N. Bohr, *Philosophical Magazine*, 26, 1–25, 1913.

15. On the Constitution of Atoms and Molecules, Part II. Systems containing only a single nucleus, N. Bohr, *Philosophical Magazine*, 26, 476–502, 1913.

16. On the Constitution of Atoms and Molecules, Part III. Systems containing several nuclei, N. Bohr, *Philosophical Magazine*, *26*, 857-75, 1913.
17. The Spectra of Helium and Hydrogen, A. Fowler, *Nature*, *92*, 95, 1913.
18. The Spectra of Helium and Hydrogen, N. Bohr, *Nature*, *92*, 231, 1913.
19. The Spectra of Helium and Hydrogen, A Fowler, *Nature*, *92*, 232, 1913.
20. *Subtle is the Lord: The Science and the Life of Albert Einstein*, Abraham Pais, Oxford University Press, 1982, p. 154.
21. The Solar Spectrum and the Earth's Crust, Henry Norris Russell, *Science*, *39*, 791-94, 1914.
22. On the Radiative Equilibrium of the Stars, A. S. Eddington, *Monthly Notices of the Royal Astronomical Society*, *77*, 16-35, 1916.
23. *Gaskugeln, Anwendungen der mechanischen Wärmetheorie auf kosmologische und meteorologische Probleme*, Robert Emden, Teubner-Verlag, 1907.
24. Further Notes on the Radiative Equilibrium of the Stars, A. S. Eddington, *Monthly Notices of the Royal Astronomical Society*, *77*, 596-613, 1916.
25. On the Conditions in the Interior of a Star, A. S. Eddington, *Astrophysical Journal*, *48*, 205-13, 1918.
26. Ibid. On the Conditions … Eddington, pp. 211-12, 1918.
27. Meghnad Saha, 1893-1956, *Biographical Memoirs, Royal Society of London*, 217-36, 1960.
28. On a Physical Theory of Stellar Spectra, M. N. Saha, *Proceedings of the Royal Society of London A*, *99*, 135-53, 1921.
29. The Properties of Matter as Illustrated by the Stars, Henry Norris Russell, *Publications of the Astronomical Society of the Pacific*, *33*, 275-90, 1921, see p. 282.
30. The Intensities of Absorption Lines in Stellar Spectra, and the Temperatures and Pressures in the Reversing Layers of Stars, R. H. Fowler & E. A. Milne, *Monthly Notices of the Royal Astronomical Society*, *83*, 403-24, 1923; see especially p. 419.
31. Astrophysical Data Bearing on the Relative Abundance of the Elements, Cecilia H. Payne, *Proceedings of the National Academy of Sciences of the USA*, *11*, 192-98, 1925.
32. Ibid., The Intensities of Absorption Lines, R. H. Fowler & E. A. Milne.
33. The Maxima of Absorption Lines in Stellar Spectra R. H. Fowler and E. A. Milne, *Monthly Notices of the Royal Astronomical Society*, *84*, 499-515, 1924.
34. Letter from Russell to Cecilia Payne, 14 January, 1925, Russell papers, cited in Ibid., Quantum Physics and the Stars (II), DeVorkin & Kenat.
35. A Possible Explanation of the Behaviour of the Hydrogen Lines in Giant Stars, H. N. Russell & K. T. Compton, *Nature*, *114*, 86-87, 1924.
36. Ibid. The Solar Spectrum and the Earth's Crust, Russell, 1914.
37. Letter from Russell to Saha cited in Ibid., Quantum Physics and the Stars (II), DeVorkin & Kenat.
38. *Stellar Atmospheres: A contribution to the observational study of matter at high temperatures*, Cecilia Helena Payne, Ph.D. thesis, Radcliffe College, 1925, see also American Doctoral Dissertations, source code L1926, 168 pp.
39. Ibid., *Stellar Atmospheres*, p. 186.
40. Ibid., *Stellar Atmospheres*, p. 188.
41. Ibid., *Stellar Atmospheres*, p. 57.
42. *Cecilia Payne-Gaposchkin: An Autobiography and Other Recollections*, edited by Katherine Haramundanis, Cambridge University Press, 1984, p. 165.
43. *The internal constitution of the stars*, Arthur S. Eddington, Cambridge University Press, 1926.
44. Ibid., *Henry Norris Russell*, DeVorkin, chapter 14.

45. On the Composition of the Sun's Atmosphere, H. N. Russell, *Astrophysical Journal*, *70*, 11–82, 1929.

46. Ibid., *Cecilia Payne-Gaposchkin*, Haramundanis, p. 203.

47. Ibid., *Cecilia Payne-Gaposchkin* , Haramundanis, p. 257.

48. New Regularities in the Spectra of the Alkaline Earths, H. N. Russell & F. A. Saunders, *Astrophysical Journal*, *61*, 38–69, 1925.

49. Ibid., *Henry Norris Russell*, DeVorkin, pp. 364–67.

5

Conclusions Based on Calculations

The logic of well-founded but far-reaching theories may at times be extended to problems so far removed from normal experience that the derived results violate intuition. Doubts then arise about the validity of applying the theory to such extreme conditions, and the theory's original inventors may become its most severe critics.[a]

A Problem Biding Its Time

When is the right time to ask a question? Some questions can be answered almost as soon as they are posed. Others cannot be answered either because they are awkwardly put or because the means for answering them are not at hand. One such question first surfaced late in the eighteenth century.

As early as 1783, the Reverend John Michell, an English natural philosopher and geologist, conceived of a star so compact and massive that its gravitational attraction would prevent light from escaping. He considered the motion of light in Newtonian terms, which suggested that the gravitational attraction of any star would slow down its emitted radiation. Although for most stars this deceleration might be minor, radiation escaping massive stars might have its speed appreciably diminished. For extremely massive compact stars, he thought the emitted light could actually come to a standstill before dropping back to the star's surface.[3]

Michell was concerned not only with whether light could actually be prevented from escaping a star, but also with how we might recognize this. Perhaps his most important realization was that the existence of such a star might be inferred from the gravitational force it would exert on an orbiting companion.[4] The French

---

[a] Events presented in this chapter have become part of a folklore of astrophysics riven by passion. Readers wishing to read further will find the paper by Michael Nauenberg and insights offered in Kip Thorne's book on black holes particularly informative.[1,2]

scholar Pierre-Simon Laplace, who considered the same problem somewhat later, approached the question along similar lines.

Although Michell's perceptions were correct, his questions had to await the arrival of Einstein's theory of general relativity before they could be answered. This much Karl Schwarzschild made clear in a seminal paper on gravity, space, and time around stars so massive and compact that they might as well be geometric points.[5]

Karl Schwarzschild and the Gravitational Attraction of Massive Bodies

As World War I started in 1914, Karl Schwarzschild, Director of the Potsdam Observatory, volunteered for military service despite having just passed his 40th birthday. At the time, he already was Germany's leading astrophysicist, a man with a wide range of interests and prodigious theoretical as well as instrumental abilities.

In the German army Schwarzschild first served in Belgium and France. Then, transferred to the Eastern front in Russia, he was struck by pemphigus, a painful skin disease, at the time known to be fatal. Recognizing his fate, he hastened to write three significant scientific papers within just a few weeks. The most important of these laid the foundations of black hole theory.[6]

Written only weeks after Einstein's completed theory of general relativity published on December 2, 1915, Schwarzschild's first paper described a massive compact star by means of mass concentrated into a geometric point. He made his calculations look easy – an adroit choice of coordinates here, a clever substitution of variables there, a transformation to show that a singularity in his mathematical solution had to reside at the point mass, and soon a beautiful analytic expression emerged that presented not only a rigorous solution conforming to the general theory of relativity, but one he also considered to be the sole possible, that is, the unique solution for a point-like mass. Indeed, as we shall see in Chapter 10, for more than four decades nobody doubted this.

However, as Schwarzschild emphasized, his efforts had yielded more. Not only did his solution reproduce and confirm Einstein's derivation of the perihelion precession of Mercury; but, where Einstein felt compelled to resort to a second-order approximation, which certainly was more than adequate for dealing with Solar System planetary orbits, Schwarzschild's exact solution enabled him to derive a variant of Kepler's laws valid much closer to the central mass. He found that the angular frequency n of a body in circular motion about the mass, no longer increases steadily with diminishing radius, as Kepler's laws demand, but rather approaches a limit of order $n = c/(\alpha\sqrt{2})$, where $\alpha = 2MG/c^2$, M is the mass of the central point, G is the gravitational constant, and c is the speed of light in vacuum. For a *solar mass*, $M_\odot$, the corresponding orbital frequency is $\sim 10^4$ revolutions per second on close approach to the central mass point.

In practice, this deviation from Kepler's prediction is too small to be measurable for Solar System planets, but for a hypothetical central mass far more compact than the Sun the difference would be significant.

From the Russian front, Schwarzschild wrote Einstein, who promptly transmitted the paper for publication in the *Sitzungsberichte der Königlich-Preussischen Akademie der Wissenschaften zu Berlin*. As Schwarzschild's paper generously phrased it, the new solution would "permit Einstein's result to radiate with enhanced purity."[b]

Apparently surprised, pleased, and impressed by the unexpectedly prompt support his work had received, Einstein wrote Schwarzschild, "I would not have expected that the rigorous treatment of the point-source problem would be so simple." [c,7,d]

For the next four decades, Schwarzschild's point-mass solution was to be included in most treatises on general relativity, and throughout this time it would torment the minds of leading astrophysicists.

The space/time separation, *ds*, between two adjacent events in the vicinity of the mass-point can be expressed in terms of their proximity in time and space, and as a function of radial distance, R. It reads:

$$ds^2 = \left(1 - \frac{\alpha}{R}\right)c^2 dt^2 - \left[\frac{dR^2}{(1-\frac{\alpha}{R})} + R^2(d\theta^2 + \sin^2\theta\, d\phi^2)\right]. \tag{5.1}$$

Here, θ is a polar and ϕ an orthogonal azimuthal angle, dt^2 is the square of the time increment separating the two events, and the term in square brackets on the right of the equation is the square of the effective increment of spatial separation in the gravitational potential of the black hole at radial distance R.

The radial distance $R = \alpha \equiv 2MG/c^2$, a critical *parameter* in black hole theory, is called the *Schwarzschild radius*. Seen by an observer far from the mass, any infalling body reaches the speed of light in crossing a hypothetical surface at this radius. For a mass comparable to that of the Sun but compressed into a volume of the order of a few cubic kilometers or less, Schwarzschild gave this radius as ~3 km; for a sufficiently compact mass of 1 gram it was ~1.5×10^{-28} cm.

Here was a rigorous solution of Einstein's general relativistic equations applied to a point mass. A simpler physical system than this was hard to conceive. Yet it was troubling.

The mathematical expression was time symmetric. It read identically whether time *t* ran forward or backward. Any body falling into a black hole should thus have a counterpart escaping the black hole. But this appeared impossible. An external observer could see particles falling toward and up to the Schwarzschild

[b] Einstein's Resultat in vermehrter Reinheit erstrahlen zu lassen.
[c] Ich hätte nicht gedacht, dass die strenge Behandlung des Puktproblems so einfach wäre.
[d] Schwarzschild had sent his paper to Einstein on December 22, 1915; Einstein replied on December 29. Released from military duty because of his fatal disease, Schwarzschild died in Potsdam on May 11, 1916.

radius, but none ever emerging. For an observer at infinity, who dropped a small mass radially toward the central mass point, the receding mass would remain in view forever, increasingly redshifted as it approached the Schwarzschild radius where clocks slowed toward a stop. From a radial distance at which the clocks might have appeared completely stopped, no light could ever escape.

In Chapter 10, we will see that it took until 1958 and beyond to reconcile some of the apparent contradictions raised by Schwarzschild's paper.

For some time, Schwarzschild's work constituted an intriguing mathematical oddity with little apparent application to the real world. All this began to change 15 years later.

Ralph Fowler and the Nature of White Dwarfs

In Chapter 4, I referred to Ralph Fowler and his work with E. A. Milne on the temperature and pressure dependence of spectral features in stellar atmospheres. Fowler is not as well known as Arthur Eddington, his colleague at Trinity College, Cambridge, but his lasting contributions to astrophysics may have been comparable.

In 1914, just as World War I was starting, Fowler had been elected to a prize fellowship for work in pure mathematics at Trinity. Aged 25 that year, he enlisted as a lieutenant in the Royal Marine Artillery and was seriously wounded at Gallipoli – the disastrous British, French, Australian, and New Zealand naval attempt to wrest control of the Dardanelles from the Turks. The losses suffered both by the Allies and the Turks before the Allies' final retreat were staggering.

After convalescence, Fowler was put to work on mathematical problems of anti-aircraft gunnery in the Experimental Section of the Munitions Inventions Department. This lit an interest in mathematical physics that continued to define his career. At war's end in 1918, he was awarded the Order of the British Empire for leading this wartime ballistic work.

Returning to Trinity, Fowler worked in many areas of mathematical physics, including statistical mechanics, thermodynamics, and astrophysics.[8] He was a particularly effective supervisor of doctoral students, many of whom went on to illustrious careers. One of his first students was Edmund Clifton Stoner, who earned his degree in 1924. By then, the 24-year-old Stoner had proposed a novel classification scheme for the electronic levels and sublevels of atoms. He noted the confirmation of his scheme by available optical and X-ray spectra and magnetic properties of atoms, and by the valences of atoms in forming chemical bonds.[9] The new scheme built on, but also replaced, a less potent enumeration of atomic sublevels due to Niels Bohr, and was quickly accepted. Within months, the 24-year-old theoretical physicist Wolfgang Pauli, then at the university of Hamburg, and one of the sharpest theoretical physicists of the twentieth century, had extended Stoner's work and generalized it into a new proposed *rule* stating that no two electrons with identical momenta and *spins* can simultaneously occupy precisely the same position in space. The product of momentum and position of nearest neighboring particles with identical spins had to differ by at least a quantity h. Recognizing that strong applied fields split atomic levels into their most basic sublevels, each designated by a unique set of quantum numbers, Pauli's rule stated that "One can never have two or more equivalent electrons in an atom for which, the values of all their quantum numbers coincide in strong applied fields."[10] This rule soon became enshrined as the *Pauli exclusion principle*. The principle appeared to hold not just for individual atoms, but also for electrons occupying any system.

Although Stoner's contributions to the Pauli principle were soon forgotten, we will encounter the promising young investigator again later, through the important contributions he made to our understanding of the fate of dying stars.

Fowler's wide-ranging interests brought him into contact with Werner Heisenberg who, on July 28, 1925, gave a lecture at Cambridge on the quantum theory of Bohr and Sommerfeld. Heisenberg was then working on the radically new quantum approach he had just discovered.

In August, a few weeks after Heisenberg's lecture, Fowler received the page proofs of Heisenberg's new paper. On going through them Fowler thought they should interest one of his students, the 24-year-old Paul Dirac, who had been working toward a doctoral degree in theoretical physics since arriving in Cambridge in 1923.[11] Fowler, one of the few theorists in Cambridge with a high interest in modern atomic and quantum theory, had earlier suggested to Dirac calculations rooted in statistical mechanics and others also dealing with the Bohr–Sommerfeld theory.

On reading through the page proofs, Dirac at first was not greatly impressed. A week later, he realized how revolutionary Heisenberg's ideas were, and soon threw himself into the task of improving on the new theory by setting Heisenberg's work into a format consistent with special relativity. This was a fundamental problem under investigation by many formidable competitors on the Continent, including Niels Bohr, Max Born, Enrico Fermi, Pascual Jordan, Wolfgang Pauli, Erwin Schrödinger, and Heisenberg himself, among others. Dirac's successes in this venture became legendary.

In 1926, while trying to place quantum mechanics on a firm relativistic footing, Dirac wrote a fundamental paper dealing with the statistical mechanics of electrons and their collective dynamics when occupying a small volume.[12] Shortly after the appearance of his paper in print, he received a note from the Italian physicist Enrico Fermi, a master of both experimental and theoretical physics, who had published an essentially identical paper a few months earlier.[13] Fermi was puzzled that Dirac had not acknowledged this paper in his article. Dirac was mortified and wrote an apology to Fermi. He recalled having come across Fermi's paper at a time when he had no interest in the subject. He had promptly forgotten all about it. Later, when he attacked the same problem motivated by a different set of concerns, he had never realized that the problem had already been solved.[14]

The theory that Fermi and Dirac had independently invented became known as Fermi–Dirac statistics. Several years later, Dirac invented the name *fermions* for particles like electrons, protons and neutrons that obey the exclusion principle. Light quanta (*photons*) and helium nuclei (α-particles), both of which could cluster as closely as they wished and obeyed what came to be known as *Bose–Einstein statistics*, Dirac dubbed *bosons*.[15]

When Dirac first invented his statistical approach, he was still working toward his PhD; so Fowler was among the first to hear about it. Fowler at once realized that Fermi–Dirac statistics might provide insight on white dwarfs.[16] His attention had been drawn to these highly compact stars by a question raised in Eddington's book 'The internal constitution of the stars,' where Eddington asked what would happen to matter in a white dwarf when the star eventually cooled to near absolute zero?[17]

The originally ionized gas in the star's interior, Eddington thought, would then necessarily tend to form an atomic solid – in his mind, the normal form of matter at low temperatures. Then, as Eddington saw it, the formation of atoms would release heat, with which the star might expand against gravity to accommodate the added space required by the newly formed atoms; but, because of the high forces of gravity in the white dwarf's compact state, the released heat would not suffice to expand the star enough to accommodate the newly formed atoms.

Did this mean that the atomic state was not the lowest energy state that matter could assume?

Figure 5.1. Ralph Fowler around the time he explained the constitution of white dwarfs. (*Courtesy of the Master and Fellows of Trinity College Cambridge*).

Fowler at once realized that the Fermi–Dirac statistics provided an answer. He applied these new quantum statistics to the highly compressed matter of white dwarfs, simplifying the discussion engagingly to point out the main features characterizing these stars, even if his approach might not provide a final rigorous analysis.

In Fermi–Dirac statistics the energy of the electrons removed from a nucleus can be very high, even at a temperature close to absolute zero. As a consequence of the Pauli exclusion principle, no two electrons with identical spins and identical momenta can simultaneously occupy an arbitrarily small spatial volume. At the very lowest temperatures, electrons compressed into a small space, like the interior of a highly compact star, are said to be *degenerate*. If a pair of electrons with identical momenta – one with positive, the other with negative spin – is squeezed into a small spatial volume whose linear dimensions are $\Delta \ell$, no other such pair can occupy the same volume unless its momentum p differs by some minimum value $\Delta p \sim h/\Delta \ell$ from the original pair's momenta. Adding further pairs of necessity raises the range of momenta, entailing ever higher particle energies as matter is progressively squeezed into smaller volumes.

Edmund Stoner and the Maximum Mass of White Dwarfs

With his early contributions to the realization of the Pauli principle, and his familiarity with work Fowler was conducting, Edmund Stoner was particularly well qualified to extend Fowler's investigations of white dwarfs. In 1929, Stoner, by then a reader at the University of Leeds, reexamined Fowler's white dwarf problem and wrote a paper asking whether a limiting density of matter might exist in these stars?[18] The question arose because as the star shrinks it releases gravitational energy. But if this released energy does not suffice to raise appropriately the energy of electrons squeezed into an ever smaller volume, the contraction should stop for lack of energy demanded by the exclusion principle. The star's temperature would then be zero, because all the lowest energy states available to electrons in the Fermi–Dirac theory would be completely filled.

In this first paper, Stoner neglected relativistic effects, just as Fowler had, and found that a star of mass M could shrink only until it reached a limiting electron density $n = 9.24 \times 10^{29}(M/M_\odot)^2$ cm^{-3}. If he took the nuclear mass per electron to be $2.5m_H$, where m_H is the *rest mass* of a hydrogen atom, this amounted to a maximum mass density, $\rho_{max} \sim 3.85 \times 10^6 (M/M_\odot)^2$ g cm^{-3}. This corresponded to a minimum radius $r_{min} = (3M/4\pi\rho_{max})^{1/3}$ which, for the white dwarf o$_2$ Eridani B, was about 60% of its observationally estimated radius $\sim 13,000$ km. This particular white dwarf apparently could still contract somewhat, but then should reach a point at which it could contract no further.

Stoner's neglect of relativistic effects was quickly noted by Wilhelm Anderson in Estonia. He pointed out that an electron's kinetic energy increases less rapidly with increasing energy than Stoner had calculated because its rest mass energy needs to be subtracted from its total energy to obtain the equivalent of a kinetic energy. Anderson recalculated the number density of electrons obtained when this is taken into account, and found that for stars of uniform density the limiting mass for Stoner's equilibrium condition could be reached only for stellar masses in a restricted range. For white dwarf masses approaching as high as $\sim 0.7 M_\odot$, the number densities of electrons at equilibrium rapidly increased by orders of magnitude, effectively becoming infinite, unless – as Anderson pointed out – other physical factors were to first intervene. What these might be Anderson could only guess, given how little about physics at these energies was known at the time.[19]

Within months of reading Anderson's work, Stoner had not only corrected his own earlier neglect, but also recalculated the stars' energetics more rigorously than Anderson. The new equilibrium condition, computed for a star again modeled as a sphere of uniform density, now raised the limiting mass at which a white dwarf could still reach equilibrium, to a value somewhat short of $\sim 1.1 M_\odot$. Above this mass, the star's density rapidly rose apparently without bounds, suggesting that the star would have to collapse.

Stoner recognized that stars modeled as uniform spheres could indicate only a likely value of the limiting mass of a white dwarf. A more rigorous calculation would involve *polytropic models*. These take into account the hydrostatic pressures that monotonically increase toward a star's center. For real gases these pressures P are related to mass density ρ through relations of the form $P = \kappa \rho^\gamma$, where κ is a constant and γ is of order 5/3 for nonrelativistic and 4/3 for relativistic ideal gases.

Given the serious possibility that a white dwarf star could only have a limiting mass, what would happen if this mass was exceeded? Would it necessarily collapse? Might the equation of state, the relation between pressure and density, then drastically change, potentially to rescue the star from total collapse? Stoner did not discuss or speculate on these questions; but in the background, about to receive closer scrutiny, loomed Karl Schwarzschild's general relativistic portrayal of a highly compact massive body.

The Young Chandrasekhar

Four years after the publication of Fowler's paper on white dwarfs, the ideas it had generated enthralled a prodigious young physics student in India. In July 1930, the 19-year-old Subrahmanyan Chandrasekhar, a fresh graduate of the Presidency College in Madras, was awarded a Government of India scholarship for graduate studies at Cambridge. By then, he was determined to carry out his doctoral studies under Fowler. On the long trip to England by sea, he combined his understanding of special relativity and the then-new quantum statistics to try to understand the structure of white dwarf stars.

In musing about the problem, Chandrasekhar was struck by the extremely high energies the electrons would reach in massive white dwarfs. Surely, the

Figure 5.2. Subrahmanyan Chandrasekhar in Cambridge while working on the collapse of white dwarf stars. He had just turned 20 when he published his first papers on the topic. (*Courtesy of the Master and Fellows of Trinity College Cambridge*).

velocities of the electrons would come close to those of light, and this meant that a relativistic form of quantum statistics needed to be applied. This was something that Fowler's paper had not done.

On his arrival in Cambridge, Chandrasekhar became a research student under the supervision of Fowler, who saw to his student's admission into Trinity College Cambridge and supervised the first two years of his studies. On the advice of Fowler's former graduate student, Paul Dirac, Chandrasekhar spent the third year of his Cambridge studies at the Institute for Theoretical Physics in Copenhagen, where Niels Bohr held sway. Thus, his entire postgraduate education was in theoretical physics, under the guidance of some of the brightest physicists of the times. At Trinity, Chandrasekhar also had continuing contact with Eddington, the leading British astrophysicist of the day and also a Trinity fellow.

By the summer of 1933, Chandrasekhar had earned his PhD at Cambridge. The same October, now aged 23, he was elected to a Prize Fellowship at Trinity, for the period 1933–37.[20]

Two years earlier, shortly after arriving in Cambridge, Chandrasekhar had already published three papers on white dwarf contraction.[21,22,23] All three articles acknowledge Stoner's prior work, albeit rather sparingly. In one of these articles, published in the *Philosophical Magazine*, Chandrasekhar also claimed that his results, based on a polytropic model of the form $P \propto \rho^{5/3}$, yielded "a much nearer approximation to the conditions *actually existent* in a white dwarf than the previous calculations of Stoner based on uniform density distribution in the star."

In one of the other papers, submitted to the *Astrophysical Journal* somewhat later, Chandrasekhar moderated his criticism. Here he considered a fully relativistic white dwarf, for which $P \propto \rho^{4/3}$. This allowed him to derive a limiting mass that such a hydrostatically supported white dwarf can attain before having to collapse under its own gravitational attraction. He calculated this maximum mass as $M = 1.822 \times 10^{33}$ grams, and compared it to Stoner's value $M = 2.2 \times 10^{33}$ grams, adding "The 'agreement' between the accurate working out, based on the theory of polytropes, and the cruder form of the theory is rather surprising." Here, the theory of polytropes referred to the assumption $P \propto \rho^{4/3}$, and the 'cruder form' to Stoner's assumption of constant density throughout the star.

Chandrasekhar's very first paper already raised the question of catastrophic collapse of the central portions of a white dwarf into a point mass of infinite density and zero radius. But he conceded that collapse could lead to a new and unknown equation of state, which could potentially prevent or at least delay total collapse into a point source.[24]

In his earlier papers Chandrasekhar could not know that the next equation of state might actually be that for matter in a neutron star. At the time the neutron had not yet been discovered. That came a year later, when the physicist James Chadwick, also at Cambridge, discovered this electrically neutral particle in his laboratory experiments.[25] We now know that the equation of state of material that follows the collapse of a white dwarf can be that of a star composed entirely of neutrons. The electrons in the core of the massive white dwarf can become so energetic that they penetrate the atomic nuclei in the star to form an electrically neutral mass consisting of neutrons. The equations that Chandrasekhar developed for white dwarfs also hold for neutron stars. The relativistic electrons simply are replaced by relativistic neutrons, although other forms of nuclear matter also are possible. Even today, we know very little about the forms of matter that can exist at the highest densities in stars.

By 1934, Chandrasekhar was more explicit, writing "It is conceivable, for instance, that at a very high critical density the atomic nuclei come so near one another that the nature of the interaction might suddenly change and be followed subsequently by a sharp alteration in the equation of state in the sense of giving a maximum density of which matter is capable." To this he added, "However, we are now entering a region of pure speculation … "[26] Neutron physics was still too poorly understood.[e]

---

[e] Even today uncertainties about neutron stars persist. Observations now indicate that neutron stars can have masses ranging up to $\sim 2 M_{\odot}$, a value that is not yet fully understood.[27]

Lev Landau, Neutron Stars, and Collapsed Stars

In 1931, independently of Chandrasekhar and Stoner, and before publication of Chandrasekhar's first paper, Lev Davidovich Landau, a 23-year-old Soviet physicist, had reached similar conclusions.[28] Landau estimated, on grounds similar to those worked out by Chandrasekhar, that a star more massive than 2.8×10^{33} grams, that is, 1.4 solar masses, must ultimately collapse. He rounded this off to $1.5 M_\odot$, but his actual value appears identical to the value to which Chandrasekhar's most sophisticated models were later converging.

Landau was particularly clear in his choice of words, "there exists in the whole quantum theory no cause preventing the system from collapsing to a point (the electrostatic forces are by great densities relatively very small)...On these general lines we can try to develop a theory of stellar structure. The central region of the star must consist of a core of highly condensed matter surrounded by matter in [an] ordinary state."

Landau's thrust differed from that of Stoner and Chandrasekhar. He was not writing about white dwarfs, but rather the ultimate structure of all stars and the source of stellar energy. His paper proposed that stars have compact central condensations onto which matter from a surrounding envelope is continually drifting with the release of energy. In Landau's view, this energy release was the source of all starlight.

Landau returned to this topic in 1938.[29] By then, James Chadwick's 1932 discovery of the neutron was well established and Landau repeated his calculations for a stellar core consisting entirely of neutrons. He first calculated the energy required to produce such a core consisting of neutrons packed to the density existing in ordinary nuclei. If he assumed that the star started with stellar material consisting entirely of atomic oxygen, he obtained this energy readily by comparing the mass of an oxygen ^{16}O atom plus 8 electrons to that of 16 neutrons. Because these neutrons are more massive than the atomic nucleus and its assigned electrons, energy would be required to form them. Formation of one neutron in this way would require 1.2×10^{-5} erg. Formation of one gram of neutronic material would require 7×10^{18} erg. On the other hand, if a homogeneous, spherical neutron core of nuclear density $\sim 10^{14}$ grams per cubic centimeter had already formed and was sufficiently massive, its (negative) potential energy would suffice to make up for this difference. Landau found that the minimum required core mass providing this *stability* was of order $1/20\ M_\odot$, and this would be even lower, $\sim 10^{-3} M_\odot$, if the neutrons behaved like a gas obeying Fermi–Dirac statistics and thus had a high kinetic energy, and also a higher relativistic mass because of the small volume into which they were squeezed.

Landau concluded, "When the mass of the body is greater than the critical mass, then in the formation of the 'neutronic' phase an enormous amount of energy is liberated and we see that the conception of a 'neutronic' state of matter gives an immediate answer to the question of the sources of stellar energy."

Landau's paper appeared in the February 19, 1938 issue of *Nature*. Aged 30, by then, he was already heading the theoretical physics division of the Institute for Physical Problems of the Academy of Sciences in Moscow. At the time, Stalin's political purges were raging in the Soviet Union, and Landau was not spared. His opposition to the ongoing barbarism brought about his arrest, just weeks later, on April 27, 1938. Incarcerated in an NKVD prison, he was released the following year, on April 29, 1939, but only after his more celebrated Soviet fellow-physicist and head of the Institute, Pyotr Kapitsa, had written a personal letter to Stalin vouching for Landau.

A Point of Contention

Chandrasekhar continued to refine and generalize his theory over the next three years, with the encouragement of Eddington. By late 1934, he had arrived at an exact equation of state capable of describing stars with degenerate cores surrounded by layers of ordinary matter.[30] Here he also referred explicitly to the work of Stoner, Landau and others, who had contributed to the theory of white dwarfs in previous years. Stoner, in particular, had produced a number of additional articles.[31,32]

Chandrasekhar submitted his paper to the Royal Astronomical Society and was invited to give a talk on his results at the Society's imminent January 1935 meeting. A few days before the meeting was to take place, Chandrasekhar learned that Eddington was to speak on relativistic degeneracy immediately after his talk. He was puzzled that Eddington had not mentioned this to him, since they saw each other at Trinity almost daily.[33]

Nevertheless, at the meeting of the Royal Astronomical Society on Friday, January 11, 1935, Chandrasekhar duly presented his latest results.[34]

Today, it is difficult to imagine Chandrasekhar's presentation on 'Stellar Configurations with Degenerate Cores,' being immediately countered by Eddington's dismissive remarks[35,36]:

> I do not know whether I shall escape from this meeting alive, but [my point] is that there is no such thing as relativistic degeneracy! …
>
> Dr. Chandrasekhar had got this result before, but he has rubbed it in, in his last paper; and when discussing it with him, I felt driven to the conclusion that this was almost a *reductio ad absurdum* of the relativistic degeneracy formula. Various accidents may intervene to save a star, but I want more protection than that. I think there should be a law of Nature to prevent a star from behaving in this absurd way! …
>
> The formula is based on a combination of relativity mechanics and non-relativity quantum theory, and I do not regard the offspring of such a union as born in lawful wedlock. I feel satisfied myself that the current formula is based on a partial relativity theory, and that if the theory is

made complete the relativity corrections are compensated, so that we come back to the "ordinary" formula.

When Eddington finished, the president immediately called on the next speaker to give his paper, giving Chandrasekhar no chance to reply.[37]

Established scientists tend to forget how devastating a jocular rejoinder can be to a young colleague. Everyone present seemed persuaded by Eddington's witty remarks. Some made signs they felt sorry for Chandrasekhar and his stumble. Over the following years, Eddington repeated such attacks on Chandrasekhar's work on other occasions. Privately, leading physicists such as Niels Bohr and Wolfgang Pauli reassured Chandrasekhar that Eddington was talking nonsense. But none of them felt it worth getting into a fight with Eddington.[38]

The 24-year-old Chandrasekhar stood almost alone facing the giant of British astrophysics. Eventually, he decided there was no sense in continuing the controversy. He wrote a book, 'An Introduction to the Study of Stellar Structure,' which became a classic, and then went on to work on other problems.[39]

* * *

Today, most astrophysicists are puzzled by Eddington's reaction. Why was he so strongly opposed to Chandrasekhar's findings?

We will see in this chapter's concluding section that legitimate questions could be asked and were asked by Einstein himself at that time. To the end of his life Einstein questioned the *completeness* of quantum mechanics and thus also quantum statistical mechanics, both of which he considered correct in application to normal problems, but failing to answers questions he considered essential. He doubted whether either theory could be legitimately extended into the relativistic regime, including the quantum statistics of particles moving at speeds approaching those of light. As late as 1939, when Eddington's confrontation of Chandrasekhar reached its final stages, Einstein wrote a major article questioning the physical reality of relativistic collapse of massive bodies.[40]

Eddington was thus not alone in worrying about the physical reality of catastrophic collapse of a star.

Questions of Priority

Eddington's attack on the young Chandrasekhar's presentation at the January 1935 Royal Astronomical Society's meeting was carefully preserved in notes on the meeting printed in the Society's publication *The Observatory*.[41] Over the years, generations of astronomers have read these for themselves and sympathized. Gradually, Chandrasekhar gained immense renown from this David and Goliath confrontation. A mythology developed in which the work of Stoner and Landau was conveniently forgotten to highlight Chandrasekhar's lonely stance confronting the astrophysical giant of the times. The limiting mass of a white dwarf that Stoner and he had worked on came to be known solely as the *Chandrasekhar Limit*.

The most authoritative examination of these events, published more than 70 years after they took place, is a careful study by Michael Nauenberg of the University of California at Santa Cruz.[42] Nauenberg leaves little doubt that Stoner published his calculations on a limiting white dwarf mass and arrived at a simple estimate of it well before Chandrasekhar. Later, both Stoner and Chandrasekhar treated more sophisticated stellar models and extended the calculations to new levels of complexity. Landau's work, which was submitted for publication before Chandrasekhar's but appeared in print later, is also given due credit. One can only hope that Nauenberg's analysis will find suitable acceptance, because both Stoner, a diabetic, and Landau, who had been badly injured in an automobile accident, passed away much earlier than Chandrasekhar, leaving him the sole survivor of those times.

No doubt can exist that Chandrasekhar was a towering astrophysicist in his own rights. But for historical accuracy, the myth that he was first to discover the limiting mass, and the sole defender of its existence, needs to be set aside. Due credit should also be given to Anderson, who first pointed out that relativistic degeneracy needs to be taken into account, and that it is this which accounts for the limiting mass.

* * *

Questions of priority aside, the style in which Chandrasekhar preferred to work still deserves mention. Toward the end of his life he emphasized how important the search for the inherent mathematical beauty of the Cosmos had been in his endeavors. In the epilogue to his monumental 'The Mathematical Theory of Black Holes,' published in 1992, he recalled a conversation with the great modern sculptor Henry Moore, who cited the sculptures of Michelangelo as examples of the greatest beauty "revealed at every scale … from the excellence of their entire proportion to the graceful delicacy of the fingernails." To this, Chandrasekhar added, "The mathematical perfectness of the black holes of Nature is, similarly, revealed at every level by some strangeness in the proportion, in conformity of parts to one another and to the whole."[43]

At various stages of his astrophysical career, Chandrasekhar exhaustively devoted himself to a single astrophysical problem for periods generally lasting a few years to a decade, contributing to a lucid clarification of its fundamentals, as much as was within his powers, before abruptly switching fields to attack an entirely different astrophysical problem. At the end of each of these intellectual journeys he wrote an authoritative text, which might be compared to the logs published by explorers and naturalists of previous centuries, except that each book covered a different branch of theoretical astrophysics and provided a new set of theoretical tools.

The problems Chandrasekhar attacked and exhaustively investigated often seemed too abstract to his contemporaries. In time, however, his papers and books were recognized to merely have been well ahead of their time. They solved problems and devised new mathematical methods that others eventually came to appreciate and adopt. His books and papers were models of craftsmanship, endurance, perfection – and beauty.

Eddington

Seen in retrospect, the most controversial character on the stage of twentieth century astrophysics may appear to have been Arthur Stanley Eddington. In his own time, he dominated theoretical astrophysics. He worked in many fields of astronomy, only a fraction of which dealt with stellar structure or cosmology and relativity. He initiated many new lines of investigation by swiftly importing into astronomy new insights from rapidly evolving areas of quantum theory and general relativity in the first quarter of the twentieth century. Others in Britain followed his thrust. Understandably, much of the work he did has been superseded, but that's the nature of all scientific progress.

Eddington's greatest triumph came from his quick realization that general relativity and its predicted deflection of light by the Sun could and should be tested expeditiously. The success of the eclipse expeditions he organized led to widespread acceptance of Einstein's general theory of relativity and public adulation of Einstein, some of which also rubbed off on Eddington, sealing his reputation as the leading European astrophysicist of his times.

In 1926, Eddington's book-length synthesis on stellar structure, 'The internal constitution of the stars', (ICS) was universally pronounced a masterpiece.[44] Ralph Fowler's paper 'On Dense Matter' only seemed to confirm this assessment.[45] But, in retrospect, the book also appears to have marked the start of decline in Eddington's powers. Others in his immediate surroundings, notably Fowler, and Fowler's graduate student Paul Dirac, quickly mastered the new quantum theory of Heisenberg, Schrödinger, and the Göttingen and Copenhagen schools with which Eddington never became as familiar as he was with the earlier work of Planck, Einstein, and Bohr. In particular, Fowler, whom Eddington thanked in the preface to ICS for having "generally been my referee in difficulties over points of theoretical physics," quickly solved the problem of the internal constitution of white dwarfs, which ICS had attempted in vain to correctly portray.

Eddington's rejection of relativistic quantum theory may seem surprising, particularly when two world experts on the subject, Ralph Fowler and Paul Dirac, were Cambridge colleagues. Perhaps Eddington was following Einstein's lead in this, as he had in his greatest success, the measurement of the solar deflection of light. Abraham Pais, Einstein's younger colleague at the Princeton Institute of Advanced Studies, and his eventual biographer, recalls, "Einstein considered quantum mechanics to be highly successful...(but) this opinion of his applied exclusively to non-relativistic quantum mechanics. I know from experience how difficult it was to discuss quantum field theory with him. He did not believe that non-relativistic quantum mechanics provided a secure enough basis for relativistic generalization. Relativistic quantum field theory was repugnant to him."[46]

Despite his early interest in mastering the advances in modern physics and adopting them for astronomy, Eddington by the late 1920s was increasingly distancing himself from contemporary physics.[47] His book 'Fundamental Theory,' completed just before his death in 1944, reflected many of his ruminations. In an attempt at quantum theory, he formulated arguments based on symmetry considerations alone, which his contemporaries found impenetrable.[48] Nothing of value appears to have emerged from the book he so painstakingly crafted.

* * *

Eddington's treatment of the much younger Chandrasekhar was certainly boorish, but Eddington also made quite different impressions on other young scientists. Cecilia Payne-Gaposchkin, whom Eddington strongly influenced when she was an undergraduate at Cambridge, recalls him as "a very quiet man . . . [A] conversation with him was punctuated by long silences. He never replied immediately to a question; he pondered it, and after a long (but not uncomfortable) interval would respond with a complete and rounded answer."[49]

J. Robert Oppenheimer and Collapsed Stars

Writing in late 1938, J. Robert Oppenheimer at the University of California at Berkeley, and his postdoctoral student, Robert Serber, found an error in Landau's calculation for the minimum required core mass of a relativistic neutron star. They concluded that the minimum mass of a stable core would be $\sim 0.1 M_\odot$, but that, at best, stars would form such a core only once they had exhausted all of their nuclear energy sources, at least in the central portions of the star.[50]

A few months later, Oppenheimer returned to the problem, this time with a graduate student, George M. Volkoff.[51] They considered whether there was a limiting mass for neutron cores above which no core could avoid collapse. The problem was similar to the one that Stoner, Chandrasekhar, and Landau had considered some years earlier, but the *relativistic particles* that would now need to be considered were neutrons rather than electrons. In addition, general relativistic effects needed to be considered. To help with these, Oppenheimer enlisted the aid of Richard C. Tolman, a professor and expert on general relativity at the California Institute of Technology, Caltech, where Oppenheimer spent half of his time each academic year. Tolman's contributions to the problem were published as a separate paper in the same issue of the *Physical Review* in which Oppenheimer's and Volkoff's paper appeared.[52]

On the assumption that static configurations exist for stars with neutron cores obeying some of the equations of state they considered, Oppenheimer and Volkoff suggested that stable cores could not exist for masses higher than $\sim 0.7 M_\odot$. But

Figure 5.3. J. Robert Oppenheimer a few years after he, his students, and one of his postdocs had developed a theory predicting the formation of neutron stars and further collapse into a singularity. The term *black hole* for these singularities had not yet been coined. (*Digital Photo Archive, Department of Energy [DOE]. Courtesy AIP Emilio Segre Visual Archives*).

as Tolman as well as Oppenheimer and Volkoff emphasized, a static solution did not guarantee that the configuration was stable. Nor was finding a static solution essential because a sufficiently slowly varying configuration could just as well be of interest.

To check on some of these considerations, Oppenheimer now embarked on a further study with yet another graduate student, Hartland Snyder, to examine how a star would behave under continued gravitational contraction.[53]

On working their way through the problem of predicting the fate of a sufficiently massive star in general-relativistic terms, Oppenheimer and Snyder concluded that "When all thermonuclear sources of energy are exhausted, a sufficiently heavy star will collapse. Unless fission due to rotation, the radiation of mass, or the blowing off of mass by radiation, reduce the star's mass to the order of that of the sun, this contraction will continue indefinitely."

By taking the pressure within the star to be negligibly low, they pointed out that the star's collapse could be quite rapid as experienced by an observer moving along with the collapsing matter as it approached the Schwarzschild radius. A distant observer would, however, see a much more gradual contraction; the time at which the star's surface appeared to shrink to the Schwarzschild radius would appear infinite, and light escaping from the star would appear progressively reddened. They believed the assumption about a negligible pressure to be in good accordance with expectations and concluded that no sufficiently massive, nonrotating star could ultimately escape total collapse.

Oppenheimer and Snyder's ideas had little immediate impact. Nothing tied their conjectures to anything astronomers were then observing. It took another quarter century for observers to find sufficient evidence for the existence of stars that actually collapsed into a geometric point.

Einstein's Doubts on Gravitational Collapse

When Oppenheimer and Snyder submitted their paper to the *Physical Review* on July 10, 1939, they appeared unaware that, two months earlier on May 10, Einstein had submitted a paper on the same subject to the *Annals of Mathematics*.[54] Neither Einstein nor Oppenheimer seemed aware of the other's work. Oppenheimer and Snyder's article did not appear in print until September 1 of that year, and Einstein's article appeared even later, in the October issue of the *Annals*.

In 1916, Einstein had enthusiastically congratulated Schwarzschild on his ingenious general relativistic derivation of the gravitational field surrounding a massive object. Over the years, however, he evidently came to doubt whether sufficiently compact massive bodies could exist in Nature. His 1939 paper attempted to theoretically construct such a body by postulating a swarm of particles in circular orbits about a common center, each particle gravitationally attracting all the others. As the radius of this swarm diminished, the particle velocities would increase in line with Schwarzschild's prediction mentioned earlier, until their velocities threatened to exceed the speed of light, a speed Einstein insisted his particles could not attain.

Einstein found no way to construct the hypothesized body from such a contracting swarm of particles, and concluded that mathematically such compact masses could be depicted, but physically they could not exist.

With this, the father of general relativity repudiated the existence of black holes!

* * *

What had gone wrong?

Half a century later, a popular account by the relativist Kip Thorne succinctly explained the apparent conflict between Oppenheimer and Snyder's views on one hand and Einstein's on the other. Einstein's 1939 considerations were correct as far as they went. He was misled by his insistence that black holes needed to be constructed through a succession of equilibrium states as the orbital radius of his swarm of particles was gradually reduced. Thorne argued that black holes cannot be created subject to such constraints. Black holes form solely through catastrophic collapse – a possibility that Einstein had evidently ignored.[55] In Chapter 10 we return to this point to see why a catastrophic collapse is essential.

Notes

1. Edmund C. Stoner and the Discovery of the Maximum Mass of White Dwarfs, Michael Nauenberg, *Journal for the History of Astronomy*, *39*, 1–16, 2008.
2. *Black Holes and Time Warps – Einstein's Outrageous Legacy*, Kip Thorne. New York: W. W. Norton, 1994.
3. On the Means of discovering the Distance, Magnitude, &c. of the Fixed Stars, in consequence of the Diminution of the Velocity of Light, in case such a Diminution should be found to take place in any of them, and such other Data should be procured from Observations, as would be farther necessary for the Purpose, John Michell, *Philosophical Transactions of the Royal Society of London*, *74*, 35–57, 1784.
4. Ibid., On the Means, Michell, paragraph 29.
5. Über das Gravitationsfeld eines Massenpunktes nach der Einsteinschen Theorie, Karl Schwarzschild, *Sitzungsberichte der Königlich-Preussischen Akademie der Wissenschaften zu Berlin*, *VII*, 189–96, 1916.
6. Ibid., *Über das Gravitationsfeld*, Schwarzschild.
7. Letter from Karl Schwarzschild to Albert Einstein, December 22, 1915, and Einstein's reply dated December 29, 1915, *The Collected Papers of Albert Einstein*, Vol. 8, *The Berlin Years*, Part A: 1914–1917, R. Schulmann, A. J. Knox, M. Janssen, J. Illy, eds. Princeton University Press, 1998, pp. 224 and 231.
8. Obituaries: Ralph Howard Fowler, E. A. Milne, *The Observatory*, *65*, 245–46, 1944.
9. The Distribution of Electrons among Atomic Levels, E. C. Stoner, *Philosophical Magazine*, *48*, 719–36, 1924.
10. Über den Zusammenhang des Abschlusses der Elektronengruppen im Atom mit der Komplexstruktur der Spektren, W. Pauli, jr. *Zeitschrift für Physik*, *31*, 765–83, 1925.
11. *Dirac, a scientific biography*, Helge Kragh. Cambridge University Press, 1990, p. 14.
12. On the Theory of Quantum Mechanics, P. A. M. Dirac, *Proceedings of the Royal Society of London A*, *112*, 661–77, 1926.
13. Zur Quantelung des idealen einatomigen Gases, E. Fermi, *Zeitschrift für Physik*, *36*, 902–12, 1926.
14. Ibid., *Dirac*, Kragh, p. 36.
15. Ibid., *Dirac*, Kragh, pp. 35–36.
16. On Dense Matter, R. H. Fowler, *Monthly Notices of the Royal Astronomical Society*, *87*, 114–22, 1926.
17. *The internal constitution of the stars*, A. S. Eddington, Cambridge University Press, 1926, Section 117.
18. The Limiting Density in White Dwarf Stars, Edmund C. Stoner, *Philosophical Magazine*, *7*, 63–70, 1929.
19. Über die Grenzdichte der Materie und der Energie, Wilhelm Anderson, *Zeitschrift für Physik*, *54*, 851–56, 1929.
20. Autobiography, S. Chandrasekhar, in *Les Prix Nobel*, The Nobel Prizes 1983, Editor Wilhelm Odelberg, Nobel Foundation, Stockholm, 1984.

21. The Highly Collapsed Configurations of a Stellar Mass, S. Chandrasekhar, *Monthly Notices of the Royal Astronomical Society*, *91*, 456–66, 1931.

22. The Density of White Dwarf Stars, S. Chandrasekhar, *Philosophical Magazine*, Series 7, *11*, 592–96, 1931.

23. The Maximum Mass of Ideal White Dwarfs, S. Chandrasekhar, *Astrophysical Journal*, *74*, 81–82, 1931.

24. Ibid. The Highly Condensed … Chandrasekhar, see p. 463.

25. Possible Existence of a Neutron, J. Chadwick, *Nature*, *129*, 312, 1932.

26. The Physical State of Matter in the Interior of Stars, S. Chandrasekhar, *The Observatory*, *57*, 93–99, 1934.

27. A two-solar-mass neutron star measured using Shapiro delay, P. B. Demorest, et al., *Nature*, *467*, 1081–83, 2010.

28. On the Theory of Stars, L. Landau, *Physikalische Zeitschrift der Sowjetunion*, *1*, 285–88, 1932.

29. Origin of Stellar Energy, L. Landau, *Nature*, *141*, 333–34, 1938.

30. The Highly Collapsed Configurations of a Stellar Mass. (Second Paper.), S. Chandrasekhar, *Monthly Notices of the Royal Astronomical Society*, *95*, 207–25, 1935.

31. The Minimum Pressure of a Degenerate Electron Gas, Edmund C. Stoner, *Monthly Notices of the Royal Astronomical Society*, *92*, 651–61, 1932.

32. Upper Limits for Densities and Temperatures in Stars, Edmund C. Stoner, *Monthly Notices of the Royal Astronomical Society*, *92*, 662–76, 1932.

33. Chandrasekhar vs. Eddington – an unanticipated confrontation, Kameshwar C. Wali, *Physics Today*, *35*, 33–40, October 1982.

34. S. Chandrasekhar, *The Observatory*, *58*, 37, 1935.

35. Stellar Configurations with Degenerate Cores, S. Chandrasekhar, *The Observatory*, *57*, 373–77, 1934.

36. Response to Chandrasekhar, Sir Arthur Eddington, *The Observatory*, *58*, 37–39, 1935.

37. *The Observatory*, *58*, 37–39, 1935.

38. Ibid., Chandrasekhar vs. Eddington, Wali, pp. 37–40.

39. *An Introduction to the Study of Stellar Structure*, S. Chandrasekhar, University of Chicago Press, 1939; Mineola, NY: Dover Publications 1957.

40. On a Stationary System with Spherical Symmetry Consisting of Many Gravitating Masses, Albert Einstein, *Annals of Mathematics*, *40*, 922–36, 1939.

41. *The Observatory*, *58*, 37–39, 1935.

42. Ibid., Edmund C. Stoner, Nauenberg.

43. *The Mathematical Theory of Black Holes*, S. Chandrasekhar, Oxford University Press, 1992.

44. Ibid., *The internal constitution of the stars*, A. S. Eddington, Cambridge University Press, 1926.

45. Ibid., On Dense Matter, Fowler, 1926.

46. *Subtle is the Lord: The Science and the Life of Albert Einstein*, Abraham Pais, Oxford University Press, 1982, p. 463.

47. Sir Arthur Stanley Eddington, O.M., F.R.S., G. Temple, *The Observatory*, *66*, 7–10, 1945.

48. *Fundamental Theory*, A. S. Eddington, Cambridge University Press, 1948.

49. *Cecilia Payne-Gaposchkin – An autobiography and other recollections* edited by Katherine Haramundanis, Cambridge, 1984, pp. 203 and 120.

50. On the Stability of Stellar Neutron Cores, J. R. Oppenheimer & Robert Serber, *Physical Review*, *54*, 540, 1938.

51. On Massive Neutron Cores, J. R. Oppenheimer & G. M. Volkoff, *Physical Review*, *55*, 374–81, 1939.

52. Static Solutions of Einstein's Field Equations for Spheres of Fluid, Richard C. Tolman, *Physical Review*, 55, 364–73, 1939.

53. On Continued Gravitational Contraction, J. R. Oppenheimer & H. Snyder, *Physical Review*, 56, 455–59, 1939.

54. Ibid., On Stationary Systems, Einstein.

55. Ibid., *Black Holes and Time Warps*, Thorne, pp. 135–37.

6

Ask the Right Questions,
Accept Limited Answers

Asking the right question and being satisfied with the answer it yields is one of the most difficult tasks of astrophysical research. Concentrate on too small a problem and you may neglect essential extrinsic factors. Attempt to answer too broad a question and you may fail across the entire front.

* * *

As the 1920s were drawing to a close, two questions were being asked with increasing urgency: "What makes the stars shine?" and "What is the origin of the Chemical Elements?" To many, the two problems appeared related.

Gravitational contraction by then appeared an unlikely source of stellar energy. A potentially abundant supply of nuclear energy appeared to exist, to keep the stars shining for billions of years, but how it might be released was unknown. The known mass defects of heavy elements suggested that a merging of light elements to form heavier elements could both release sufficient energy and account for the existence of heavy elements. The emerging realization that hydrogen was the most abundant element in the Universe made this notion particularly attractive.

If we only understood the nature of nuclear reactions we might account for both the source of stellar energy and the relative abundances of the chemical elements. The prospects were heady![a]

Robert Atkinson and His Two-Pronged Search

By 1929, two observational findings were beginning to influence the discussion. The expansion of the Universe had gradually become accepted as real.

---

[a] The ongoing debate on this question at the start of the twentieth century has been documented by Karl Hufbauer in a number of his publications, particularly in his 'Astronomers take up the stellar energy problem, 1917–1920.'[1] I base much of the present chapter on his findings in that paper as well as those in three of his others, the last two of which have not yet been published.[2,3,4]

And, if its rate of expansion was assumed constant, an extrapolation backward in time suggested a cosmic age of the order of billions of years. Meanwhile, Henry Norris Russell's reluctant recognition that the Sun and the stars were largely composed of hydrogen indicated that the conversion of hydrogen into helium might be a serious contender for powering the radiation emitted by the Sun. The energy that could be liberated in whatever conversion process one might consider was directly proportional to the mass difference – the *mass defect* – between four hydrogen atoms and an atom of ^{4}He. This difference was by then well known, and amounts roughly to $\Delta m = 7 \times 10^{-3}$ gram for every gram of hydrogen consumed. Multiplied by the speed of light squared, the energy potentially liberated per consumed gram of hydrogen would be of order $c^2 \Delta m \sim 6 \times 10^{18}$ erg per gram – enough energy to keep the Sun shining for billions of years on its reserves of hydrogen.

Even before the publication of Russell's 1929 estimate of hydrogen abundance in the Sun, Robert d'Escourt Atkinson, an English physicist in his early thirties working on his PhD thesis at the University of Göttingen, was beginning to consider the build-up of elements in stars. He was interested in simultaneously attacking both of the questions on the minds of contemporary astronomers, "What makes the stars shine?" and "What is the origin of the heavy elements?"[5,6] These questions appeared related because the mass defect of heavy elements fitfully increases up to the mass of the iron isotope ^{56}Fe. By building a succession of heavier elements out of hydrogen, one might be able to account simultaneously both for the sources of energy that make stars radiate and the abundances of the heavy elements.

Joined by a younger experimental physicist at Göttingen, Friedrich Georg (Fritz) Houtermans, born near the city of Danzig in West Prussia, Atkinson cast around for a suitable starting point for analyzing these possibilities. They found it in a strikingly new theoretical approach that George Gamow had developed the previous year.

George Gamow and the Nuclear Physics of Stars

For the summer term of 1928, the 24-year-old Gamow, a graduate student in theoretical physics at Leningrad University, had been sent to study at Göttingen, one of the main centers in which quantum physics was being developed. In Göttingen that summer the buzz was all about developing a quantum theory of atomic and molecular structure. Gamow deliberately chose not to join this effort. Instead, as he later wrote, "I decided to see what the new quantum theory could do in the case of the atomic nucleus."[7]

On arrival in Göttingen Gamow chanced on a recent article by Ernest Rutherford, and found he could explain a set of puzzling experimental results Rutherford had obtained. Years earlier, in 1919, Rutherford had bombarded nitrogen nuclei with α-particles – helium nuclei – and noticed that they formed the oxygen isotope

^{17}O with the emission of a proton, p, the nucleus of a hydrogen atom,

$$^{14}\text{N} + ^4\text{He} \rightarrow\ ^{17}\text{O} + p .\tag{6.1}$$

However, Rutherford now was finding that even extremely energetic α-particles directed at a uranium nucleus were scattered, unable to penetrate to produce a nuclear reaction. This contrasted to the ability of uranium to emit α-particles of quite low energy. Rutherford gave an explanation for this which Gamow found quite unpersuasive.[8,9]

Gamow quickly solved this problem. Using whatever quantum theory he already had absorbed, he proposed a startlingly novel, totally counterintuitive explanation, based on an occasional tunneling through the electrostatic repulsive barrier set up by the positive charge of an atomic nucleus.[10] He showed that quantum theory permitted an energetic atomic nucleus, like an α-particle, occasionally to penetrate another atomic nucleus, even if it did not have sufficient energy to overcome fully the mutual repulsion between the two positively charged nuclei. There remained a small but finite probability that the α-particle would "tunnel" through this repulsive barrier. It was as though you could drive a very fast car against a wall and have it emerge unscathed on the other side, without breaking the wall or vaulting over it.

This was a new aspect of quantum theory that had no counterpart in classical physics. The tunneling permitted α-particles to enter a moderately charged nitrogen nucleus, but not a more highly charged uranium nucleus. For uranium, however, the tunneling permitted the occasional emission of an α-particle, in radioactive decay.

A second paper written jointly with Houtermans showed that Gamow's novel approach also explained an experimentally determined relation that had been known for more than a decade. This law, first formulated by the physicists Hans Geiger and John Mitchell Nuttall in 1911, reflected a relation between the half-life for emitting an α-particle from an atomic nucleus and the energy with which the particle was emitted. The higher the energy of the emitted particle, the shorter the half-life of the emitting nucleus.[11,12]

An Invitation from Niels Bohr

At the end of his summer's stay in Göttingen, of 1928, and almost out of money, Gamow thought he'd still visit Copenhagen and Niels Bohr's Institute. He arrived unannounced, but Bohr's secretary arranged for him to meet the famous physicist. Gamow told Bohr about his quantum theory of radioactive α-decay. Bohr was evidently impressed. "My secretary told me that you have only enough money to stay here for one day. If I arrange for you a Carslberg Fellowship at the Royal Danish Academy of Sciences, would you stay here for one year?"[13] Gamow enthusiastically accepted. It was hard to believe his good fortune.

Papers that followed led to invitations for even-longer visits to Rutherford's laboratories in Cambridge that Bohr helped to arrange with support of fellowships from the Rockefeller Foundation.[14]

Atkinson and Houtermans' Calculations

Atkinson and Houtermans were impressed by Gamow's tunneling theory. They conjectured it might permit protons and electrons to tunnel into atomic nuclei at the temperatures of about 40 million degrees Kelvin that Eddington had calculated to prevail at the center of the Sun.[b,15] If a succession of four protons and two electrons were to impinge on and somehow stick to a ^{4}He helium nucleus, a beryllium ^{8}Be nucleus might form. But ^{8}Be nuclei are unstable and quickly break up to form two helium nuclei. Helium would thus be autocatalytic, and the conversion of hydrogen into helium would provide the requisite energy for stars to shine. If the breakup of beryllium could be avoided through the addition of a further proton before the breakup occurred, then even heavier elements might be produced. The two physicists recognized that their theory was based on a number of tenuous assumptions, and that it did not account for many anomalies, but nevertheless they felt it was a step in the right direction.

Figure 6.1. George Gamow in a photograph appearing to date back to 1929, the year he was a visitor at Bohr's institute in Copenhagen. In 1928, aged 24, Gamow had developed the theory of quantum mechanical tunneling, a process that proved essential to understanding energy generation at the center of stars. (*American Institute of Physics Emilio Segre Visual Archives, Margrethe Bohr Collection*).

---

[b] Eddington's estimate was based on his expectation that stars were composed of heavy elements, which led to a temperature too high by a factor of ~2.

The approach, however, soon ran into difficulty. Houtermans himself noticed that the isotopes ^{5}Li and ^{5}He, both of which were intermediate products in the buildup chain they had postulated, are absent on Earth and thus clearly unstable. This threatened the validity of the process they had described. Their work also ran into two other barriers as Hufbauer has pointed out.[16] First, astrophysicists were too unfamiliar with the theoretical tools of nuclear physics; second, Milne soon was attacking Eddington's theories on stellar structure, declaring instead that the central temperatures of stars should be in the billions of degrees Kelvin, a temperature at which the tunneling process would be irrelevant.[17]

Atkinson, nevertheless, continued to work on the theory. By late 1929, he had been appointed an assistant professor at Rutgers University in New Jersey. His proximity there to Princeton provided him access to, and encouragement from, Henry Norris Russell after Russell's return in 1930 from a year-long vacation. By 1931, however, Atkinson had gone about as far as he felt he could proceed with his approach. He turned his attention to other problems.

Others took over Atkinson's general thrust. Advances in experimental nuclear physics were continually shifting the details. Most notably the discovery of deuterium by Harold Urey with the help of Ferdinand G. Brickwedde and George M. Murphy in 1931 needed to be taken into account, as did the discovery of neutrons by James Chadwick in 1932.[18,19,20,21] Nevertheless, by 1935, many leading astrophysicists, including Eddington, believed that Atkinson's efforts at least were on the right track.[22]

George Gamow's Investigations into Compact Cores

A melding of the theory of stellar structure and nuclear physics, however, was still lacking. George Gamow, although a strong early supporter of the efforts of Atkinson and Houtermans, could still entertain quite different ideas as late as 1935. Following up ideas on the existence of collapsed cores at the centers of stars that Lev Landau, his friend from student days in Leningrad, had suggested three years earlier, Gamow now rejuvenated the stellar contraction theory in a new guise.[23] Landau had proposed that stars might contain a dense collapsed core, which would permit stars to shine for far longer periods than the contraction of less dense stars permitted. As matter from an outer shell fell onto the core it would liberate energy in vastly greater amounts. Gamow speculated that, if the surface of the core were occasionally to erupt and toss nuclear matter into the star's surface layers, disintegration of this material could give rise to the heavy elements found in Nature.

Basing himself on this model, Gamow published an article in the *Ohio Journal of Science*.[24] Why he should have chosen this obscure journal where his work was unlikely to ever be found is not obvious; but, whatever his reasons, his primary interest now was to paint a broad-brush image not only of how matter falling onto a compact stellar core could be expected to provide sufficient energy for a star to shine, but that specific nuclear reactions could also account for the production of heavy elements in stars.

Gamow referred both to the work of Rutherford, who had shown how light elements could be transmuted into heavier elements through α-particle bombardment, and to the work of Enrico Fermi's group in Rome, which had found that bombarding heavy elements with neutrons could transform them into their heavier isotopes. These could then *beta decay* to yield nuclei of greater mass and charge than the original element.[25] As a specific example Gamow cited the two-step transformation of iodine into xenon with the emission of an electron e^- and a photon with energy $h\nu$:

$$^{127}\text{I} + \text{n} \rightarrow {}^{128}\text{I} + h\nu, \quad {}^{128}\text{I} \rightarrow {}^{128}\text{Xe} + e^- . \tag{6.2}$$

Weizsäcker's Influential Papers

The 25-year-old German astrophysicist Carl Friedrich von Weizsäcker was pursuing a similar research program. Following the lead proposed by Robert Atkinson in 1929, he too believed that solving the problem of energy generation in stars might also account for the abundances of the various chemical elements. By 1937 he had identified several potential processes by which the Sun might generate sufficient energy to continue radiating at roughly current rates for at least 2×10^9 years, a time span that appeared required by geological data.[26]

Weizsäcker was also said to be working on a second article on nuclear transmutation that was expected to be ready for submission to a journal within a few months. In a paper published two years earlier, at age 23, he had shown how the masses and mass defects of nuclei and their isotopes could be quantum mechanically calculated to arrive at improved estimates.[27] Now, he was working on nuclear transmutation in an attempt to also understand the relative abundances of the chemical elements.[28]

His two papers may best be seen in this context. They read like tutorials on nuclear reactions that could generate the energy to make stars shine while also building up heavy elements.

Weizsäcker's first paper, 'Über Elementumwandlungen im Innern der Sterne. I.' (On the Transmutation of Elements in the Interior of Stars) began with a section titled 'Die Aufbauhypothese' (The Hypothesis of Assembly) reflecting Weizsäcker's preoccupation with the buildup of the chemical elements through assembly. But further considerations led him to a more specialized formulation of the hypothesis, namely that the temperature in the interior of a star adjusts itself so that nuclear transmutation of the lightest elements through reactions with hydrogen become possible. The star radiates away the energy generated in this transmutation at a rate that just keeps the stellar temperature constant. Weizsäcker thus mused that a star may be a "machine which through the release of nuclear energy automatically maintains the conditions required for its functioning." He added "it may also be the only possible machine of this kind," an interesting remark considering the work on nuclear reactors on which he would soon embark as part of Germany's military efforts in World War II.[c]

The paper is divided into eight sections. In Section 3 of the paper, Weizsäcker considered reaction chains that could lead to the formation of ^{4}He. But because of his preoccupation with forming heavy elements, he concentrated on processes that release neutrons needed to form those elements.

Section 4 briefly sketched the thermal structure of the star and came to the strange and erroneous conclusion that the high radiation pressure at the center of a star "makes the density of matter near the star's center extraordinarily small; it vanishes at the very center."[d] Weizsäcker seemed more familiar with nuclear processes than with stellar structure.

In Section 5, Weizsäcker turned to the formation of successive heavy elements through absorption of neutrons. An especially important and novel insight was his recognition that, for the heaviest elements to be formed, an extremely rapid sequence of neutron absorptions, at intervals of no more than a minute, are needed to counter the short lifetimes of some of the intermediate nuclei in a sequence leading to the heaviest elements, thorium or uranium.

By the following year, 1938, when Weizsäcker published his second paper, he was keenly aware of errors in his first paper, and wrote "Further examination has shown that some of the hypotheses assumed there cannot be sustained."[30] He recognized that the transmutation of hydrogen to generate energy could not lead to the formation of heavy elements. He now believed it altogether possible that the formation of heavy elements could have taken place before the stars formed, during a phase of the Cosmos that inherently differed from that of today.[e]

In Section 6 of this paper, Weizsäcker finally specified the energy generating reactions in stars. He mentioned both the proton–proton reaction to form deuterium and the CNO cycle, both of which he apparently had considered on his own; but he conceded, in separate footnotes for each of these reactions, that he had learned from recent conversations with Gamow that Hans Bethe at Cornell University had by then quantitatively studied both.

Although the two papers by Weizsäcker undoubtedly were influential at the time in alerting astronomers to the myriad reactions that could be of astronomical interest, for understanding nuclear processes both in stars and in the early Cosmos, they lacked the specificity and assurance that Bethe, as we will see, was about to provide in two papers on stellar energy generation.

[c] Wenn diese Vorstellung richtig ist, so ist der Stern eine Maschine, welche mit Hilfe der freigemachten Kernenergien die zu ihrer Freimachung notwendingen äusseren Bedingungen automatisch gleichförmig aufrechterhält.[29]

[d] Erstens wird infolge des hohen Strahlungsdruckes die Materiedichte in der Umgebung des Sternmittelpunktes ausserordentlich klein; sie verschwindet im Mittelpunkt selbst.

[e] Es ist durchaus möglich, dass die Bildung der Elemente vor der Entstehung der Sterne, in einem vom heutigen wesentich verschiedenen Zustand des Kosmos stattgefunden hat.

Independently of Weizsäcker, Ernst Öpik, by then working at the Dorpat Observatory in Tartu, Estonia, had come up with similar ideas. In a 115-page long article on 'Stellar Structure, Source of Energy, and Evolution,' appearing in 1938 in the *Publications of the Astronomical Observatory of the University of Tartu*, he noted that "The starting reaction of the atomic synthesis is most probably the not yet observed direct synthesis of the deuteron from protons (with the expulsion of a positron)" and also "The observed synthesis $^{12}C + ^1H$, with the observed amount of carbon in the sun, is able to supply the radiation of the sun for 500 million years, at [a central temperature of] $T_c \sim 2 \times 10^7$." Öpik did not anticipate the carbon cycle that Bethe would describe the following year, but the stellar structure considerations he took into account, in many ways, make his work more impressive than the articles Weizsäcker was publishing about the same time.[31] Öpik's work, however, remained almost unnoticed, because few astrophysicists read the *Publications of the Astronomical Observatory of the University of Tartu*, where Öpik published much of his work.

* * *

With the crucial determination in the late 1920s of an expansion pervading the Universe to the furthest reaches one could observe, astronomers concluded that the Universe had been expanding for billions of years. Expansion cools any closed system, while compression heats it. If the Universe started out in a far more compact state than today, primeval gases and radiation could have been extremely hot long before the Cosmos had sufficiently expanded and cooled for stars to form.

In the third part of his second paper, Weizsäcker accordingly took up the question of whether the chemical elements of Nature could have been formed in a hot early Universe.[32] He noted that application of the Saha equation to neutrons in thermal equilibrium with more massive nuclei could correctly predict the relative abundances of isotopes of individual elements once their mass defects were known.

In Saha's original equation, the number of electrons tied to an atomic nucleus defined the different levels of ionization of an atom and related these to the temperature. In Weizsäcker's new application, the Saha equation promised to relate the temperature to the number of nuclear neutrons defining the different isotopes of an element. But because the mass defects were known only for some of the lighter elements, this approach did not yield as much information at the time as might have been hoped.

Weizsäcker also knew that the heaviest elements could be formed only at extremely high temperatures and pressures. He concluded that two epochs of nucleosynthesis might have been required to produce the distribution of elements observed today. A first intensely hot and dense phase could have produced the heaviest elements, while a later, cooler, more tenuous phase of expansion might have led to the formation of the more abundant lighter elements, such as oxygen.

But he realized that the data in hand were insufficient to pursue such calculations further.

Gamow, Teller, and the Washington Conference of 1938

In 1934, the 30-year-old Gamow had been appointed professor of physics at the George Washington University in Washington, DC. One of Gamow's two conditions for coming to George Washington had been that the university would allow him to attract another physicist to the university. Gamow chose the Hungarian-born theorist Edward Teller for this position. The other condition to which the university had agreed was to commit funds to support an annual Washington Conference on Theoretical Physics.

In 1934, Enrico Fermi had developed a novel theory of β-decay.[33] On investigating this theory in greater depth, Gamow and Teller realized that Fermi's theory was overly simplified. It had failed to note that the spins of the reacting particles had to be taken into account in arriving at *cross sections* for β-decay.[34] They modified the theory to correct this omission, and found their results to better agree with available experimental evidence.

By the beginning of 1938, as Gamow was arranging for the fourth of his conferences on theoretical physics, he was aware that Weizsäcker in Germany was concentrating on understanding nuclear processes in stars. This, as well as Gamow's own work with Teller, suggested to him that nuclear physics appeared far enough advanced to solve the problem of energy generation in the Sun. He chose "Nuclear Processes as the Source of Stellar Energy" for the topic to be discussed that year. The conference was to be held at the Carnegie Institution of Washington's department of Terrestrial Magnetism on March 21–23.

In Gamow's mind, the undoubted expert on theoretical nuclear physics, at the time, was the 31-year-old Hans Bethe, born in Germany, but by then a full professor at Cornell University. Despite his youth, he had shown himself the master of nuclear theory through a blaze of calculations and three review papers published in the influential journal *Reviews of Modern Physics*.[35,36,37] This triad of papers jointly accounting for 447 pages in the journal soon was dubbed *Bethe's Bible*, an acknowledgment of Bethe's mastery of nuclear physics.

Gamow felt that Bethe's presence at the conference would be essential.[f]

Bethe was reluctant to attend this fourth conference in the series of the by-then well-established annual Washington Conferences. He had been an enthusiastic attendee at the three previous conferences, which afforded an opportunity to exchange ideas with fellow theoretical physicists in a congenial setting; but he had little at stake in energy generation in stars, and would have preferred to

---

[f] Karl Hufbauer has written an insightful chapter on the astonishingly complete and persuasive theory on energy generation in stars that Hans Bethe developed over a period of weeks in the spring of 1938.[38] Here, I recount only a few of the salient features surrounding Bethe's insights and calculations.

Figure 6.2. Hans Bethe in 1938, the year he developed his insights on the nuclear processes that power main-sequence stars. (*Hans Bethe papers #14-22-976, Division of Rare and Manuscript Collections, Cornell University Library*).

devote himself to the many fundamental problems in pure nuclear physics that were captivating him.

Gamow, however, felt the Conference could not do without Bethe's participation, and asked Teller, a close friend of Bethe's, to persuade him to attend. In his insistence on inviting Bethe, Gamow may have recognized his own preference not to get involved in messy detailed problems. Writing in his autobiography 'My World Line,' of his original decision in 1928 to apply the quantum theory to nuclear rather than atomic or molecular problems, he recalled the prevailing atmosphere in Göttingen, where[39]

> Seminar rooms and cafés were crowded with physicists, old and young, arguing about the consequences which [the new developments] in the quantum theory would have on our understanding of atomic and molecular structure. But somehow I was not engulfed in this whirlpool of feverish activity. One reason was that ... I always preferred to work in less crowded fields. Another reason was that whereas any new theory is almost always expressed originally in very simple form, within only a few years it usually

grows into an extremely complicated mathematical structure … Of course I knew that it is absolutely necessary for the solution of complicated scientific and engineering problems, but I just did not like it.

By 1938, Gamow may have recognized that solving the stellar energy problem was going to be complicated, that he himself was temperamentally unsuited to tackle its complexities, and that it would take an expert like Hans Bethe to sort it out.

Resolution of the Stellar Energy Problem

Also invited to the Washington Conference and attending were the 30-year-old Danish astrophysicist Bengt Strömgren and the 28-year-old Subrahmanyan Chandrasekhar, both of whom had recently been attracted to the University of Chicago. Strömgren had just written a comprehensive review article on the theory of the stellar interior and stellar evolution, 'Die Theorie des Sterninnern und die Entwicklung der Sterne,' and Chandrasekhar was completing his magisterial book 'An Introduction to the Study of Stellar Structure' that he would send to his publishers later that year.[40,41] The presence of these two full-time astrophysicists at this conference provided a sense of perspective for the attending theoretical physicists on questions regarding established astronomical observations, reasonably understood astrophysical theories, and ideas that were still largely speculative.

Particularly significant in Strömgren's talk was his reporting that astronomers now were agreed that hydrogen was a major constituent of stars, amounting to about 35% by weight according to estimates of the times. Also, the Sun's central temperatures could be on the order of 19 million degrees Kelvin, with densities reaching 76 grams per cubic centimeter. The new temperature estimate was significantly lower than Eddington's earlier estimate of 40 million degrees Kelvin.

In the ensuing discussions, Bethe suggested that, given the high abundances of hydrogen in stars, proton interactions with other protons might well provide the required source of stellar energy. At this, Gamow and Teller felt compelled to tell Bethe that one of Teller's graduate students, Charles L. Critchfield, had been trying to persuade them to let him study this same process. Always fair minded, Bethe suggested that he and Critchfield try to examine this process in detail to see where their calculations would lead. This soon gave rise to a joint paper submitted to the *Physical Review* only three months later, on June 23, 1938.

The paper is only seven pages long and divides its attention between two calculations. The first concerns the likelihood or *cross section* for two hydrogen nuclei, H, to interact to form a deuterium nucleus – the *deuteron*, D – plus a positron e^+,

$$\mathrm{H} + \mathrm{H} \rightarrow \mathrm{D} + e^+ . \tag{6.3}$$

Bethe and Critchfield calculated this quantum mechanically, to take into account the relative spins of the hydrogen nuclei and the deuteron.[42] In this they were employing Fermi's theory as extended by Gamow and Teller.[43,44]

Estimating the likelihood of this process was the most difficult part of their considerations, mainly because production of a positron in a nuclear reaction involves *weak* nuclear forces that are far less likely to take place than reactions that produce gamma photons – high-energy electromagnetic radiation. But once a deuteron is formed, it quickly reacts with a proton to form the helium isotope ^{3}He with emission of a photon, γ,

$$\mathrm{D} + \mathrm{H} \rightarrow {}^3\mathrm{He} + \gamma \;. \tag{6.4}$$

Bethe and Critchfield estimated that this process was 10^{18} times faster than the production of deuterons, so that hydrogen in stars would be about 10^{18} times more abundant than deuterium.

One more process was required to form the abundant, stable helium isotope ^{4}He, either directly or indirectly through the addition of a proton to the helium isotope ^{3}He. Bethe and Critchfield suggested two different pathways, between which it was difficult to choose – either

$$^3\mathrm{He} + {}^4\mathrm{He} \rightarrow {}^7\mathrm{Be} + \gamma; \quad {}^7\mathrm{Be} \rightarrow {}^7\mathrm{Li} + e^+; \quad {}^7\mathrm{Li} + \mathrm{H} \rightarrow 2\,{}^4\mathrm{He}\;, \tag{6.5}$$

or

$$^3\mathrm{He} + e^- \rightarrow {}^3\mathrm{H} \quad; \quad {}^3\mathrm{H} + \mathrm{H} \rightarrow {}^4\mathrm{He} + \gamma \;. \tag{6.6}$$

As it turns out, there is yet a third reaction, which they did not consider but which is now recognized to be dominant

$$2\,{}^3\mathrm{He} \rightarrow {}^4\mathrm{He} + 2\mathrm{H} \;. \tag{6.7}$$

As their paper clearly stated, the net effect was to convert four hydrogen nuclei plus two electrons into a helium ^{4}He nucleus. The difference in mass between the consumed particles and the helium nucleus produced they took to be roughly equivalent to $\Delta m = 0.0286$ atomic masses of hydrogen yielding $c^2 \Delta m = 4.3 \times 10^{-5}$ erg per helium nucleus formed.

The likelihood of the reactions they envisaged also depended on the temperature and density of the ionized gases at the center of the star. These two quantities determined how often two protons would collide with energies sufficiently high to overcome the nuclear repulsion of their positive electric charges and interact to produce a deuteron. This was a problem in quantum statistics that was relatively simple. Following Strömgren, Bethe and Critchfield assumed the central temperature of the Sun to be 20 million degrees Kelvin, its density 80 grams per cubic centimeter, and its hydrogen abundance 35%. The energy yield of the Sun would then be about 2.2 ergs per second for each gram of solar material.

They concluded that (their italics) "the *proton–proton combination gives an energy evolution of the right order of magnitude for the sun.*"[45] This compared favorably with Strömgren's estimates for the entire Sun; but, as the two authors emphasized, the density and temperature in the Sun rapidly diminishes from the center outward, so that the average energy production might be an order of magnitude lower. They added "it seems that there must be another process contributing somewhat more to the energy evolution in the sun. This is probably the capture of protons by carbon." Here they referred to a forthcoming paper by Bethe alone.

* * *

Before turning to this second process two comments may be in order.

The first is that, whereas the proton–proton reaction seemed to satisfy the energy production in stars considerably less luminous than the Sun, it could not account for those that were hundreds of thousands times more luminous. This was because the proton–proton reaction rate increased only as rapidly as the 3.5^{th} power of temperature, that is, in proportion to $T^{3.5}$, and standard hydrostatic and thermodynamic calculations showed that the central temperature of even the most massive stars did not rise sufficiently for the proton–proton reaction to generate the observed rates of energy production, that is, the extremely high luminosities of these stars.

The second comment is that, by the time the Bethe–Critchfield paper was being submitted for publication, Bethe had realized that another process that begins with the capture of protons by carbon nuclei was considerably more important than the proton–proton reaction, not only for the Sun, but also for all stars that are more massive. This was both a considerably more complex process, and one that Critchfield had not anticipated. So Bethe, who had given it considerable thought, arranged with Critchfield that this was a paper that Bethe would work out and publish by himself. The arrangement was not unreasonable, given that the style of the Bethe–Critchfield paper strongly suggests that most of the novel insights and the guiding information based on the work of others had been contributed by Bethe.

Bethe's Paper of 1939

The most significant points of Bethe's 1939 paper can already be gleaned from its abstract, which emphasized that[46]:

> [T]he *most important source of energy in ordinary stars is the reaction of carbon and nitrogen with protons.* These reactions form a cycle in which the original nucleus is reproduced. . . . Thus carbon and nitrogen merely serve as catalysts for the combination of four protons (and two electrons) into an α-particle.

The italics are Bethe's.

The abstract also announced the uniqueness of the carbon–nitrogen cycle. The paper went to great lengths to show that no other reactions existed, or that they were unimportant if they did. It then went on to discuss the theory's fit to astronomical data and found the agreement of the carbon–nitrogen reactions with observational data to be excellent for all bright stars of the main sequence but not for red giants.

Another scientist might have been bothered by the theory's difficulties in explaining the energy consumption of the red giants. Bethe was willing to leave good enough alone. For him, it evidently was sufficient that he could explain how stars on the main sequence shine. The red giants were another matter. Although the red giant Capella had a much lower density and temperature than the Sun, it was far more luminous. He felt this could not possibly be explained by the same mechanism that accounted for the luminosity of main sequence stars. Bethe dismissed some nuclear processes that appeared unlikely to him – and today can be outright rejected – and thought that the only other known source of energy was gravitational, which would require a core model for giants. Two accompanying footnotes referred, respectively, to Lev Landau's paper 'Origin of Stellar Energy,' and to a suggestion made to Bethe by Gamow, probably along the lines of Gamow's 1935 article in the *Ohio Journal of Science*.[47] Both of these alternatives dealt with a massive neutron core onto which matter from the star's outer regions deposits itself in a process that radiates large amounts of energy as the impact of infalling matter heats the core.

Landau's paper suggested that this process holds even in the Sun, a view that Bethe would by now have been inclined to reject on the basis of the good agreement he was finding for *main sequence stars* with the two energy generating processes he had proposed. But for red giants, a neutron core model seemed credible, although Bethe remarked that "any core model seems to give small rather than large stellar radii."[g]

Bethe's abstract also pointed out that his proposed mechanism of energy production in main sequence stars also led to conclusions about the mass–luminosity relation for these stars, their stability against temperature changes, and their evolution. The paper devoted a separate section to each of these topics.

For the mass–luminosity relation, Bethe returned to Strömgren's 1937 paper to make use of the equations of hydrostatic equilibrium, gas pressure, radiative equilibrium, and Srömgren's numerical results for opacity derived from quantum mechanics, to show that the luminosity L could be expressed as[50]:

$$L \sim M\rho z T^{\gamma} , \tag{6.8}$$

---

[g] This point, however, continued to be debated. As late as 1975, Kip S. Thorne and Anna N. Żytkow resurrected the idea in a paper on 'Red Giants and Supergiants with Degenerate Neutron Cores.'[48] Their model was further elaborated at least until 1992, and may still be under investigation even today.[49]

where M is the star's mass, ρ its density, z the product of the concentrations of hydrogen and nitrogen, T the star's central temperature, and γ a numerical factor of the order of 18 that indicated the high dependence of luminosity on temperature, a feature that had been lacking in the proton–proton reaction, where Bethe and Critchfield had estimated $\gamma \sim 3.5$. Bethe now recalculated the minimum values that γ could attain. For stars a hundred times fainter than the Sun, where the proton-proton reaction would dominate, he found that γ could drop as low as ~ 4.5 at a central temperature around 11 million degrees Kelvin but would rise both at lower and higher temperatures.

Expressing the star's central temperature and density in terms of mass M, Bethe was able to show that, in the range where the carbon–nitrogen cycle dominated, the luminosity should increase slightly more rapidly than the fourth power of the star's mass – considerably less rapidly than the customarily cited value of order 5.5, but in better agreement with observations. By melding the proton–proton reaction with the carbon–nitrogen cycle, he further showed how the energy production, that is, the luminosity of stars, should depend on central temperature, with a crossover from the proton–proton reaction to the carbon-nitrogen cycle occurring in the central temperature range around 15 million degrees Kelvin.

In a succeeding section, Bethe addressed a question that had already been raised by the British astrophysicist Thomas George Cowling, namely, whether or not stars should be stable or become unstable and start oscillating. When the radiation pressure is negligible, the star should be highly stable. But when radiation pressures became high, as in very massive stars, oscillations could be expected to arise. Bethe nevertheless showed that for energy generation proportional to T^{17}, the ratio of radiation to gas pressure remains satisfactorily low. A further consideration, however, arose from the variable ratio of carbon to nitrogen that could change in the star as its central temperature changed. Here too, Bethe was able to show that the nuclear reactions involved would keep stars within reasonable stability bounds.

In a final section Bethe still considered questions of evolution. He found that for stars like the Sun, evolutionary ages were rather longer than the age of the Universe, which at the time was estimated to be as low as 2×10^9 years, while his estimate for the total lifetime of the Sun was of order 12×10^9 years. But he foresaw that when all the star's hydrogen had been exhausted, the star would be able to keep radiating only by contraction. For low-mass stars, he presumed that this would continue until the star reached the white dwarf stage, as was already known from the work of Ralph Fowler, Stoner, and Chandrasekhar. For massive stars, the end-point was more likely to be the formation of a neutron core.

With these considerations, he stopped. He had covered a breathtaking range of astrophysical topics!

* * *

By early September 1938 Bethe's paper 'Energy Production in Stars' was ready for submission to the *Physical Review*, which recorded receiving it on September 7. Initially the editor, John Torrence Tate, suggested that the *Reviews of Modern Physics* would be a more appropriate journal in which to publish this broadly–based paper.[51] Perhaps Tate mistook the grand sweep of Bethe's paper as representing a review. But Bethe argued that the paper's originality warranted its publication in the *Physical Review*. He also asked for a delay in publication. One of Bethe's graduate students, Robert Marshak, who later was to become a leading nuclear physicist at the University of Rochester, had called Bethe's attention to a $500 prize being offered by the New York Academy of Sciences for the best original paper on energy production in stars; the Academy stipulated that the paper should not have been previously published.

The delay in publication in the *Physical Review* permitted Bethe to first submit his paper to the prize contest, which he won. When his paper eventually appeared in the *Physical Review* issue of March 1, 1939, a footnote on its first page proudly announced that the paper had been awarded an A. Cressy Morrison Prize in 1938, by the New York Academy of Sciences.

But there was more significance to it than a winner's pride. Of the $500 prize he received, Bethe later recalled,[52] "I gave $50 to Marshak as a 'finder's fee.' The reason he had known about all these sources of money was his own poverty at the time, and he found this fee very welcome. Another $250 was used as a 'donation' to the German government to secure the release of my mother's goods because she had finally decided to emigrate." These were difficult times for Bethe, a recent immigrant in the United States worried about a parent reluctant to leave her native Germany. And $500 was a lot of money during those Depression years.

* * *

Bethe's explanation, not just of the energy generation in stars but also of many potentially related stellar problems, had an immediate effect. As David DeVorkin noted[53]:

> [In the winter of 1939], Bethe came to Princeton, had an extended discussion with Russell about his work, and left his manuscript with him…Russell was convinced by the breadth of Bethe's argument and, most definitely, the expertise he demonstrated in manipulating both nuclear physics and astrophysical evidence…Russell was now a convert…And after Bethe's original paper had appeared in the *Physical Review*, Russell proclaimed it 'The Most Notable Achievement of Theoretical Astrophysics of the Last Fifteen Years.'[54]

With Russell's high praise, Bethe's depiction of energy generation in main sequence stars was accepted essentially at once. The new theoretical tools

Bethe had provided astrophysicists now would allow them to pursue the subject further.[h]

The words *tour de force* are widely overworked, but Bethe's paper cannot well be described as anything else. What he wrote with Critchfield in 1938, and in the more comprehensive second paper by himself in 1939, is essentially what textbooks teach our students today. In the second of these papers he even took into account that a small fraction of the energy generated in the transformation of hydrogen into helium is carried away by neutrinos that escape the star and do not contribute to the observed starlight. In its major features, the combination of the Bethe and Bethe/Critchfield papers was self-contained and complete; only small differences from their original theories have emerged over the years.

Baade, Zwicky, and the Energy Generated in Supernova Explosions

Not all stars are main sequence stars, of course; nor are they necessarily red giants, or *pulsating variables*. In the early 1930s, Walter Baade and Fritz Zwicky of the Mount Wilson Observatory and California Institute of Technology had noted that one class of explosively erupting stars, the *novae*, all have roughly the same luminosity. In our nearest neighboring large galaxy, the Andromeda Nebula, 20 or 30 novae are observed each year. All of these reach an apparent visual magnitude of roughly 17. Taking the great distance of the Nebula into account, this makes them about a hundred thousand times as luminous as the Sun.

In 1885 an enormous explosion, ten thousand times brighter even than a nova, had been observed in the Andromeda Nebula. Baade and Zwicky called this a *super-nova*, today contracted to *supernova*. They noted that a similarly powerful explosion had been observed by Tycho Brahe in our own Milky Way in 1572, and wrote, "The extensive investigations of extragalactic systems during recent years have brought to light the remarkable fact that there exist two well-defined types of new stars or novae which might be distinguished as *common novae* and *super-novae*. No intermediate objects have so far been observed."[57]

On the supposition that these explosions represented the brightening of a single star perhaps no more than 50 times as massive as the Sun, they calculated the total energy liberated in the explosion, both in the form of light and as kinetic energy

---

[h] Bethe later recalled an invitation, from the physicist John Hasbrouck van Vleck, to speak at the Harvard physics colloquium which, Bethe wrote, "I accepted happily. In the front row at the colloquium sat Henry Norris Russell of Princeton, the acknowledged leader of American astrophysics. When I had finished my presentation he asked a few searching questions. He was convinced and became my most effective propagandist." [55] Karl Hufbauer has kindly pointed out to me that Bethe's recollection, 65 years after the events, must have been faulty. Hufbauer's research shows that Bethe did give a colloquium talk at Harvard in May 1938; but that, as David DeVorkin has recorded, Bethe and Russell did not meet until early February 1939.[56]

in the ejection of the star's surface layers. Those layers had to expand rapidly for the star to brighten suddenly to the extent observed. They estimated that the explosion might convert a fraction of somewhere between 0.1% and 5% of a solar mass–energy, $M_\odot c^2$, into energy liberated in the explosion. This is a far larger fractional conversion of mass into energy than stars convert in the course of their entire lives on the main sequence.

As far as astronomers of the 1930s were concerned, however, supernovae created a further energy puzzle. No longer did one have to account solely for the energy expenditure of stars shining steadily. Supernovae posed similar energy supply problems.

Gamow, Schoenberg, and Supernova Explosions

Early in 1941, Gamow and the Brazilian theoretical physicist Mário Schoenberg, a visiting fellow at George Washington University at the time, tackled the problem of how the energy of a supernova might be generated.[58]

They first considered how the collapse of a star might take place. They postulated that the collapse must create extremely high temperatures, at which electrons would penetrate atomic nuclei, lowering the nuclear electric charge and inducing the emission of an *antineutrino* $\bar{\nu}$. Alternatively, a nucleus containing a combination of N *nucleons* and having a nuclear charge z might emit an electron plus a neutrino, ν,

$$^{z}\mathrm{N} + e^{-} \rightarrow \; ^{(z-1)}\mathrm{N} + \bar{\nu}; \quad ^{z}\mathrm{N} \rightarrow \; ^{z+1}\mathrm{N} + e^{-} + \nu \, . \tag{6.9}$$

Either way, the neutrino or antineutrino would escape the star because of the transparency of ordinary atomic matter to neutrinos. This would lead to rapid cooling as the neutrino or antineutrino carried off much of the energy of gravitational collapse. In the meantime, however, radiation and atoms caught up in the collapse would become exceedingly hot, and the radiation pressure would explosively accelerate the outer layers of the star, hurling them out into space at speeds of thousands of kilometers per second. The rapid expansion of these layers would also give rise to the observed steep rise in the star's luminosity.

The energy liberated in this fashion, Gamow and Schoenberg postulated, would be supplied by the rapid gravitational collapse of a massive central condensation of matter cooled by the emission of the neutrinos. Referring to Chandrasekhar's limiting mass, they wrote "stars possessing a mass larger than the critical mass will undergo a much more extensive collapse and their ever-increasing radiation will drive away more and more material from their surface. The process will probably not stop until the expelled material brings the mass of the remaining star below the critical value. This process may be compared with the supernovae explosions, in which case the expelled gases form extensive nebulosities [such as the] Crab-Nebula."[59]

Despite their confidence in their calculations on the rate of neutrino emission, the two authors were conscious of the complexity of the dynamics of the collapse, which they knew represented "very serious mathematical difficulties."[i]

Gamow and Schoenberg called their neutrino process the *urca process*. In his autobiography, Gamow explains that "We called it the Urca Process, partially to commemorate the casino in which we first met, and partially because the Urca Process results in a rapid disappearance of thermal energy from the interior of a star, similar to the rapid disappearance of money from the pockets of the gamblers on the Casino da Urca [in Rio de Janeiro]."[60]

Although equation (6.9) showed how heavy elements could be converted into other heavy elements in supernova collapse, this did not solve the problem of the origin of heavy elements. Even there, the barriers at mass 5 and 8 would first have to be overcome. A solution to this problem still did not exist.

* * *

With Bethe's successful solution of the stellar energy problem, it became clear that insight into the origin of the heavy chemical elements was still lacking. Contrary to the earlier expectations of Atkinson and Gamow, the stellar energy and elemental abundance problems had not been simultaneously solved. Bethe had made clear that although the conversion of hydrogen into helium explained energy generation in main sequence stars, the formation of heavier elements than helium faced the problem of transcending nuclear masses 5 and 8, both of which are unstable. He provided no solution to this conundrum.

Weizsäcker's two papers continued to be cited for some time, but their lasting contributions were less to an understanding of energy generation in stars, than to the buildup of elements. His realization, that extremely rapid neutron absorption would be required for building the heaviest elements in stars, and his appreciation that the relatively level abundances of the heaviest elements required their formation at extraordinarily high temperatures found only in very massive stars about to explode, is still accepted today.

Although the origin of the chemical elements thus remained unknown, a clear set of requirements for establishing a successful theory was also emerging, and those prospects looked daunting!

More important than it had been for solving the stellar energy problem would be the need for assembling a vast compendium of nuclear interaction cross sections for all conceivable astrophysical settings. These would have to be fed into detailed calculations of likely processes to determine whether they had any chance of successfully explaining the observed abundances. As Weizsäcker had already noted,

---

[i] Current theories agree that neutrinos play a fundamental role in the collapse; but difficulties persist in fully accounting for the complex dynamics of the collapse and explosion.

for the formation of the heaviest elements astrophysical processes taking place on time scales of mere minutes would then have to be found and matched to compatible nuclear reactions. Settling these problems would require close collaboration among theoretical astrophysicists, experimental nuclear physicists, and experts in massive computations.

Notes

1. Astronomers take up the stellar energy problem, 1917–1920, Karl Hufbauer, *Historical Studies in the Physical Sciences*, 11: part 2, 277–303, 1981.
2. Stellar Structure and Evolution, 1924–1939, Karl Hufbauer, *Journal for the History of Astronomy*, 37, 203–27, 2006.
3. A Physicist's Solution to the Stellar-Energy Problem, 1928–1935, Karl Hufbauer, unpublished manuscript. I thank Karl Hufbauer for making this available to me.
4. Hans Bethe's breakthrough on the stellar-energy problem, Karl Hufbauer, unpublished manuscript dated 1997. I thank Karl Hufbauer for making this available to me.
5. Transmutation of the Lighter Elements in Stars, R. d'E. Atkinson & F. G. Houtermans, *Nature*, 123, 567–68, 1929.
6. Zur Frage der Aufbaumöglichkeit der Elemente in Sternen, R d'E. Atkinson & F. G. Houtermans, *Zeitschrift für Physik*, 54, 656–65, 1929.
7. *My World Line*, George Gamow. New York: Viking Press, 1970, p. 59.
8. Structure of the Radioactive Atom and Origin of the α-Rays, Ernest Rutherford, *Philosophical Magazine*, Series 7, 4, 580–605, 1927.
9. Ibid., *My World Line*, Gamow, pp. 59ff.
10. Zur Quantentheorie des Atomkerns (On the Quantum Theory of the Atomic Nucleus) G. Gamow, *Zeitschrift für Physik*, 51, 204–12, 1928.
11. Zur Quantentheorie des radioaktiven Kerns. (On the Quantum Theory of the Radioactive Nucleus), G. Gamow & F. G. Houtermans, *Zeitschrift für Physik*, 52, 496–509, 1928.
12. Ibid., *My World Line*, Gamow, p. 61.
13. Ibid., *My World Line*, Gamow, p. 64.
14. *Ibid. My World Line*, Gamow, pp. 64 ff.
15. *The internal constitution of the stars*, Arthur S. Eddington. Cambridge University Press, 1926, p. 151.
16. Ibid., A Physicist's Solution, Hufbauer.
17. The Analysis of Stellar Structure, E. A. Milne, *Monthly Notices of the Royal Astronomical Society*, 91, 4–55, 1930.
18. Harold Urey and the discovery of deuterium, F. G. Brickwedde, *Physics Today*, 35, 34–39, September 1982.
19. A Hydrogen Isotope of Mass 2, Harold C. Urey, F. G. Brickwedde & G. M. Murphy, *Physical Review*, 39, 164–65, 1932.
20. A Hydrogen Isotope of Mass 2 and its Concentration, Harold C. Urey, F. G. Brickwedde & G. M. Murphy, *Physical Review*, 40, 1–15, 1932.
21. Possible Existence of a Neutron, J. Chadwick, *Nature*, 129, 312, 1932.
22. Ibid., A Physicist's Solution . . . , Hufbauer.
23. On the Theory of Stars, L. Landau, *Physikalische Zeitschrift der Sowjetunion*, 1, 285–88, 1932.
24. Nuclear Transformations and the Origin of the Chemical Elements, G. Gamow, *Ohio Journal of Science*, 35, 406–13, 1935.

25. Artificial Radioactivity produced by Neutron Bombardment, E. Fermi, et al., *Proceedings of the Royal Society of London A*, *146*, 483–500, 1934.
26. Über Elementumwandlungen im Innern der Sterne. I. (On Transmutation of Elements in Stars. I.), C. F. v. Weizsäcker, *Physikalische Zeitschrift*, *38*, 176–91, 1937.
27. Zur Theorie der Kernmassen, C. F. v. Weizsäcker, *Zeitschrift für Physik*, *96*, 431–58, 1935.
28. Über Elementumwandlungen im Innern der Sterne. II. C. F. v. Weizsäcker *Physikalische Zeitschrift*, *39*, 633–46, 1938.
29. Ibid., Über Elementumwandlungen I., Weizsäcker, 1937, p. 178.
30. Ibid., Über Elementumwandlungen II., Weizsäcker, 1938.
31. Stellar Structure, Source of Energy, and Evolution, Ernst Öpik, *Publications de L'Obervatoire Astronomique de L'Université de Tartu*, *xxx*, (3), 1–115, 1938.
32. Ibid., Über Elementumwandlungen II., Weizsäcker, 1938.
33. Versuch einer Theorie der β-Strahlen. I., E. Fermi, *Zeitschrift für Physik*, *88*, 161–77, 1934.
34. Selection Rules for the β-Disintegration, G. Gamow & E. Teller, *Physical Review*, *49*, 895–99, 1936.
35. Nuclear Physics A. Stationary States of Nuclei, H. A. Bethe & R. F. Bacher, *Reviews of Modern Physics*, *8*, 82–229, 1936.
36. Nuclear Physics B. Nuclear Dynamics, Theoretical, H. A. Bethe, *Reviews of Modern Physics*, *9*, 69–224, 1937.
37. Nuclear Physics C. Nuclear Dynamics, Experimental, M. Stanley Livingston & H. A. Bethe, *Reviews of Modern Physics*, *9*, 245–390, 1937.
38. Ibid., Hans Bethe's Breakthrough Solution, Hufbauer.
39. Ibid., *My World Line*, Gamow, p. 58.
40. Die Theorie des Sterninnern und die Entwicklung der Sterne B. Strömgren, *Ergebnisse der Exakten Naturwissenschaften*, *16*, 465–534, 1937.
41. *An Introduction to the Study of Stellar Structure*, S. Chandrasekhar, University of Chicago Press, 1939; Dover edition, 1961.
42. The Formation of Deuterons by Proton Combination, H. A. Bethe and C. L. Critchfield, *Physical Review*, *54*, 248–54, 1938.
43. Ibid. Versuch einer Theorie…Fermi, 1934.
44. Ibid., Selection Rules, G. Gamow & E. Teller.
45. Ibid. The Formation of Deuterons…Bethe and Critchfield, 1938, p. 254.
46. Energy Production in Stars, H. A. Bethe, *Physical Review*, *55*, 434–56, 1939.
47. Origin of Stellar Energy, L. Landau, *Nature*, *141*, 333–34, 1938.
48. Red Giants and Supergiants with Degenerate Neutron Cores, Kip S. Thorne and Anna. N. Żytkow, *Astrophysical Journal*, *199*, L19–24, 1975.
49. R. C. Cannon, et al., "The Structure and Evolution of Thorne-Żytkow Objects," *Astrophysical Journal*, *386*, 206–14, 1992.
50. Ibid., Theorie des Sterninnern, Strömgren, 1937.
51. Ibid., Hans Bethe's Breakthrough, Hufbauer.
52. My Life in Astrophysics, Hans A. Bethe, *Annual Reviews of Astronomy and Astrophysics*, *41*, 1–14, 2003.
53. *Henry Norris Russell – Dean of American Astronomers*, David H. DeVorkin, Princeton University Press, 2000 pp. 254–55.
54. What Keeps the Stars Shining?, Henry Norris Russell, *Scientific American*, *161*, 18–19, July 1939.
55. Ibid., My Life in Astrophysics, Bethe
56. Karl Hufbauer email to the author, September 8, 2012.

57. On Super-novae, W. Baade & F. Zwicky, *Proceedings of the National Academy of Sciences of the USA*, *20*, 254–59, 1934.

58. Neutrino Theory of Stellar Collapse, G. Gamow and M. Schoenberg, *Physical Review*, *59*, 539–47, 1941

59. Ibid., Neutrino Theory of Stellar Collapse, Gamow and Schoenberg, pp. 546–47.

60. Ibid., *My World Line*, Gamow, p. 137.

Part II A National Plan Shaping the Universe We Perceive

7

A New Order and the New Universe It Produced

July 5, 1945 marked a break in the way science and engineering were to be conducted in the United States in the postwar era. The government would take a leading role in the transition. The future was to belong to scientists and engineers working as teams partly dedicated to basic research, but more importantly working for the benefit of the nation, its security needs, its ability to feed its people, and the health of its children. This chapter recounts in some depth how and why the new program came to be initiated and adopted.[a]

In the new arrangement, the future of astrophysics was nowhere mentioned, but time would show astronomy's emphasis on surveys and surveillance to be most closely aligned with military priorities. It took less than three decades to show that the alliance between the military and astronomers was leading to formidable advances no astronomer had anticipated. Soon other nations began to emulate the U.S. lead, and astronomy started advancing at a dizzying pace.

The contrast with astronomy before World War II could not have been greater.

Support for Basic Research in the Prewar Years

When Niels Bohr, during an impromptu visit from the young George Gamow late in the summer of 1928, assured him a year's Carlsberg Fellowship at the Royal Danish Foundation starting the next day, no senior physicist would have been surprised. When Bohr and Ernest Rutherford, the following year, asked the Rockefeller Foundation to provide Gamow a fellowship so he could work with Rutherford at Cambridge University, this too would have been expected. The world of physics and astronomy was small. Leading scientists knew each other personally and had close contacts with the few organizations that supported basic research.

---

[a] Readers wishing to examine further the broader context of the rise of postwar science in the United States will find the investigations of Daniel J. Kevles, G. Pascal Zachary, and David Dickson informative on a broad range of topics.[1,2,3]

In the United States the Rockefeller Foundation and the John Simon Guggenheim Foundation were two of the leading sources supporting promising young scholars.[4]

The structure of astronomy in the United States may have been even more intimate than the organization of physics. Beginning in the late nineteenth century, philanthropy had played an indispensable role in the construction of the major U.S. astronomical observatories. The Yerkes Observatory of the University of Chicago was financed by the Chicago industrialist Charles T. Yerkes and completed in 1897.[5] The Mount Wilson Solar Observatory was completed in 1904 under the auspices of the philanthropically founded Carnegie Institution of Washington; its 100-inch telescope was added in 1919 and was paid for through a philanthropic donation from John D. Hooker of Los Angeles, after whom the telescope was named.[6] In 1928 the Rockefeller Foundation pledged $6M to Caltech for the construction of the 200-inch telescope on Mt. Palomar; this complex undertaking was not completed until 1949.[7] All three observatories were the brainchildren of George Ellery Hale an enormously talented astronomer/entrepreneur. The dream of constructing the Yerkes telescope at Williams Bay, Wisconsin, had captivated him in 1892 when he was only 24. By the time the 200-inch telescope was completed, half a century later, he had long passed away.

Although both the Rockefeller and Guggenheim Foundation fellows were selected through the National Research Council (NRC) of the National Academy of Sciences, this sole organization in Washington straddling the interests of all scientific disciplines, lacked federal authorization and the government was reluctant to have the NRC provide federally funded research fellowships. At the end of World War I, the military services had cut back drastically on all areas of research and also on their interactions with the NRC. The historian of science Daniel Kevles attributes this hands-off approach to the support of basic research to a "disjunction between, on the one hand, the freedom demanded by civilian scientists and, on the other, the insistence of the armed services on controlling their research" as well as the military's "necessity of secrecy."[8]

In the United States the conduct of science in the 1920s and 1930s had a decidedly pragmatic bent. The 1920s had seen industry opening its own research laboratories, to the point where industry was investing twice as much in research as the federal government. The government's research was geared largely to promoting military and industrial needs. Basic research in physics and astronomy was supported mainly through private resources.

The few leading U.S. universities, including Caltech, Chicago, Cornell, Harvard, Princeton, Rochester, Stanford, and Vanderbilt, dedicated to conducting basic research competitive at international levels, might be supported by the Rockefeller Foundation, the Carnegie Institution, or by wealthy industrialists such as George Eastman or T. Coleman Du Pont. Leading state universities like Wisconsin, Michigan, and Berkeley expected research support from their state legislatures. And Wisconsin profited from its Wisconsin Alumni Research Foundation, funded in part by patents taken out on processes the university's researchers had discovered.[9]

Basic research in prewar days was a venture on which neither the government, nor industry was willing to embark, though virtually everyone in science was aware of its potential.

A National Research Program

The plan, *Science – The Endless Frontier*, submitted that July 1945 day to President Harry S Truman, was the brainchild of Vannevar Bush, the visionary former Vice President and Dean of Engineering at MIT. As Director of the wartime *Office of Scientific Research and Development, OSRD*, the agency charged with mobilizing civilian science, Bush had been instrumental in setting up the Manhattan Project to build the atomic bomb; organizing improvements in military *radar* at the MIT Radiation Laboratories; and overseeing the production of the insecticide DDT, penicillin, sulfa drugs, and better vaccines, which had dramatically reduced military death rates in World War II.

In these successes of large-scale science and engineering efforts, Bush foresaw a new way to conduct postwar science. It entailed a communal effort that would make the United States a world leader. Scientists and engineers working hand-in-hand with the military would make America a military superpower; basic research conducted in the universities would educate a first-rate generation of young scientists and engineers; medical researchers would seek to raise levels of health and longevity; and agricultural research would ensure the nation's ability to feed itself.

Military funding of basic and applied research in the quarter century after World War II dramatically increased scientific productivity in the United States. Astronomy in particular obtained infusions that initiated radio, infrared, X-ray, and gamma-ray observations. A side effect of the new order was a heightened influence of outside forces on astronomy; some were beneficial, others irksome. No comparable incentives or intrusions had existed in astronomy or astrophysics before World War II. The transformation initially was gradual. But with the escalating Cold War, U.S. governmental research support of astrophysics rapidly escalated to make astronomy part of a burgeoning national research effort.

Vannevar Bush, Creator of the New Order

Vannevar Bush, born in Everett, Massachusetts in 1890, graduated from Tufts College in 1913.[b] In 1916, he obtained a doctorate in engineering jointly from MIT and Harvard and then returned to Tufts as an assistant professor.

By 1919, Bush had joined the faculty of MIT, where one of his greatest successes was the construction of a *differential analyzer*, a complex mechanical array of shafts

---

[b] The name *Vannevar* rhymes with the word *beaver*.

Figure 7.1. Vannevar Bush about the time he arrived in Washington, just before World War II. As Director of the Office of Scientific Research and Development (OSRD). Bush became the chief organizer of wartime scientific developments in the United States. At war's end, he conceived the visionary document *Science – The Endless Frontier*, the blueprint governing the development of science in America in the second half of the twentieth century. It foresaw a loosely knit alliance in which basic science would always be ready to respond to evolving national priorities. (*Courtesy of the Carnegie Institution for Science*).

and gears designed to solve differential equations. Becoming Dean of Engineering in 1932, he helped transform MIT into an institution dedicated not only to advanced engineering but also to research in basic science. He left the Institute in 1938 to head the Carnegie Institution of Washington, at the time one of the largest foundations supporting science in the United States.

Once in Washington, Bush sought a way to present a plan for improving scientific research in the United States to President Franklin D. Roosevelt. Bush realized, as few others, that the nation's universities could offer powerful leadership on research involving military scientific problems that no other group of institutions could readily duplicate. In a meeting that lasted only 10 to 15 minutes, he handed Roosevelt a single-page memorandum that proposed the creation of a new research agency to harness this intellectual resource for the benefit of the nation at a time of rising global tensions. Roosevelt wasted no time in approving the document, saying, "That's okay. Put 'OK, FDR' on it."[10]

Changing how Washington's institutions work is never easy, even with the backing of a president. Bush now had to struggle to place his newly formed *National Defense Research Committee* and its successor, the Office of Scientific Research and

Development (OSRD), on an equal footing with the U.S. Army and Navy in shaping military research and a global war strategy.

By nature blunt, Bush soon found he needed to concentrate on forging good relationships with Roosevelt, the President's trusted advisor Harry Hopkins, the prevailing leadership at lower levels in Washington, and ultimately also Roosevelt's successor, President Harry S Truman.[11]

In late 1944, as the war was approaching an end, Bush increasingly turned to planning the future. He sensed that the United States could maintain ascendancy in peace only as it had during the war, as a world leader in science and technology.

Although he held no such formal title, Bush in effect had become the first science advisor to a president of the United States. Once again he turned to Roosevelt for supportive instructions.[12]

On November 17, 1944, only a few months before his death, Roosevelt directed Bush to prepare a report to answer four questions of national priority suggested by the enormous success of U.S. science in helping to win the war.

> (1) What can be done, consistent with military security, and with the prior approval of the military authorities, to make known to the world as soon as possible, the contributions which have been made during our war effort to scientific knowledge?
>
> (2) With particular reference to the war of science against disease, what can be done now to organize a program for continuing in the future the work which has been done in medicine and related sciences?
>
> (3) What can the Government do now and in the future to aid research activities by public and private organizations?
>
> (4) Can an effective program be proposed for discovering and developing scientific talent in American youth so that the continuing future of scientific research in this country may be assured on a level comparable to what has been done during the war?

But, as if to emphasize the fourth of these questions, Roosevelt's letter (which Bush had helped to craft) began with:

> The Office of Scientific Research and Development, of which you are the Director, represents a unique experiment of team-work and cooperation in coordinating scientific research and in applying existing scientific knowledge to the solution of technical problems in the war. Its work has been conducted in the utmost secrecy and carried out without public recognition of any kind; but its tangible results can be found in the communiques coming in from the battlefronts all over the war. Some day the full story of its achievement can be told.
>
> There is, however, no reason why the lessons to be found in this experiment cannot be profitably employed in times of peace. The information, the techniques, and the research experience developed by the Office of

Scientific Research and Development and by the thousands of scientists in the universities and in private industry, should be used in the days of peace ahead for the improvement of the national health, the creation of new enterprises bringing new jobs, and the betterment of the national standard of living.

It is with that objective in mind that I would like to have your recommendations ...

It is this introductory statement that perhaps best encapsulated the difference between prewar and postwar astronomy. Its importance cannot be overstressed.

* * *

In response to the president's directive, Bush submitted a report on July 5, 1945 on which some of the finest scientific, technical, and legal minds in the United States had worked over the winter and spring. Its title was 'Science – The Endless Frontier.'[13]

The report had been completed in only seven months, but by then Roosevelt had died, and Bush submitted the document to President Truman. The summary of the report was a visionary statement written largely by Bush himself. It proposed that the nation's future depended on continuing technological progress across a broad front, including medicine, agriculture, engineering, and the basic sciences on which these technologies would depend. As far as the physical sciences were concerned, the opening statement, in part, read:

Scientific Progress Is Essential

... New products, new industries, and more jobs require continuous additions to knowledge of the laws of nature, and the application of that knowledge to practical purposes. Similarly, our defense against aggression demands new knowledge so that we can develop new and improved weapons. This essential new knowledge can be obtained only through basic scientific research.

Science can be effective in the national welfare only as a member of a team, whether the conditions be peace or war. But without scientific progress no amount of achievement in other directions can insure our health, prosperity, and security as a nation in the modern world.

Sentences that followed made clear that Bush saw science as a key element in a seamless effort to improve the welfare and security of the nation:

There must be more – and more adequate – military research in peacetime. It is essential that the civilian scientists continue in peacetime some portion of those contributions to national security which they have made so effectively during the war. This can best be done through a civilian-controlled organization with close liaison with the Army and Navy ...

Basic scientific research is scientific capital … For science to serve as a powerful factor in our national welfare, applied research both in Government and in industry must be vigorous. …

The training of a scientist is a long and expensive process. Studies clearly show that there are talented individuals in every part of the population, but with few exceptions, those without the means of buying higher education go without it. If ability, and not the circumstance of family fortune, determines who shall receive higher education in science, then we shall be assured of constantly improving quality at every level of scientific activity. The government should provide a reasonable number of undergraduate scholarships and graduate fellowships in order to … attract into science only that proportion of youthful talent appropriate to the needs of science in relation to the other needs of the Nation for high abilities. …

The effective discharge of these new responsibilities will require the full attention of some over-all agency devoted to that purpose. … Such an agency should be composed of persons of broad interest and experience, having an understanding of the peculiarities of scientific research and scientific education. It should have stability of funds so that long-range programs may be undertaken. It should recognize that freedom of inquiry must be preserved and should leave internal control of policy, personnel, and the method and scope of research to the institutions in which it is carried on. It should be fully responsible to the President and through him to the Congress for its program.

On the wisdom with which we bring science to bear in the war against disease, the creation of new industries, and in the strengthening of our Armed Forces depends in large measure our future as a nation.

I have quoted these passages so extensively because they show how deeply Bush was convinced that the nation's military needs and the health, welfare, and prosperity of its people depended on an integrated, government-sponsored program of technology, science, and science education. Much of this chapter will show the extent to which this vision, inspired largely by America's exploitation of science and technology in concluding World War II, came to shape U.S. science in the second half of the twentieth century.

* * *

Congressional action is seldom rapid. Five years were to elapse before Congress passed a bill in March 1950 establishing the National Science Foundation (NSF) as the institution Bush had envisioned to carry out his recommendations. Not everything that Bush had requested in his report was granted. But most of the crucial recommendations of 'Science – the Endless Frontier' came through unscathed.

David DeVorkin has written about the decidedly unenthusiastic initial reception the U.S. astronomical leadership, perhaps predictably, gave the new policy[14]:

They all worried about autonomy, loss of control, and a loss of traditional forms of support. Home institutions might lessen their support when they saw that sciences like astronomy could attract outside funding, and, as Struve and Shapley and other observatory directors knew, government patronage in any form, if not channeled through traditional lines of authority, also held out the possibility that hierarchical patterns would change. For astronomy, it could mean the loss of the autonomy, power, and authority of the observatory director... Men like Struve and Shapley would not allow their autonomy as directors to be weakened to the point where they could not control the course of research... [they] dominated much of American astronomy in a manner that would be hard to appreciate today.[c]

Younger astronomers, in contrast, quickly grasped the new opportunities the changed conditions offered.

* * *

In 1990, the historian of science Daniel J. Kevles wrote a comprehensive 'Appreciation of the Bush report' in conjunction with a reissue of 'Science – the Endless Frontier' on the 45th anniversary of the report's submission and the 40th anniversary of the creation of the NSF.[15]

As Kevles recalled, in the five years it took to establish the NSF, the nascent Cold War had focused research and development (R&D) on national security. In 1949–50, the federal R&D budget was approximately $1B, 90% of which was shared between the Department of Defense (DoD) and the Atomic Energy Commission (AEC), which was largely concerned with nuclear weaponry. Much of the applied research was carried out by industry; more basic technology research was frequently assigned to universities. Through the Office of Naval Research (ONR) the military also was the main supporter of basic scientific research in academic institutions, a responsibility the Navy had assumed during the five years it took to agree on the establishment and functions of the NSF. Figures cited by Kevles indicate that in 1953, during the Korean War, military R&D was funded at a level of $3.1B, while Congress provided the NSF with a mere $4.75M, apparently in the belief that basic research was a luxury in time of war.

The launch of the Soviet satellite *Sputnik* in 1957 shocked a bewildered United States into investing more heavily in research. Under President Dwight D. Eisenhower, the National Aeronautics and Space Administration (NASA) was established to conduct a civil space program, separate from military space efforts. The NSF budget also grew. Figures provided by Kevles show that, by 1967, the federal R&D budget had grown to nearly $15B, only about half of which was related to

---

[c] Otto Struve and Harlow Shapley, respectively, were the powerful directors of two of the largest observatories, the University of Chicago's Yerkes observatory in Williams Bay, Wisconsin, and the Harvard College Observatory.

defense. The NSF budget that year approached $500 M, part of which enabled the Foundation to provide more than 13% of the federal support for academic research. DoD supplied 19%, NASA about 8%, and the AEC roughly 6%.[d] U.S. universities awarded nearly 13,000 doctoral degrees in science and engineering that year, more than twice the number awarded in 1960.[17]

In keeping with the recommendations of 'Science – The Endless Frontier,' both military and civilian agencies respected intellectual self-determination and gave academic scientists great freedom in how research funding was spent in accord with their professional judgment.

This was the atmosphere in which astronomy and space science was conducted in the United States during the 1960s. The national security shock delivered by the Soviet's launch of the *Sputnik* satellite in 1957 totally transformed astronomy. Many astrophysicists remember those years as a golden era, when grant applications and reports were easy to write and researchers could concentrate on what they did best.

Research funding was abundant; capable academic scientists and engineers were often given free rein in deciding what they would study, as well as what and where they would publish. Fellowships and research assistantships for graduate students and postdocs were readily obtained in astronomy, particularly in the 1960s, as the space race was heating up.

Commenting on these changes, the British pioneer of radio astronomy, Sir Bernard Lovell, noted that whereas 1957 saw only 168 graduate students in astronomy enrolled in 28 U.S. universities, nine years later there were 793 – almost a fivefold increase. This annual growth rate of 19% was twice the growth rate in related sciences.[18]

Astronomy in Britain before and after World War II

In Britain the contrast in the field of astronomy between pre and post–World War II years also was striking. Three decades after the War, Lovell, by then knighted for his postwar astronomical contributions, provided an insider's account of what he called 'The Effects of Defence Science on the Advance in Astronomy.'[19] The thrust of his paper is summarized in its first few lines:

> The period of 22 years from 1935 to 1957 will surely be regarded in history as an epoch when defence science not only influenced but actually revolutionized observational astronomy. The centre piece of this period is World War II but there are three separate events which are substantially localized in the longer period and which can now be seen to be the critical factors determining the nature of astronomical research today. They are

---

[d] The remaining 54% not mentioned by Kevles came mainly from the Department of Health and Human Services, the Department of Agriculture, and other U.S. funding agencies, and thus were not directly related to the physical sciences. The total federal obligation to R&D in academic institutions was about $1.45B in 1967.[16]

the development of radar, the launching of Sputnik One and the growth of scientific manpower stimulated by the success of Allied Science in World War II. . . .

Lovell's remark about manpower needs an added comment. He refers not just to sheer numbers. Those in themselves were dramatic, but postwar expenditures for science in Britain grew rapidly as well, rising from an average of £778 K for the five prewar years to £6.68M by 1954, enabling the Department of Scientific and Industrial Research to increase the number of research grants by a factor of 38 and the number of students supported by a factor of nearly 10.[20]

When the young university research staff were drafted into the establishments in 1939 they found that a completely different technical level applied. At the end of the war they were themselves acquainted with the most sophisticated and advanced techniques of the day. They had learned how to collaborate with a wide range of scientific and non-scientific disciplines, and they had learnt the way of committees and how to use Government machinery. In . . . the critical decade after the war the new generation of scientists were taught by individuals whose outlook on research was utterly different from that of the pre-war environment. The involvement with massive operations had conditioned them to think and behave in ways which would have shocked the pre-war university administrators. All these factors were critical in the large-scale development of astronomy.

This statement represents the crucial difference between prewar and postwar astronomy. Lovell elucidated it nicely in the British context with which he was familiar. But the same message could have been heard from young scientists who had left their positions at U.S. universities to participate in wartime efforts at Los Alamos or at the MIT Radiation Laboratories and had then returned to academe after the war.

Astronomy before World War II

Popular books written before 1939 and the outbreak of World War II depicted the Universe in majestic terms. Displayed in the light of serenely shining stars, the Cosmos inexorably expanded over the eons. A few stars known to pulse with great precision lent them the aspect of beacons shining far across the expanses of space. Occasionally a nova or an overpoweringly larger supernova explosion punctuated the proceedings in a random display of distant fireworks; but the overall impression was one of infinite serenity.

Because some of the books in this genre were written by leading astronomers and astrophysicists, such as Sir James Jeans, we may surmise that they largely reflected the views of professional astronomers.

World War II and the subsequent Cold War changed all this with the new tools they introduced into astronomy. These revealed an entirely different landscape and demanded changes in how the Universe should be perceived!

Widespread governmental support also changed the social structures of astronomy and astrophysics. This had, at least in part, been anticipated in 'Science – The Endless Frontier'; but the transformation of the entire field through the release of previously secret military technologies had not. Major changes in astronomy resulted, often accidentally, when uses of military devices revealed something curiously unexpected in the heavens. Time and again, unanticipated findings initiated new lines of inquiry channeling the attention of astronomers into previously unimagined directions. By the late 1960s, a series of such abrupt breaks with traditional prewar procedures had transformed the field.

A New Sociology of Astronomy

Historians of science have recorded these changes; but it is the ethnographers and sociologists of science who can most clearly correlate what came to be learned about the Universe with the changes in how astronomy was organized, funded, and retooled through the import of military technologies.

Three particularly thoughtful treatises come to mind.

The first to appear was the book 'Astronomy Transformed – The Emergence of Radio Astronomy in Britain,' jointly written by David O. Edge, a former radio astronomer turned sociologist, and by the professional sociologist Michael J. Mulkay.[21] It describes how radio techniques ushered in by World War II radar changed astronomy. Most of the practitioners of the novel field of radio astronomy were physicists or engineers who were familiar with radio techniques and eager to apply them to new scientific investigations. As Edge has pointed out, most optical astronomers welcomed the influx of these outsiders.[22]

The second investigation is one of the earliest historical/sociological treatments tracing the birth and growth of X-ray astronomy, 'Glimpsing an Invisible Universe – The Emergence of X-Ray Astronomy,' by Richard E. Hirsh. Here, similar relationships between military and astronomical interests found a confluence.[23]

The third study is a book-length review on 'Infrared Sky Surveys' by Stefan D. Price who, over several decades, worked as a U.S. Air Force researcher.[24] Price documents the Air Force rationale for pursuing infrared sky surveys throughout the Cold War.

The military was concerned with rapidly detecting an enemy launch of ballistic missiles and the emplacement of Earth-orbiting military satellites in space. Thermal emission from satellites, rockets, rocket exhausts, as well as aircraft and their engine exhausts might readily be detected by sensitive infrared techniques, but so also would be the infrared radiation from celestial sources; and this was largely unknown.

The military needed infrared maps of the sky to tell how well or how badly a defense system based on infrared detection would work. For this purpose the U.S. Air Force began to conduct its own sky surveys, in which Price and Russell G. Walker, his colleague and one-time supervisor at the Air Force Cambridge Research Laboratories, played leading roles. Simultaneously, however, the Air Force began funding the research of industrial and academic engineers and physicists specializing in infrared optics and willing to pursue astronomical investigations with sensitive infrared detectors.

Throughout the postwar era, the military's capabilities in surveillance increased in power, and as the armed forces outgrew a technique, a capability, or a device and no longer needed it, the military establishment passed it on to colleagues in astronomy. Aside from radio techniques, X-ray, gamma-ray, and infrared devices were leading to the discovery of entirely new astronomical phenomena – stars and even entire galaxies that emitted more power at radio or X-ray wavelengths than at all other wavelengths combined. No less significant were observations of dusty star-forming regions that shone mainly in the infrared where heated *dust* grains radiate most powerfully.

A new sense of excitement and purpose swept over astronomy, leading to revolutionary new views on the origins and evolution of the Universe, the formation of galaxies, and the birth and death of stars.

A trait shared by the discovery instruments the military bequeathed was an ability to look at the Universe through entirely new eyes. Though blind to visible light, these instruments sensed radiations that astronomers had never tapped before.

The next few sections summarize some of the most striking discoveries the new technologies brought about. In recalling these, I will refrain from repeatedly emphasizing how each of the discoveries was made with equipment handed down by the military-industrial complex; but the reader should keep in mind that low-noise electronic components and circuitry, receivers and detectors sensitive at the most varied wavelengths, and the ubiquitous technical know-how developed throughout World War II and the Cold War consistently played a central role. Use of this magical equipment became so natural that astronomers no longer even noticed. Only after the passage of nearly half a century, when participation on large international projects became a priority, would U.S. astronomers become aware of how dependent they had become on military technology. At the end of this chapter we return to this topic to see how the realization then sank in.

The Birth of Radio Astronomy

In retrospect, the serene astronomical landscape holding sway before World War II had begun to show fault lines even before the war – although few seemed to notice. The first signs came from a puzzling observation of radio noise emanating from the Milky Way's center that Karl Guthe Jansky, a radio engineer at the Bell Telephone Laboratories in Holmdel, New Jersey, had discovered in 1932 as he attempted to track down spurious noise interfering with radio telephone transmission.[25] At the time, the astronomical community did not know what to do with this information and largely ignored it. The sole exception appears to have been a paper by Fred L. Whipple and Jesse G. Greenstein at Harvard, who tried but did not succeed in accounting for the radiation.[26]

World War II introduced powerful *radar* systems to track enemy airplanes. Invented just before the onset of the war and further developed throughout the conflict, radar enabled British intelligence to see the approach of German aircraft long before they had crossed the English Channel. Fighter planes could be scrambled in time and the approaching aircraft shot down. Later, Allied bombers were equipped with radar so their crews might see their targets through cloud cover and drop their bombs with greater accuracy.

Wartime radar operated at radio wavelengths. Though not designed for the purpose, the most sensitive radar receivers became the earliest radioastronomical telescopes. The first celestial source of any kind found to emit radio waves was the Sun, a radio outburst from whose surface was initially noted by James Stanley Hey in Britain.

Although Hey's surprising findings dated back to early 1942, military secrecy delayed publication of his data until after the war. In a 1946 letter to *Nature*, Hey reported[27]:

> It is now possible to disclose that, on one occasion during the War, Army equipments observed solar radiation of the order of [a hundred thousand] times the power expected from the sun. ... This abnormally high intensity of solar radiation occurred on February 27 and 28, 1942. ... The main evidence that the disturbance was caused by electromagnetic radiations of solar origin was obtained by the bearings and elevations measured independently by the [4–6 meter band radar] receiving sets, sited in widely separated parts of Great Britain (for example Hull, Bristol, Southampton, Yarmouth). ...

At the time of these observations strong solar flares were observed at the French observatory at Meudon, and it soon became clear that solar radio emission is enhanced when the Sun is active. Independently from Hey, Grote Reber, an electrical engineer who had constructed a radio telescope in his backyard in Wheaton, Illinois, and G. C. Southworth, a research engineer at the Bell Telephone Laboratories, also were detecting radio emission from the Sun during the war. Southworth's observations were carried out at shorter wavelengths, in the 1–10 cm range, and provided measures of what he initially believed to be the Sun's quiescent thermal emission at these wavelengths.[28] Reber's observations were conducted in the 1.83–1.92 m wavelength band, detecting the weak emission from the Sun at times when it apparently was not flaring.[29]

Shortly after the war, Martin Ryle, whose research group would later make ground-breaking radio-astronomical discoveries, began using discarded British military radar equipment to set up a radio astronomy program at Cambridge. A. C. Bernard Lovell established a similar group at Manchester, and in the Netherlands Jan Oort used former German radar equipment to start a Dutch radio astronomy program. An active program in radio astronomy also took root in Australia.[30]

These are the groups that pioneered radio astronomy at its outset.

Sketches of how some of the most significant discoveries came about confirm the central role that novel instrumentation played.

Radio Galaxies

Shortly after the war, Hey and his co-workers discovered a strong, time variable radio flux from the Cygnus region of the sky, now known to be occupied by the strongest radio source observable in the sky, a galaxy designated as Cygnus A.[31] The time variability was later attributed to passage of the arriving radiation through turbulent regions in *interplanetary space* or Earth's magnetosphere.

By 1954, Walter Baade, working with his colleague Rudolph Minkowski at the 200-inch telescope at Mount Palomar, had succeeded with the optical identification. By then, this was easy. Francis Graham Smith in Cambridge had previously used a radio *interferometer* to locate the radio position to within about a minute of arc.[32] Once the 200-inch telescope was pointed at this region a curious object emerged that, as Baade and Minkowski pointed out, appeared to be "a rare case of two galaxies in actual collision."[33] As improved angular resolution at radio frequencies was achieved, a progression of powerful extragalactic radio sources was being identified.

Quasars

In the late 1950s, the Manchester radio astronomer Cyril Hazard began measuring the angular diameters of radio sources listed in the third catalog compiled by Martin Ryle's Cambridge radio astronomy group. One of these, the source 3C 273, appeared to be highly compact. Available radio interferometers did not have adequate *spatial resolving power* to measure the apparent diameter of this or other sources that appeared to be *stellar*, in the sense of appearing to be point-like sources. These came to be named *quasi-stellar objects* (QSOs), a name soon contracted to *quasars*.

3C 273 lies in a part of the sky occasionally traversed by the Moon. A compact source dims rapidly as the limb of the Moon passes in front of it. A similarly rapid brightening occurs as the Moon moves on and the occulted radio source reemerges. Precise timing of these two events can provide an improved position of the source, as well as a better estimate of its angular dimensions.[34] Hazard and his team thus used lunar occultations to study 3C 273. The positional accuracy was sufficiently high, about 1 arc second, that an optical identification was assured.

The optical astronomer Maarten Schmidt at the California Institute of Technology obtained a photographic plate of the region with the Mount Palomar 200-inch telescope, as well as an optical spectrum that appeared quite unusual. Soon he realized that the spectrum indicated a 47,400 kilometer per second redshift, implying an enormous distance across the Universe and a fantastically high luminosity if the redshift was due to the cosmic expansion![35] An article in *Nature* reporting the findings of Cyril Hazard and his colleagues was immediately followed by Maarten Schmidt's article in the journal – an exemplary collaboration between radio and optical astronomers.

The Microwave Background Radiation

In the late spring of 1964, Arno A. Penzias and Robert W. Wilson were calibrating the radio noise temperature of a 7.3 cm wavelength system in the 20-foot horn-reflector antenna at the Bell Telephone Laboratories' Crawford Hill facility in Holmdel, New Jersey. The antenna had originally been designed for signal detection from communications satellites. Despite every care they had taken, Penzias and Wilson found an excess noise temperature in the zenith, amounting to about 3.5 K. This excess constituted an unanticipated signal once allowance had been made for atmospheric effects and antenna characteristics, as well as *diffraction* responsible for a modest thermal back-lobe signal emanating from the ground.[36]

The excess signal they observed was, within the limits they could set, isotropic, unpolarized, and unaffected by seasonal variations between July 1964 and April 1965. They were puzzled by this signal until they heard that Robert H. Dicke and three younger colleagues at nearby Princeton University in New Jersey had just predicted the potential existence of a cosmic thermal background signal approximately of the strength Penzias and Wilson were observing – a proposed vestige of radiation that had pervaded the Universe at early epochs when the Cosmos was highly compact and very hot.[37]

The same prediction, however, had already been anticipated 17 years earlier by two young theorists, Ralph A. Alpher and Robert C. Herman, who had estimated that the radiation temperature should be in the range of 1–5 K, remarkably close to the temperature determined by Penzias and Wilson.[38,39]

This may have been the most momentous astrophysical prediction and discovery of the second half of the twentieth century. It made crystal clear that the Universe had at one time been intensely hot!

Cosmic Masers

The *maser*, a precursor of the laser but operating in the microwave radio region, had been discovered by Charles Townes and his co-workers at Columbia University in 1954. In 1964, cosmic masers were accidentally discovered, but their true nature took several years to unravel. It took until 1967 for them to be properly recognized and understood.[40,41,42]

Pulsars

Earlier, I had mentioned the discovery of pulsars by Jocelyn Bell, a graduate student working under the supervision of Antony Hewish at the Mullard Radio Astronomy Observatory of Cambridge University in the late 1960s.[43]

Gravitational Radiation

The accuracy and stability of the pulsars' pulse rates later made possible a wide variety of important astrophysical timing studies. One of the most significant of these provided new insights on general relativity by establishing the emission of *gravitational radiation* by a pair of tightly bound pulsars, whose orbital period gradually declined as the pulsars slowly drew closer to each other.[44,45] In pulsars, Nature had bequeathed us a set of highly reliable astronomical clocks.

Superluminal Radio Sources

In 1966, Martin J. Rees, a 24-year-old graduate student at Cambridge University, suggested that light variations in quasars might be due to explosions in which fragments flying apart at extremely high velocities would appear to be receding from each other at speeds greater than the speed of light, c, even though their actual velocities of mutual recession remained less than c. This incongruous appearance would be due to a difference in the respective times at which radiation from the receding and approaching fragments had been emitted. Radiation received from the nearer fragment would have been emitted some time after radiation from the more distant fragment.

Several years later, in 1970, a team led by MIT physicist Irwin Shapiro used a radio interferometer with a 3900-km baseline reaching from California to Massachusetts, and radio wavelengths as short as 3.8 cm, to observe the quasar 3C 279 at an angular resolution of about 10^{-3} arc seconds. Over a period of four months, they found the separation between two fragments shot out of the quasar to have increased by 10%. Given the quasar's high redshift and thus great distance, this seemed to indicate a *superluminal* relative velocity between the fragments – a mutual recession apparently faster than the speed of light and thus in violation of the laws of special relativity.[46] At this point Shapiro became aware of the paper by Rees and realized that the situation was more complex but had a ready-made explanation in apparent superluminal expansion.[47]

Interstellar Magnetic Fields

In 1949 the physicist Enrico Fermi suggested that cosmic-ray particles reaching us from beyond the Solar System might be accelerated to their high observed energies through a succession of reflections off magnetic fields embedded in moving clouds of interstellar gas. Fermi calculated that the required magnetic field strength in the clouds would have to be about 5 microgauss (5×10^{-6} gauss), roughly a hundred thousand times lower than the strength of the magnetic field at Earth's surface.[48]

Such a weak magnetic field is difficult to measure. But 20 years after Fermi's predictions, two entirely different radio astronomical measurements, one in 1969 by Gerrit L. Verschuur, and another, three years later, in 1972, by Richard Manchester, inferred an interstellar magnetic field strength of the same order as Fermi had predicted. Both observations were obtained at the National Radio Astronomy Observatory in Green Bank, West Virginia.[49,50]

The Introduction of Powerful Rockets

Astronomers and the military share an interest in surveillance techniques that often can be used by either community. Even the astronomical sources observed can be of importance to both, although for the military an astronomical source may merely be a nuisance that interferes with detecting enemy activities in space.

Obtaining the clearest view of the sky often requires launching a telescope into space to avoid the limitations of observing through the atmosphere. Here also, the military paved the way.

Over a period of a dozen years starting in the early 1930s, Wernher von Braun and an army of technical experts in Germany had painstakingly learned how to build the powerful military V-2 rockets with which Hitler later hoped to turn the tide of war. These missiles could be guided with reasonable accuracy to such targets as London or Antwerp.

After the War, von Braun and a team of his experts were brought to the United States, along with captured V-2 rockets. The captured rockets were to be tested to help the United States design its own missiles, and as part of the test series, scientists were offered an opportunity to build payloads to be taken to altitudes above the atmosphere.

X-Ray Observations from Above the Atmosphere

In 1948 a group of researchers at the U.S. Naval Research Laboratory (NRL) began to place ultraviolet- and X-ray-sensing detectors aboard the captured rockets. The ultraviolet and X-rays they sought to detect are strongly absorbed by atmospheric gases and could be observed only with telescopes at high altitudes. On the earliest flights, the physicist T. R. Burnight used simple detectors consisting of photographic films covered by thin metal plates. With a rocket launched at White Sands, New Mexico, on August 5, 1948, he observed solar X-rays that could penetrate through beryllium plates three-quarters of a millimeter thick.[51]

This flight provided evidence for X-radiation in unexpected intensities. A few months later Burnight's colleagues, Richard Tousey, J. D. Purcell, and K. Watanabe, confirmed these observations.[52] And on September 29, 1949, Herbert Friedman, S. W. Lichtman, and E. T. Byram, also of NRL, launched a V-2 payload containing modern photon counters that ushered in a whole new era of X-ray astronomy.[53]

X-Ray Stars

With U.S. Air Force support and a payload designed at least in part to search for X-rays generated at the surface of the Moon by energetic radiation arriving from the Sun, Riccardo Giacconi, Herbert Gursky, Frank Paolini, and Bruno Rossi of the American Science and Engineering Corporation, in June 1962 detected the first signals from an X-ray source outside the Solar System![54]

Soon, the NRL team, led by Herbert Friedman, accurately located the position of this source in the constellation Scorpius,[55] and, within a year, a group at the Lawrence Radiation Laboratory in Livermore, California, obtained a first coarse spectrum of the Scorpius source in work conducted under the auspices of the AEC. Its spectrum indicated the emission of far more energetic X-rays than emitted by the Sun.[56]

Some of the stellar X-ray sources seemed particularly intriguing. Could they be the neutron stars or even the black holes long postulated by theorists since the late 1930s? If such compact bodies were members of binary systems in which the companion was a red giant, tidally removed outer layers of the companion crashing down on the surface of the compact star could account for the emission of X-rays. These were tantalizing conjectures, but it would take time to determine whether or not they were correct.

X-Ray Galaxies

In the ensuing years, further discoveries in X-ray astronomy rapidly followed. Increasingly they involved sophisticated X-ray sensors that ultimately revealed the existence of a variety of *X-ray stars*, galaxies, and a uniform-appearing *X-ray background*. Possibly the most astonishing among the discoveries was the NRL group's identification of a galaxy more luminous in X-rays than in all other energy ranges combined. It was Cygnus A, the same galaxy that, 20 years earlier, Hey had found to emit strongly in the radio wavelength regime. But Cygnus A was not the only such source. The galaxy Messier 87, also a well-known radio emitter, was highly luminous in its X-ray emission as well.[57]

The Infrared Sky

Since the early 1930s, Edgar W. Kutzschner, in the Department of Physics at the University of Berlin, had been developing and evaluating a series of infrared-sensing detectors. Such detectors appeared promising for many military purposes, including the night-time detection of ships at sea. Around 1933 Kutzschner's research obtained support from the German Army, which financed his efforts to perfect lead sulfide (PbS) infrared detectors.[58]

The significant promise of infrared surveillance led Germany, Britain, and the United States to concentrate efforts during World War II on rapidly developing infrared-sensing lead sulfide detectors. This classified research often remained unknown even to researchers from one and the same country.

In the fall of 1945, the Dutch-born U.S. astronomer Gerard P. Kuiper found out about the German efforts through interviews he conducted with German scientists as part of the U.S. postwar assessment of German technologies. He then discovered that similar efforts had been underway for some time in the United States. In late 1941, the OSRD had given Robert J. Cashman at Northwestern University in Illinois a contract for developing infrared sensors. On becoming aware of Cashman's efforts, Kuiper proposed their teaming up to initiate a program of astronomical observations of stars and planets at the McDonald Observatory in Texas.[59,60,61]

Wartime work had also been conducted on lead sulfide detectors in England.[62] Peter Fellgett, a junior member of the team dedicated to this effort, completed his doctoral studies at Cambridge University in 1951 with a thesis on the near infrared magnitudes of 51 stars obtained using lead sulfide detectors – one of the first attempts to study systematically the infrared emission from stars.[63]

During the 1950s and early 1960s, a number of astronomers in the United States also began to explore the near- and mid-infrared sky. Often they employed infrared detectors developed for the military. By the mid-1960s, these efforts had revealed the existence of extremely luminous *infrared stars* and galaxies.[64,65,66] Balloon, airborne, and rocket observations soon also began to explore the infrared sky at longer infrared wavelengths, which can be studied only from very high altitudes or from beyond Earth's absorbing atmosphere.

Two particularly surprising advances brought about by the infrared detection techniques came to the fore in 1965 and 1970.

Infrared Stars

The first of these advances, published in 1965 by Gerald Neugebauer, Dowell E. Martz, and Robert B. Leighton at the California Institute of Technology, resulted from a sky survey conducted with sensitive PbS detectors that led to the discovery of a number of stars highly luminous in the 2.0 to 2.4 μm band. Most were either undetectable or only barely detectable at visible wavelengths even with the largest existing telescopes. Judging from their infrared colors, these stars had surface temperatures as low as 1000 K, a temperature about six times lower than the surface temperature of the Sun. For these stars to be as luminous as observed, their radii had to be orders of magnitude greater than that of the Sun. Here was a new class of stars – a set of stars that shone solely in the infrared![67]

Infrared Galaxies

The second astounding advance came with the discovery of galaxies extremely luminous in the mid-infrared around 25 μm, that is, at wavelengths a factor of 10 longer than those of the luminous infrared stars. Measurements conducted by Frank J. Low and his Rice University student Douglas E. Kleinmann, at wavelengths ranging from 2 to 25 μm, revealed a most startling result[68]:

> Observations . . . support the view that all galaxies emit far-infrared radiation from their nuclei by a common physical process. More than a factor

of 10^7 separates the weakest from the most luminous sources, and there are five galaxies which radiate about 1000 times the normal output of an entire spiral galaxy.

The longer wavelengths at which these galaxies were discovered were detected with a highly sensitive bolometer that Low had developed some years earlier while working as a research physicist at the Texas Instruments Corporation.[69]

Two years later, in 1972, Low and George H. Rieke, his colleague at the University of Arizona, published further mid-infrared observations of extragalactic sources. This effort was supported in part by the Air Force Cambridge Research Laboratory. The Air Force, at the time, was initiating efforts to map the mid-infrared sky, with the aim of developing techniques that could single out Soviet ballistic missiles or Earth-orbiting satellites against a background of naturally occurring celestial sources. For this purpose, an improved understanding of mid-infrared astronomical sources was essential. Without a good map of the infrared sky, it would not be possible to identify quickly enemy ballistic missiles, Earth-orbiting satellites, or even aircraft emitting infrared radiation.

Rieke and Low's observations now showed that the infrared luminosity of the nearby quasar 3C 273 was roughly 4×10^{39} Watt, out to wavelengths of 25 μm. This exceeded the quasar's luminosity in all other wavelength ranges by a factor of nearly 10. They estimated that this was more than a thousand times higher than the rate at which our own Galaxy emits visual radiation.[70]

Some of these highly luminous sources are by now understood to be galaxies in which young highly luminous stars are forming at formidable rates. Others are powerfully emitting quasars. The giant black holes now known to power the quasars generate vast amounts of energy absorbed in surrounding gas and dust and re-emitted at infrared wavelengths. The relationship between massive star formation and black hole activity has still not been satisfactorily sorted out.

Gamma-Ray Astronomy

The United States was not the only beneficiary of German rocket techniques. At the end of the war, the Soviet Union had assembled its own set of German experts and had similarly begun to develop a powerful rocket industry. With the launch of the first Earth-orbiting satellite *Sputnik* in 1957, they exhibited the impressive capabilities their rockets and guiding mechanisms had attained in the dozen years since the war. The accuracies with which the Soviet Union was able to place satellites into Earth orbit showed that their rockets could now reach any place on Earth with great precision and presumably with significant nuclear warheads. Moreover, the high ground of space gave them the ultimate means of surveillance of military installations anywhere on Earth.

To counter this ascendancy, the United States created a crash program to develop both more powerful rockets and more incisive surveillance techniques from space. The Space Race of the 1960s had begun! It would culminate in the Moon landings of 1969 and then level off.

The Cold War efforts to land men on the Moon, though in the public's mind a race between the United States and the Soviet Union to reach the Moon first, was actually far more desperate. It entailed an all-out effort to gain military ascendancy in space, through precision launches and maneuvers, often with massive payloads. Perhaps the most phenomenal astronomical discovery this struggle revealed came in 1969. So secret was this finding and others that succeeded it that they remained classified until 1973.

In an era of mistrust among nations, extreme measures are taken to keep track of weapons tests carried out by other countries. The *Limited Test Ban Treaty* signed in the Kennedy/Khrushchev era of 1963 forbade the testing of nuclear weaponry in the atmosphere, in outer space, and under water, and directed nations to carry out their tests underground. The United States was worried that the Soviet Union might violate this agreement by exploding its test devices not underground, but at great distances out in space. Were the Soviet military to do this, it would be possible to sense the burst of *gamma rays* released in such highly energetic explosions. To detect these potential bursts, the United States designed the *Vela* project, which placed a series of γ-ray sensing satellites into Earth-circling orbits. They were stationed so at least one satellite would always be positioned to detect a nuclear burst no matter where it exploded in space.

Simultaneous, precisely timed observations of a burst of γ rays by two of these satellites provided partial information on the direction from which the burst had arrived. If three of the Vela satellites simultaneously observed the source, it had to lie in one of two possible directions perpendicular to the plane in which the three satellites were located. This orbital emplacement provided at least partial information on where a γ-ray burst had occurred. Gradually, it became clear that the observed bursts were not originating within the Solar System.[71]

γ-Ray Bursts, GRBs

In mid-1973 Ray W. Klebesadel, Ian B. Strong, and R. A. Olson at the Los Alamos Scientific Laboratory, New Mexico, surprised astronomers everywhere by announcing that, ever since 1969, occasional short bursts of γ rays lasting only a few seconds had been observed by several of the spacecraft and appeared to arrive from beyond the Solar System and from quite different portions of the sky, roughly at intervals of a few months.[72]

It took another two decades to arrive at some measure of understanding of these bursts. During this time, NASA had launched more powerful γ-ray observatories into space. These had accumulated sufficient systematic information by 1991 to lead Bohdan Paczyński at Princeton to surmise that γ-ray bursts (GRBs) might be extragalactic explosions.[73] Eventually, near-simultaneous observations at X-ray and γ-ray energies by telescopes launched into space, and by specially constructed ground-based optical telescopes, which enabled rapid follow-up before visible light from the bursts had a chance to decay appreciably, revealed the precise location of the explosions and permitted systematic studies of their aftermath. Many γ-ray bursts are by now identified as *hypernovae* – a class of extremely powerful supernovae – exploding at enormous distances reaching out to redshifts $z \sim 8$. Others may arise in the merger of two neutron stars, which also can lead to supernova explosions.

Military Contributions to Theoretical Work

Astronomical observers were not alone in working both sides of the curtain of secrecy that largely partitioned the worlds of academic science and military technology but occasionally opened enough, throughout the Cold War, to hand academic researchers a novel tool with which to embark on a new line of research. Theorists may have been just as involved in this schizophrenic arrangement.

One of the outstanding contemporary American theorists, Kip Thorne, reminisces in his book, 'Black Holes & Time Warps – Einstein's Outrageous Legacy,' about conversations that he and other American astrophysicists occasionally had with leading Soviet colleagues at the height of the Cold War.[74] Most of the top astrophysicists on both sides had been involved in the development of atomic and hydrogen bombs. And so, at some stage the discussions on an intricate problem of astrophysics would grind to an embarrassing halt, as both sides realized they were getting uncomfortably close to revealing a military secret, though each side probably already knew as much as the other. The physics of nuclear bombs is closely linked to the physics of supernova explosions, except that the astronomers' explosions are incomparably more powerful than the military's.

Astrophysicists who straddled the fence between academic and military work read like a Who's Who of prominent theorists: By 1939, J. Robert Oppenheimer, who later led the Manhattan Project to develop the atomic bomb, had done seminal work on neutron stars and black holes. Hans Bethe that same year had explained how stars shine by converting hydrogen into helium. During the war, he became the chief theorist of the Manhattan Project. Late in life, Bethe returned to astrophysics to work on the physics of supernova explosions. Edward Teller, father of the American hydrogen bomb during the Cold War, similarly contributed to astrophysics both before and after World War II. John Archibald Wheeler, one of the pioneers of black hole theory and general relativistic investigations, was also involved in nuclear bomb projects, both during World War II and the Cold War.

In Germany, the young astrophysicist Carl Friedrich von Weizsäcker worked on the aborted German World War II atomic bomb project under Werner Heisenberg, inventor of quantum theory. After the war, Weizsäcker turned to philosophy and became a leading advocate of world peace.

In the Soviet Union, Yakov Borisovich Zel'dovich, among the most ingenious theoretical astrophysicists of the postwar era, was deeply involved in the design of the Soviet atomic and hydrogen bombs. And the father of the Soviet hydrogen bomb, Andrei Dmitrievich Sakharov, who later was celebrated for his efforts at establishing world peace, also made fundamental contributions to cosmology and our understanding of the origins of the Universe.

Accompanying these giants came an army of less well known theorists, who worked on γ-ray bursts, supernovae, and other outbursts of interest to astronomers, as well as to the military. The classified sphere funded a good fraction of the basic theory as well as experiments to determine nuclear cross sections needed to calculate stellar opacities and nuclear reaction rates in stars, parameters that also are of importance for the design of nuclear weapons.

Computers

The military provided much of the instrumentation that theorists required for their ever-expanding need for complex simulations. During World War II computers were designed to carry out increasingly complicated calculations in ballistics, optimization of aerodynamics in aircraft design, and the construction and optimization of atomic bombs.[75] The calculations applicable to bomb design quickly found use in simulating nuclear reaction sequences in stars and, more globally, in studies of stellar evolution. Calculations far too complex to be made with pencil and paper were being replaced by computations and simulations carried out with increasingly sophisticated machines. Computer simulations of chains of nuclear reactions in supernova explosions continued at Los Alamos throughout the Cold War, and are still ongoing today. Simulations of this type are critical both to weapons design and astrophysics.

As World War II gave way to the Cold War, the military developed fast, light-weight computers to control precisely the flight of rockets and guide missiles to their final destinations. For surveillance from space, computers had to place payloads accurately in their intended orbits. Simultaneously, industry was rapidly discovering markets not only for supercomputers but also for small portable or desktop machines. These smaller personal computers developed in the mid-1970s and early 1980s quickly came into scientific use as well and, as their capabilities expanded and prices dropped, they became indispensable tools used by scientists worldwide, in all fields, including astronomy and astrophysics.

The use of computers has become so ubiquitous by now that it is difficult to disentangle where the military stimulated a new development or advance and where industry was quick to realize and grasp new opportunities. Nor is it easy to describe the speed with which computers have changed the lives and habits not only of scientists but even of kindergarten children. For scientists, the ease and speed with which networks of colleagues, worldwide, can jointly work on a major problem, without leaving their home institutions, has also resulted in global collaborations on complex problems.

Even Vannevar Bush would have been impressed by the range and speed of the interplay.

Looking Back

Between the end of World War II and the early 1970s, astronomy was enriched through the discovery of some 14 new and unanticipated major phenomena listed below. Only two of the discoveries resulted from observations in the optical regime. Both could well have been made with instrumentation available before World War II, or with marginally improved techniques. These two involved:

Magnetic Variable stars, 1947, and
Flare stars, 1949

But twelve other major discoveries became possible only through the use of techniques and instrumentation initially designed for military purposes during World War II or the Cold War. This list of discoveries included:

Radio galaxies, 1946–54
X-ray stars, 1962
Quasars, 1963
The Cosmic microwave background, 1965
Infrared stars, 1965
X-ray galaxies, 1966
Cosmic masers, 1967
Pulsars, 1967
Superluminal radio sources, 1971
Infrared galaxies, 1970–72
Interstellar magnetic fields, 1972
Gamma-ray bursts, 1973

Elsewhere, I have described these discoveries in considerably more detail than in the current chapter, which has dealt largely with the technologies bequeathed to astronomy by the military and occasionally by industry.[76] There, I also identified the professional backgrounds of those who made the discoveries. By a wide margin, the discoverers mostly were physicists, sometimes helped by astronomers and electrical engineers.[77]

The sociologist of science David O. Edge has summarized the radio astronomical contributions succinctly. But a quote from his article 'The Sociology of Innovation in Modern Astronomy' may resonate also with most of the other major astronomical discoveries made with post–World War II or Cold War instrumentation[78]:

> [T]he innovation consisted of the systematic, state-of-the art exploitation of a set of observational *techniques* by those whose background and training equipped them with the necessary skills and competences, to produce results for an audience many of whom were not so equipped.

From the context of his paper, Edge clearly meant the word *audience* to refer to the community of traditional optical astronomers. As he was also quick to point out, however, this community largely appreciated and welcomed the rapid developments in radio astronomy – a welcome that optical astronomers subsequently also extended to discoveries made with infrared, X-ray, and γ-ray instrumentation.

By the end of the 1970s, only a quarter of a century after World War II, the sky had changed unrecognizably!

Leading astrophysicists of the prewar years, such as Arthur Stanley Eddington, James Jeans, or Henry Norris Russell, would have found their world views largely outdated. By the 1970s and 1980s, we needed a new consolidating view of the Universe.

The Mansfield Amendment

In the United States, the era of easy acquisition of military instrumentation for astronomical purposes, so prevalent in the two decades immediately after World War II, partly declined with the passage of the *Mansfield Amendment* to the Military Procurement Authorization Act for FY 1970.

More than four decades ago, at the height of the Vietnam War, which he ardently opposed, Senator Michael Joseph (Mike) Mansfield, Democratic Party Majority Leader in the U.S. Senate, a man of vision and sagacity, and one of the most trusted leaders of the U.S. Senate in the postWar era, became alarmed by the prospect of the military co-opting the nation's science without any authority to do so. The policies advocated by 'Science – The Endless Frontier,' as well as legislation passed by Congress, were clear and had always anticipated that the NSF would set directions for the nation's basic scientific research.

In late 1969, Mansfield inserted Section 203, now commonly referred to as the *Mansfield Amendment*, into the Military Procurement Authorization Act for FY 1970, Public Law 91-121, to put an end to the practice.[79] The amendment forbade the Defense Department the use of appropriated funds "to carry out any research project or study unless [it had] a direct and apparent relationship to a specific military function."[80]

In a December 5, 1969 letter to William D. McElroy, Director of the NSF, Mansfield clarified his intent.[81] "In essence, it [the amendment] emphasizes the responsibility of the civilian agencies for the long-term, basic research. It limits the research sponsored by the Defense Department to studies and projects that directly and apparently relate to defense needs."

The implementation of the Mansfield Amendment may have been regrettably clumsy: The military found it difficult to divest itself of all its newly proscribed activities presumably because, over the years, basic and applied research had become so thoroughly interwoven. And, with the huge cost of the Vietnam War, the NSF and other civilian agencies could not be adequately funded to fully support the research they, rather than the military, were now expected to manage. Nobody was happy.

The amendment, which raised alarms throughout a civil research community that had become dependent on military support, was dropped from the military authorization bill the following year, but in practice, the military took this as a more permanent restriction.[82] Moreover, in the late 1960s, many U.S. universities began to prohibit classified research on their premises following student uprisings to protest against the war in Vietnam.

The idea of separating military and civilian research may have had sound reasons; but it eliminated easy communication between military and civilian researchers, leading to a decline in basic research on both sides of the secrecy barrier, but probably more so in military research.

By the late1970s, the military argued that its basic research efforts were falling behind those of the Soviet Union. Under President Jimmy Carter and his Secretary of Defense Harold Brown, who had been Secretary of the Air Force under President Lyndon B. Johnson before becoming president of the California Institute of Technology, funding for basic research began to increase substantially. The flow of research funds into the universities once again rose. Purely from a defense perspective, these investments made a great deal of sense. The land armies of the United States were always going to be small compared to those the Soviet Union could field. To counter those, the United States would have to maintain a clear technological advantage.[83]

An Era of Surveys

The late 1970s saw the introduction of new, more focussed, more systematic approaches to search and discovery. Large surveys conducted at wavelengths that had been inaccessible to astronomers before World War II became a significant part of this effort. Other surveys involved new spectral or precise timing capabilities designed to look for features that previously might have been missed. Optical astronomers were more heavily engaged in these surveys, often using new photoelectric devices that were gradually displacing photographic plates.

Many of these surveys were *blind*, in the sense that it was not clear what they would unveil. But the preceding two decades had shown how searches with new instruments often resulted in the most unexpected findings. As a result, the decade of the 1970s had focused on plans for future surveys that would systematically lead to further recognition of phenomena previously missed, while also elucidating how rare or how common some of the recently discovered phenomena might be or how they might fit into evolutionary patterns that prevailing theory could explain.

Some of these surveys, no doubt, would be far costlier than any previously contemplated, and would involve far larger consortia of astronomers than had ever been assembled before. Surveys conducted with telescopes launched into space would be particularly costly and would make sense only if large segments of the astronomical community agreed on their importance. For, the total funding allotted to astronomy was certainly limited, and funding an expensive survey would inevitably require the curtailing of large numbers of small investigations.

Most astronomers' lists of the striking discoveries of the 1975–2010 era would probably include a good fraction of the 19 listed below. The dating of the discoveries is approximate, because many required some time to explore carefully and establish:

The Intergalactic Medium, 1971–80
Microquasars; stellar-mass black holes, 1978–79
Gravitational lenses, 1979
Dark matter pervading galaxies, but detectable, to date, only through its gravitational attraction, 1979–80
Gravitational radiation emitted by a binary pulsar, detected indirectly through observations of the loss of orbital energy, 1982
Evolutionary merging of galaxies, 1984–96
Prestellar objects and circumstellar disks, 1984–99
Antineutrinos and neutrinos from the 1987 supernova explosion in the Large Magellanic Cloud, 1987
Extragalactic filaments, sheets, and voids, 1989
Planets around other stars, 1992–95
Solar neutrinos and their oscillations among the different neutrino species, 1994
Brown Dwarfs, 1995–98
Cosmic chemical abundance evolution, 1996–2005
The highest-energy cosmic rays, 1997–2005
Dark energy pervading all space and providing the dominant mass of the Universe, 1998
Cosmic reionization, 2001
Supermassive black holes, 2004
Large-scale cosmic flows, 2008
Microwave background fluctuations, remnants of early cosmic structure, 2011

Directly or indirectly, all of these resulted from surveys dedicated to probing more deeply to find more meaningful ways to understand the Universe.

The discovery of dark matter and dark energy made a particularly deep impression. Together, these two components accounted for roughly 96% of the mass-energy of the Universe. And yet we knew nothing much else about them. What were they? Where had they arisen? Their cumulative mass–energy vastly exceeded the cumulative mass–energy of all galaxies, stars, planets, and the gases that pervaded them, which jointly accounted for a mere 4% of the overall mass–energy. The entire history of astronomical research had concentrated on studying just this *baryonic* component consisting of atoms, electrons and nuclei. Yet, here we were suddenly confronted with far more prevalent components about which we knew virtually nothing!

A Continuing Military Presence

Most astronomers remain unaware of the extent to which military tools have been incorporated in their work. Europeans, in particular, tend to be under the impression that military influences are largely an American phenomenon. But the Infrared Space Observatory (ISO), the highly successful infrared astronomical mission launched by the European Space Agency in 1995, is a clear counter example. The infrared camera on board, constructed primarily in France, had a detector array especially constructed for the mission by the French defense establishment. The Short-Wavelength Spectrometer, built primarily in the Netherlands, incorporated previously classified infrared detectors provided through the efforts of a co-investigator at the U.S. Air Force. Yet most of the many hundreds of astronomers who carried out observations with these instruments were unaware of those contributions, and would deny that this space mission had military ties. The U.S. infrared astronomical mission *Spitzer* had vastly greater *sensitivity* than ISO and incomparably larger detector arrays – all military hand-me-down devices.[e]

The Extent of Military Restrictions

In the opening paragraphs of this chapter I dwelled so extensively on the quotes from the summary of 'Science – the Endless Frontier,' to show how strongly Vannevar Bush felt the need for seamlessly weaving basic science into every aspect of the nation's future.

During the decades following World War II, the integration of military and scientific priorities became aligned to such an extent, and military influences came to pervade astronomy so thoroughly that astronomers themselves no longer noticed. Most of them would argue vigorously that nothing in their work depended on military influences, whereas almost everything they did was dependent in one way or another on military contributions.

They were so common, they had become invisible!

Often this became apparent only in later years, as half a century had passed under the postwar arrangements Bush had advocated. Then, as international collaborations became increasingly essential because the costs of many large projects could no longer be borne by any one nation alone, a previously unremarked set of U.S. export regulations surfaced with increasing frequency. These proscribed

---

[e] Because a complete picture is still elusive, it is difficult to document fully the strong ties of science to the military in the postwar Soviet Union. However, all available evidence indicates that the Soviet physical sciences were entirely dominated by governmental and military priorities, and that military-industrial secrecy shrouded every effort. Roald Sagdeev, Director of the Soviet Space Research Institute IKI for more than 15 years, from the early 1970s to the late 1980s, has provided a revealing picture of the conduct of Soviet space science in his times. It paints a picture of a scientific program strongly tied to, and supported by, military priorities, but also greatly hindered by secrecy and fierce political and military-industrial infighting.[84]

the potential sharing – the import and export – of large classes of defense-related information and articles and services identified on the United States Munitions List.

All of a sudden astronomers were confronted with a painful set of *International Traffic in Arms Regulations (ITAR)* of the U.S. Department of State, which strictly limit the *export* of military equipment and knowledge. This meant that militarily sensitive technical information or devices could not be shared readily with international partners, prohibiting what appeared to be the most innocent, mutually beneficial scientific collaborations – projects from which all sides would benefit.

ITAR had been around since the late 1970s.[85] But because international collaborations on large space missions became common only in the 1980s and 1990s, most astronomers had not been previously aware of the regulations' existence. Only then did U.S. astronomers begin to realize the extent to which they had become used to accepting military support in the form of critically sensitive detectors and other powerful hardware and software. Only then did it become apparent that the extent to which Bush had advocated the melding of military and civil research had a down side that would need to be reviewed and revised in the context of science conducted on a global scale.

Sometimes ways are found to make sensible arrangements that meet the purposes both of scientific investigations and those of military secrecy and arms control. But often the length of time it takes to work through all the necessary bureaucratic procedures prevents international cooperation; some less developed observational techniques may then be implemented on an international astronomical mission, even where more powerful but classified techniques actually exist and would significantly improve the mission's capabilities.

If Vannevar Bush and Bernard Lovell were concerned primarily with efficiency, they were certainly right. Sharing the scientific insights that academic research offered to expedite invention or perfection of novel weapons in time of war had been a crucial factor in winning World War II. The organizational structures that fast-moving military and industrial projects demanded had taught a generation of young scientists that a great deal more can be accomplished through teamwork than by individual researchers or small groups vying against each other to arrive at a new result and be first to publish.

This much was clear.

Not as easily disposed of are two interrelated questions: (1) How well does science advance when it is so strongly dependent on military subvention? (2) Does this form of funding distort our resulting views?

The first of these questions needs some amplification: Making military equipment available to astronomy, to deliberately or perhaps innocently seed a postwar revolution in astronomy, was not at all in accord with developments 'Science – the Endless Frontier' had anticipated. Wartime research on radar and the atomic bomb had shown that basic research could lead to powerful new defensive and offensive weaponry. Vannevar Bush now wanted to show that the United States would be well served by a comparable peacetime program of federally funded basic research whose discoveries could quickly be assimilated to solve military problems.

Although the wording of 'Science – the Endless Frontier' was deferential to the military, throughout, Bush had expected that the postwar years would see military research supervised by a successor organization similar to the wartime Office of Science Research and Development. The flow of information under this arrangement would be from basic research to military developments. Basic research would seldom, if ever, have to be classified secret.

Instead, in postwar astronomy the flow of information had been reversed, flowing from military development to basic science. Although the seamless sharing of information that Bush had encouraged had been preserved, the flow of information had been turned on its head. Astronomical equipment and methods could now be classified secret.

Bush's other expectation was also dashed. As soon as the war was over, the military adamantly resisted any further civilian control. There would be no civilian supervision of military research. Bush, never a man to be easily dissuaded, continued to battle for his vision of control. He used all the support in Washington that his wartime successes had won him. But eventually he had to concede defeat by a rarely united Army and Navy. Deeply disappointed, he resigned from public commitments and retired to private life.[86]

The Broad Interplay of Military and Basic Research

Obtaining military support for astronomical ventures can be intoxicating. Military budgets in all nations still are vastly greater than any purely astronomical priority could command.

Hard numbers for military expenditures can be difficult to obtain. But in a 1995 lecture and article on 'Ballistic Missile Defense and Infrared Technology,' approved for public release, John A. Jamieson, a leader in the development of military infrared technologies, provided a detailed review of President Ronald Reagan's Strategic Defense Initiative (SDI).[87] The thrust of this initiative was to provide the United States with a protective umbrella against Soviet ballistic missile attacks.

After a full account of all the activities SDI entailed, Jamieson provided the program's costs for the years from 1985 to 1992. The whole SDI defense development, he wrote, "was achieved for only about $3.4 billion per year for seven years. The infrared related portions amounted to less than $7.5 billion. (Probably $1.5 billion of this was spent on infrared technology of optics, focal plane cryocoolers, signal processors and studies and analysis . . .)"

This fraction of $1.5B spent on just this particular defense effort by 1992 most likely exceeded the amounts spent on all of infrared astronomy, world wide, throughout all previous history. Moreover, the military's expenditures on infrared devices, and its support of infrared astronomy, had not just started with the SDI project. Military interest in infrared sensing had continued seamlessly since the end of World War II. This makes clear how much the infrared detectors and optical systems made available to astronomers by the military, at no cost to the astronomical community, contributed to the rapid advances in infrared astronomy in the United States.

With this pronounced emphasis on the military's contributions to astronomical instrumentation, I do not wish, in any way, to minimize the impressive instrumental advances that astronomers themselves brought about as they adopted and adapted the technologies handed down to them by the military. With funding levels an order of magnitude lower than those available for military technological developments, by far the most effective strategy for advancing their science was to make the best use of the sensitive equipment the military could offer. Reconfiguring it to open entirely new astronomical capabilities enabled them literally to reshape the field in a new light, and reveal an incredibly richer universe than could have ever been imagined!

Astronomy is an observational science. Progress comes through detailed observations often conducted over years, decades, and sometimes centuries. Military surveillance has a similar character. Astronomers also have a deep and abiding interest in explosions, as does the military. The principles on which these are based are universal, which is why a significant fraction of research on supernovae and the modeling of supernova explosions is carried out by scientists working in weapons laboratories such as Los Alamos. Here too an interdependence has developed. Although it has served astronomy well, it raises a serious question:

What happens when astronomy becomes as dependent on military largesse as it has? Does this dependence merely advance, here and there, what we are able to observe, calculate, and model? Or has it permeated the field to such an extent that we selectively perceive the Universe through a distorting lens that warps the cosmologies we construct?

How Far Are Our Cosmic Views Dependent on Available Technologies?

Neither Bush nor Lovell discussed the extent to which the exchange of ideas, the flow of instrumentation and technologies, and the organizational structures of military, industrial, and academic institutions could change the ways scientists conceive Nature!

It seems almost obvious that a sensitive detection system perfected for military purposes should be made available to advance astronomical research as long as this does not threaten national or global security. Indeed, as this chapter has attempted to show, this is how many of the most striking postwar astronomical advances came about. But although it may be expeditious to adopt novel military tools to advance astronomical discovery, we need to keep in mind that those technologies, and the discoveries they enable, inexorably also steer the future of the discipline. They lead to new world views and raise new questions that call for further exploration along paths the new technologies have opened.

The technologies made available by World War II, and later by Cold War military crash programs, have led to ever expanding astronomical research programs driven largely by a handful of distinct classes of research tools. Most astronomers will argue that these tools do not influence how our community now portrays the origin, structure, and evolution of the Universe. Yet, if asked about what they do, these same individuals may well identify themselves as optical astronomers, X-ray astronomers, radio astronomers, or infrared astronomers – without giving the question of how these compartmentalizations have arisen or how the respective tools they inherited from the military might affect their insights on the Universe.

Something must be awry: How can we trust what we learn about the Universe when it so clearly depends on how researchers organize themselves, how they are funded, and the tools they may be offered by a military or industrial establishment flush with money? Does a cosmology, an understanding of the Universe, exist that does not depend on how scientists are funded, conduct their daily work, organize themselves, have access to tools, and develop a theoretical framework that encapsulates their findings, including perhaps a work ethic that depicts and even prescribes how they go about their work and formulate their thoughts?

It is hardly conceivable that our understanding of the Universe is not driven by the tools we employ. We know this because we were there once before!

In the years leading up to World War II, astronomy was driven by just one set of tools, those of optical astronomy. The Universe as viewed then was quite different from the one we portray today. Yet, what we understand now comes from piecing together the observational results delivered by our newer tools, just as the portrayal of the Universe by prewar astronomers was based on piecing together and trying to understand what their optical tools were telling them. And, no doubt can exist, those prewar conceptions of the Universe differed dramatically from the views we hold today!

To be sure, thanks to the wealth of data provided by our more powerful observing tools, as well as the fragmentary neutrino observations obtained so far, we do know much more today than we did seven decades ago, just before the onset of World War II.

Nonetheless, we should remain deeply aware that only about 4% of the current energy content of the Universe comprises baryons, leptons, and electromagnetic radiation, and that these particles and radiation components are all we are able to observe directly today. The other 96% of the energy is dominated by dark matter and dark energy, about which we know nothing except for the gravitational influences they exert.

How much would our insights on the Universe differ today if military or industrial efforts of the postwar era had focused on exploring dark matter and dark energy, rather than electromagnetic radiation, and had handed down to astrophysicists just the tools developed for those searches? It is difficult to believe that our views of the Universe then would be anything like those we take for granted today.

Notes

1. *The Physicists – The History of a Scientific Community in Modern America*, Daniel J. Kevles. Cambridge, MA: Harvard University Press, 1987.
2. *Endless Frontier – Vannevar Bush, Engineer of the American Century*, G. Pascal Zachary. New York: The Free Press, 1997.
3. *The New Politics of Science*, David Dickson. New York: Pantheon Books, 1984.
4. Ibid., *The Physicists*, Kevles, pp. 191–98.
5. http://astro.uchicago.edu/yerkes/index.html
6. A 100-Inch Mirror for the Solar Observatory, George E. Hale, *Astrophysical Journal*, 24, 214–18, 1906.
7. Ibid., *The Physicists*, Kevles, p. 285.
8. Ibid., *The Physicists*, Kevles, p. 148.
9. Ibid., *The Physicists*, pp. 190–99 and 268.
10. Ibid., *The Physicists*, Kevles p. 297.
11. The General of Physics, Peter Dizikes, *(MIT) Technology Review*, May/June 2011, M20–M23.
12. *Science – The Endless Frontier*, Vannevar Bush. Reprinted by the National Science Foundation on its 40th Anniversary 1950–1990, National Science Foundation, 1990.
13. Ibid., *Science – The Endless Frontier*, Bush, 1990 reprint.
14. The Post-War Society: Responding to New Patterns of Patronage, David H. DeVorkin, in *The American Astronomical Society's First Century*, ed. David H. DeVorkin, American Astronomical Society and American Institute of Physics, 1999, pp. 109 & 111.
15. Ibid., *Science – The Endless Frontier*, Daniel J. Kevles, 1990 reprint, pp. ix–xxxiii.
16. Science Indicators 1982, National Science Board, 1983, Figure 2–14, *Federal R&D budget authority for national defense*, p. 52; and Appendix Table 5–13: Federal obligations for R&D to university and college performers for selected agencies: 1967–83, p. 308.
17. Ibid., *Science – the Endless Frontier*, Kevles, pp. xvii–xix.
18. The Effects of Defence Science on the Advance of Astronomy, Sir Bernard Lovell, *Journal for the History of Astronomy*, 8, p. 168, 1977.
19. Ibid., The Effects of Defence Science, Lovell, pp. 151–73.
20. Ibid., The Effects of Defence Science, Lovell p. 167.
21. *Astronomy Transformed – The Emergence of Radio Astronomy in Britain*, David O. Edge & Michael J. Mulkay. New York: John Wiley & Sons, 1976.
22. The Sociology of Innovation in Modern Astronomy, David Edge, *Quarterly Journal of the Royal Astronomical Society*, 18, pp. 334–35, 1977.
23. *Glimpsing an Invisible Universe – The Emergence of X-Ray Astronomy*, Richard E. Hirsh. Cambridge University Press, 1983.
24. Infrared Sky Surveys, Stephan D. Price, *Space Science Reviews*, 142, 233–321, 2009.
25. Electrical Disturbances Apparently of Extraterrestrial Origin, Karl G. Jansky, *Proceedings of the Institute of Radio Engineers*, 21, 1387–98, 1933.
26. On the Origin of Interstellar Radio Disturbances, F. L. Whipple & J. L. Greenstein, *Publications of the National Academy of Sciences*, 23, 177–81, 1937.
27. Solar Radiations in the 4–6 Meter Radio Wavelength Band, J. S. Hey, *Nature*, 157, 47–48, 1946.

28. Microwave Radiation from the Sun, G. C. Southworth, *Journal of the Franklin Institute*, *239*, 285–97, 1945.

29. Cosmic Static, Grote Reber, *Astrophysical Journal*, *100*, 279–87, 1944.

30. Ibid., *Astronomy Transformed*, Edge & Mulkay.

31. Fluctuations in Cosmic Radiation at Radio Frequencies, J. S. Hey, S. J. Parsons, & J. W. Phillips, *Nature*, *158*, 234, 1946.

32. An Accurate Determination of the Positions of Four Radio Stars, F. G. Smith, *Nature*, *168*, 555, 1951.

33. Identification of Radio Sources in Cassiopeia, Cygnus A, and Puppis A, W. Baade & R. Minkowski, *Astrophysical Journal*, *119*, 206–14, 1954.

34. Investigations of the Radio Source 3C 273 by the Method of Lunar Occultations, C. Hazard, M. B. Mackey, & A. J. Shimmins, *Nature*, *197*, 1037–39, 1963.

35. 3C 273: A Star-like Object with Large Red-Shift, M. Schmidt, *Nature*, *197*, 1040, 1963.

36. A Measurement of Excess Antenna Temperature at 4080 Mc/s, A. A. Penzias & R. W. Wilson, *Astrophysical Journal*, *75*, 419–21, 1965.

37. Cosmic Black-body Radiation, R. H. Dicke, P. J. E. Peebles, P. G. Roll, & D. T. Wilkinson, *Astrophysical Journal*, *142*, 414–19, 1965.

38. Evolution of the Universe, Ralph A. Alpher & Robert Herman, *Nature*, *162*, 774–75, 1948.

39. Remarks on the Evolution of the Expanding Universe, Ralph A. Alpher & Robert C. Herman, *Physical Review*, *75*, 1089–1095, 1949, specifically see p. 1093.

40. Observations of a Strong Unidentified Microwave Line and of Emission from the OH Molecule, H. Weaver, D. R. W. Williams, N. H. Dieter, and W. T. Lum, *Nature*, *208*, 29–31,1965.

41. Observations of Polarized OH Emission, S. Weinreb, et al., *Nature*, *208*, 440–41, 1965.

42. Measurements of OH Emission Sources with an Interferometer of High Resolution, R. D. Davies, et al., *Nature*, *213*, 1109–10, 1967.

43. Observations of a Rapidly Pulsating Radio Source, A. Hewish, et al., *Nature*, *217*, 709–13, 1968.

44. Discovery of a Pulsar in a Binary System, R. A. Hulse & J. H. Taylor, *Astrophysical Journal*, *195*, L51–L53, 1975.

45. Observations of Post-Newtonian Timing Effects in the Binary Pulsar PSR 1913+16, J. M. Weisberg & J. H. Taylor, *Physical Review Letters*, *15*, 1348–1350, 1948.

46. Quasars: Millisecond-of-Arc Structure Revealed by Very-Long-Baseline Interferometry, C. A. Knight, et al., *Science*, *172*, 52–54, 1971.

47. A telephone conversation between Irwin Shapiro and Martin Harwit, July 19, 1979.

48. On the Origin of the Cosmic Radiation, Enrico Fermi, *Physical Review*, *75*, 1169–74, 1949.

49. Further Measurements of Magnetic Fields in Interstellar Clouds of Neutral Hydrogen, G. L. Verschuur, *Nature*, *223*, 140–42, 1969.

50. Pulsar Rotation and Dispersion Measures and the Galactic Magnetic Field, R. N. Manchester, *Astrophysical Journal*, *172*, 43–52, 1972.

51. Soft X-Radiation in the Upper Atmosphere, T. R. Burnight, *Physical Review*, *76*, 165, 1949.

52. Observations at High Altitudes of Extreme Ultraviolet and X-Rays from the Sun, J. D. Purcell, R. Tousey, & K. Watanabe, *Physical Review*, *76*, 165–66, 1949.

53. Photon Counter Measurements of Solar X-Rays and Extreme Ultraviolet Light, H. Friedman, S. W. Lichtman & E. T. Byram, *Physical Review*, *83*, 1025–30, 1951.

54. Evidence for X-Rays from Sources Outside the Solar System, R. Giacconi, H. Gursky, F. Paolini, & B. Rossi, *Physical Review Letters*, *9*, 439–43, 1962.

55. X-Ray Sources in the Galaxy, S. Bowyer, E. Byram, T. Chubb & H. Friedman, *Nature*, *201*, 1307–08, 1964.

56. X-ray Spectra from Scorpius (SCO-XR-1) and the Sun, Observed from Above the Atmosphere, G. Chodil, et al., *Physical Review Letters*, *15*, 605–07, 1965.

57. Cosmic X-ray Sources, Galactic and Extragalactic, E. T. Byram, T. A. Chubb, & H. Friedman, *Science*, *152*, 66–71, 1966.

58. The Development of Lead Salt Detectors, D. J. Lovell, *American Journal of Physics*, *37*, 467–78, 1969.

59. Ibid., The Effects of Defence Science, Lovell.

60. Infrared Sky Surveys, Stefan D. Price, *Space Science Reviews*, *142*, 233–321, 2009.

61. An Infrared Stellar Spectrometer, G. P. Kuiper, W. Wilson, & R. J. Cashman, *Astrophysical Journal*, *106*, 243–51, 1947.

62. Use of Lead Sulphide Cells in Infra-red Spectroscopy, G. B. B. M. Sutherland, D. E. Blackwell, & P. B. Fellgett, *Nature*, *158*, 873–74, 1946.

63. An Exploration of Infra-Red Stellar Magnitudes Using the Photo-Conductivity of Lead Sulphide, P. B. Fellgett, *Monthly Notices of the Royal Astronomical Society*, *111*, 537–59, 1951.

64. Infrared Photometry of Galaxies, Harold L. Johnson, *Astrophysical Journal*, *143*, 187–91, 1966.

65. Two-Micron Sky Survey, G. Neugebauer & R. B. Leighton, National Aeronautics and Space Administration, NASA SP–3047, Washington, DC, 1969.

66. Observations of Infrared Galaxies, D. E. Kleinmann & F. J. Low, *Astrophysical Journal*, *159*, L165–L72, 1970.

67. Observations of Extremely Cool Stars, G. Neugebauer, D. E. Martz, & R. B. Leighton, *Astrophysical Journal*, *142*, L399–L401, 1965.

68. Ibid., Observations of Infrared Galaxies, Kleinmann & Low.

69. Low-Temperature Germanium Bolometer, Frank J. Low, *Journal of the Optical Society of America*, *51*, 1300–04, 1961.

70. Infrared Photometry of Extragalactic Sources, G. H. Rieke & F. J. Low, *Astrophysical Journal*, *176*, L95–L100, 1972.

71. Letter from Ray W. Klebesadel to Martin Harwit, October 24, 1978. A large portion of this letter is reproduced in *Cosmic Discovery – The Search, Scope & Heritage of Astronomy*, Martin Harwit, Basic Books, New York, 1981.

72. Observations of Gamma-Ray Bursts of Cosmic Origin, Ray W. Klebesadel, Ian B. Strong, & Roy A. Olson, *Astrophysical Journal*, *182*, L85–L88, 1973.

73. Cosmological Gamma-Ray Bursts, Bohdan Paczyński, *Acta Astronomica*, *41*, 257–67, 1991.

74. *Black Holes & Time Warps – Einstein's Outrageous Legacy*, Kip Thorne,. New York: W. W. Norton, 1994.

75. *Landmarks in Digital Computing*, Washington, DC: Peggy A. Kidwell & Paul E. Ceruzzi, Smithsonian Institution Press, 1994.

76. *Cosmic Discovery – The Search, Scope & Heritage of Astronomy*, Martin Harwit. New York: Basic Books, 1981, chapter 2.

77. Ibid., *Cosmic Discovery*, Harwit, pp. 234–39.

78. Ibid., The Sociology of Innovation, Edge, pp. 326–39.

79. Military Procurement Authorization Act for FY 1970, Public Law 91–121, Section 203.

80. http://www.nsf.gov/nsb/documents/2000/nsb00215/nsb50/1970/mansfield.html

81. Renewing U. S. Mathematics Critical Resource for the Future, http://www.archive.org/stream/ Renewing_US_Mathematics_Critical_Resource_For_The_Future/TXT/00000128/txt._p._112

82. Ibid., *Science – The Endless Frontier*, Kevles, 1990 reprint, p. xxii.

83. Ibid., *The New Politics of Science*, Dickson, p. 125 ff.

84. *The Making of a Soviet Scientist – My Adventures in Nuclear Fusion and Space from Stalin to Star Wars*, Roald Z. Sagdeev. New York: John Wiley & Sons, 1994.

85. Ibid., *The New Politics of Science*, Dickson, pp. 143–53.

86. Ibid., *Endless Frontier*, Zachary, 1997.

87. Ballistic Missile defense and Infrared Technology, John A. Jamieson, *Proceedings of the Infrared Instrumentation Society IRIS*, 40 No. 1, 1995, pp.13–39.

8

Where Did the Chemical Elements Arise?

Among the immediate beneficiaries of 'Science – the Endless Frontier,' the new postwar policy of closely meshing basic research with applied scientific efforts of national interest, were astrophysicists studying the origins of the chemical elements. Cosmology, in particular, underwent a profound resurgence; a field that had previously been restricted to exploring arcane mathematical models suddenly found itself anchored to real-world nuclear physics. Increasingly detailed studies of nuclear interactions also began to shed light on nuclear processes in evolved stars. A sense of renewed excitement swept through the field![a]

Chandrasekhar's Early Venture into Cosmology

In 1942, Chandrasekhar and his University of Chicago doctoral student, Louis R. Henrich, had postulated that the Universe at some epoch could have been extremely dense and at a temperature of a few billion degrees Kelvin. They calculated the expected thermodynamic equilibrium distribution of nuclear species at different temperatures, and sought a range in which the chemical abundances of heavy elements came close to those observed in Nature. The conditions for this to happen, they estimated, were a temperature of order 8×10^9 K and a density $\rho = 10^7$ g/cm^3.[1] Although their work constituted a valiant attempt, they concluded that their paper, "should be regarded as of a purely exploratory nature and that such 'agreements' as may have been obtained should not be overstressed."

Four years later, in 1946, Gamow pointed out that, at the high densities Chandrasekhar and Henrich had postulated, general relativity predicted an expansion so rapid that the temperatures they had postulated could last for no longer than

---

[a] In this chapter, I begin using the centimeter-gram-second notation more extensively. Density is measured in grams per cubic centimeter, g/cm^3, and seconds are denoted by *sec*. Temperatures are given in degrees Kelvin denoted by K. Additional information on units can be found in the Notes on Usage in the book's front matter and is further elaborated in the appendix on symbols, glossary, and units.

a matter of seconds. Although he too believed "that the relative abundances of various chemical elements were determined by physical conditions existing in the universe during the early stages of its expansion, when the temperature and density were sufficiently high to secure appreciable reaction-rates for the light as well as for the heavy nuclei," he thought that this could have produced the observed abundances only if neutrons had been the dominant form of matter at earliest times.

Because of the high cosmic expansion rates, most of the neutrons would not have time to decay into protons. As the Universe cooled, they might, by an unspecified process, coagulate "into larger and larger neutral complexes which later turned into various atomic species by subsequent processes of β-emission."[2] As Gamow probably realized, this vague proposal was not particularly informative.

By early 1948, he was ready to send another Letter to the Editor of the *Physical Review*, this time with his George Washington University doctoral student Ralph Alpher. Although Hans Bethe had not been involved in Alpher and Gamow's investigations, Gamow invited him to co-author this paper, so that the list of authors – Alpher, Bethe, Gamow – might read more like the first three letters of the Greek alphabet.[3] As a result of this bit of buffoonery, the letter gained considerable notoriety.

The novel feature of Alpher, Bethe, and Gamow's approach was the incorporation of newly available neutron capture cross sections provided by Donald J. Hughes, a nuclear physicist and participant on the Manhattan Project.[b] The data showed that, at low energies of $\sim$1 MeV, neutron capture cross sections increased exponentially with atomic number halfway up the periodic table and then remained constant for heavier elements. Calculations then showed that obtaining the best fit to the known abundances of heavy chemical elements required a neutron mass density of order $\rho \sim 10^6/t^2$ g/cm^3-sec^2. Its dependence on time, $1/t^2$, was dictated by the general relativistic expansion rate of the Universe.

Alpher, Bethe, and Gamow postulated that the elemental abundance distribution was built up through successive neutron captures at rates given by the cross sections Hughes had provided. At their assumed density, a build-up begun as soon as 20 seconds after the birth of the Universe – by which time the density would have dropped to $\rho \sim 2.5 \times 10^3$ g/cm^3 – produced a good fit to the best available elemental abundance curve. The abundances of heavy elements they sought to simulate were those of a distribution in Nature that had been compiled by the geochemist Victor Moritz Goldschmidt in Norway just before the war.[4] Once formed, these neutron-rich species might β-decay to become the atomic nuclei currently populating the Universe.

---

[b] That the requisite nuclear cross sections, gathered for the design of atomic bombs, should have been made available to Alpher and Gamow right after cessation of hostilities may be an indication of the ease with which such information could be exchanged between defense-related and basic researchers, even so soon after the War.

Gamow's Cosmology

The Alpher, Bethe, Gamow letter had been submitted on February 18, 1948. On June 21 that same year, Gamow sent in one more letter, not much more than a journal page in length. But, whether he realized it or not, this short paper was to lay the foundations of modern cosmology![5]

His most important realization was that heavy element production could proceed only through the formation of helium, presumably along the lines Bethe had already indicated in 1939. Accordingly, he now examined how this might come about. A second factor that Gamow might not have noted in his two previous cosmological forays was that the radiation density at early times could far exceed the density of matter, and might have dominated the rate at which the Universe expanded:

> [T]he building-up process [of elements] must have started with the formation of deuterons from primordial neutrons and the protons into which some of these neutrons have decayed. . . . [T]he temperature at that time must have been of the order of $T_0 \sim 10^9$ K (which corresponds to the dissociation energy of deuterium nuclei), so that the density of radiation $\sigma T^4/c^2$ was of the order of magnitude of water density.

Here, $\sigma = 7.6 \times 10^{-15}$ erg cm^{-3} K^{-4} was the radiation density constant in cgs units, and $c = 3.0 \times 10^{10}$ cm/sec was the speed of light, yielding $\sigma T^4/c^2 = 8.4$ g/cm^3, that is, rather higher, but of the order of the density of water, as Gamow concluded.

Using the expansion rate the equations of general relativity provided at this density, Gamow determined that the interval during which deuterium or heavier elements might have formed could not have lasted more than about 100 seconds. If he then assumed that about half of the original particles combined to produce deuterium or heavier particles during this time, he could calculate the particle density that must have prevailed during that brief interval. This depended only on the particles' thermal velocity at 10^9 K, namely $v \sim 5 \times 10^8$ cm/sec, and the neutron capture cross section by hydrogen, $\sim 10^{-29}$ cm^2, yielding an atomic matter density of 10^{-6} g/cm^3. This density was 250 million times smaller than that estimated four months earlier by Alpher, Bethe, and Gamow.

But there were more important conclusions yet to come: As the Universe expands, the mass density of nonrelativistic particles drops in proportion to ℓ^3, where ℓ is an arbitrary scale representing the degree to which the Universe has expanded by a given time. In contrast, the mass density of radiation drops in proportion to ℓ^4 – the more rapid decline representing the photons' drop in energy as the Universe expands and cools. Gamow was thus able to calculate the approximate time at which the mass density of matter would equal, and thereafter forever exceed, the mass density of radiation. He estimated that this crossover occurred when the Universe was $\tau \sim 10^7$ years old, its temperature was $\sim 10^3$ K, and its density was 10^{-24} g/cm^3.[c]

Gamow then pointed out that this crossover point was important because it could initiate a gravitational instability predicted by the British astrophysicist James Jeans, enabling the uniformly distributed matter in the Universe to begin to clump into clouds whose dimensions would be of the order of 10^3 light years. Because the density of the clouds at the time was of the order of the density in the Galaxy today, he thought these clouds would be the precursors of today's galaxies. He suggested that the heavier chemical elements might then form molecules and dust, which would lead to star formation.

Although not correct in its details, the Universe and its evolution that Gamow had just sketched and semiquantitatively justified, in barely more than one page of the journal, was breathtaking in its audacity and foresight.

Shortly thereafter, Gamow wrote a somewhat longer article in *Nature*. It still was no more than three pages long, but indicated much of the progress that he and Alpher had made. The paper's title was 'The Evolution of the Universe.'[6] It appeared in the October 30, 1948 issue of *Nature*, summarized much of the work of his previous three letters, and came up with a 10-times larger estimate for the sizes of the galaxy-like condensations, which he pointed out were still smaller than many galaxies, but not significantly so, considering the simplifications he had assumed in his calculations.

Two weeks later, on November 13, Ralph Alpher and his colleague at the Johns Hopkins Applied Physics Laboratory, Robert Herman, published a short letter in *Nature* in which they thanked Gamow for his "constant encouragement during the process of error-hunting," but pointed out three rather elementary errors in Gamow's *Nature* paper.[7]

Most of the changes required in Gamow's conclusions were relatively minor, but one phrase in this short letter stands out. It reads "the temperature in the universe at the present time is found to be about 5 K." This was the first prediction of the radiation temperature characterizing today's Universe! It led to considerable recrimination when the cosmic microwave radiation was accidentally discovered 17 years later, by which time Alpher and Herman's short paper had long been forgotten. Their audacious prediction was seldom acknowledged. When they did get credit, it tended to be grudging.

Gamow still published a review article 'On Relativistic Cosmology' the following year, 1949.[8] In this paper, he (i) was aware that the decay of free neutrons in the early universe needed to be taken into account, (ii) recognized that the mass density of the early Universe was entirely dominated by thermal photons, (iii) reported that Enrico Fermi and Anthony Turkevich at the University of Chicago

---

[c] Given the information available to Gamow at the time, this was remarkably close. Current best estimates for the age and temperature at crossover of mass densities are $\tau \sim 70,000$ years and $T \sim 8000$ K.

had examined potential ways in which massive nuclei might be built up, but had found no way of building the heavier known elements due to the absence of a stable isotope of mass 5 – recalling a similar conclusion on heavy element formation in stars that Bethe had reached in 1939 – but (iv) suggested that "in order to turn one-half of [the primordial] material into deuterons and heavier nuclei leaving another half in the form of pure hydrogen, one must assume a [matter density at age 400 seconds] $\rho_0 = 5 \times 10^{-4}$ g/cm^3." Taken together, these statements promoted a picture remarkably similar to that still envisioned today, except that Gamow still clung to the hope that all heavy elements might be created primordially, whereas later work described below shows that only a few of the lighter elements beyond helium could be produced primordially.

There was one difficulty that Gamow felt he still needed to overcome. When extrapolated backward to earlier epochs, the Hubble constant ~540 km/sec-Mpc, current around 1949, yielded a cosmic age of about 1.8×10^9 years, considerably shorter than the age of Earth. To overcome this difficulty, Gamow incorporated a cosmological constant in his relativistic equations. It did not alter his conclusions about the early Universe, and was sufficiently small to avoid detection through its dynamic influence on Solar System bodies, but it accelerated the cosmic expansion at later times to make the age of the Universe appear shorter than it actually was. Some years later, a reevaluation of the actual expansion rate showed that the Universe actually appeared older and that this correction had not been necessary.

* * *

Gamow's approach to astrophysics, although highly influential and productive, was almost the complete antithesis of Chandrasekhar's. Where Chandrasekhar sought mathematical rigor and beauty, Gamow was more interested in grand insights. He was more inclined to import into astronomy novel aspects of modern physics of which most astronomers were unlikely to be aware, and thus to initiate new lines of research in which his contributions could make a seminal impact. His papers usually were short, sometimes not much longer than a single journal page but sufficiently explicit to introduce a new idea or approach – a single encompassing point of view.

Gamow preferred simplicity. He realized that a complete analysis would eventually require a complicated mathematical treatment that included a large number of minor contributing factors. He knew these calculations were necessary, but recognized that this was not the kind of work at which he excelled.

There is no doubt that Gamow's deep physical intuition permitted him to cut to the core of an astrophysical problem, despite his aversion to detail. This was what he enjoyed, the way he made his major contributions.

As Ralph Alpher, Gamow's PhD student in the late 1940s, would later recall,[9]

> George was absolutely terrible when it came to doing a detailed calculation. He knew it and he reveled in it. . . . His enthusiasm was for

ideas, for back of the envelope calculations that came within some factor which maybe some people wouldn't accept, which he found amusing. . . . I can't imagine Hans Bethe doing a calculation unless it was correct, so to speak, without question . . . but with George . . . it's perfectly clear, he [was] a man of marvelous ideas and his physics was almost always correct. But when it came to doing the actual calculations, that was not his thing. . . . I think there were some occasions on which errors – a numerator and denominator canceled – and it was kind of amusing later to come across these things and find that the answer was right or nearly right, but there were embedded errors which cancelled . . . and he really didn't care whether he had a square root of two out in front or two to the five-thirds. . . .

Teller always grumbled because they would discuss something, and Gamow would have these ideas, and they would do some preliminary calculations, and Teller would end up having to do the detailed calculations, . . . to make sure that the result was above criticism. . . .

Going back to this paper that he did on the evolution of the universe, in which he wrote down this magnificent expression . . . the diameter or the mass of the galaxy in terms of the binding energy of the deuteron.[d] You know, I think he probably just regarded that as [a] marvelous thing to be able to do.

Alpher and Herman and the Cosmic Background Radiation

At about the time Gamow was reaching his cosmological conclusions, in 1949, Alpher and Herman were carrying out similar calculations in greater detail – detail that Gamow never enjoyed.

In the work that Alpher and Herman now pursued they made a number of explicit assumptions.[10] (i) The primordial substance from which the elements were formed consisted of neutrons at high density and temperature. (ii) Protons were formed by neutron decay, and the successive capture of neutrons led to the formation of the elements. (iii) The expanding universe is filled with an isotropic mixture of radiation and matter assumed to be non-interconverting – that is, the Universe knows no preferred direction, and radiation and matter evolve without either converting into the other. (iv) The equations of general relativity with a cosmological constant $\Lambda = 0$ apply throughout cosmic expansion; here, they deviated from Gamow's assumption of a small but finite cosmological constant. (v) Like Gamow, they assumed that the expansion of the Universe was *adiabatic*, that is, that no energy was fed into, or extracted from matter or radiation in the

---

[d] Here Alpher was referring to Gamow's 1948 paper in *Nature*, in which he related the density of matter required for primordial deuterium production to the size of galaxies the Universe would eventually produce.

expansion. This led to a ratio

$$\rho_m^4/\rho_r^3 = \text{constant} \tag{8.1}$$

throughout cosmic expansion. Here ρ_m and ρ_r, respectively, were the mass density of matter and of radiation at an arbitrary cosmic age t. Under these assumptions, Alpher and Herman pointed out, the temperature T must drop in inverse proportion to the scale of the Universe, ℓ, and "If one assumes that the universe contains blackbody radiation, then $\rho_r \ell^4 = B = \text{constant}$."

With these assumptions, they considered two cosmological models. In both of them, they took the radiation density to be $\rho_r' \sim 1\ \text{g/cm}^3$ at an age of a few hundred seconds; but in one model they assumed the matter density at that time to have been $\rho_m' \sim 10^{-4}\ \text{g/cm}^3$, while in the other, shown in Figure 8.1, $\rho_m' \sim 10^{-6}\ \text{g/cm}^3$. These led, respectively, to a radiation temperature today of ~ 5 and ~ 1 K, "to be interpreted as the background temperature which would result from the universal expansion alone". Informative displays such as Figure 8.1 later became classic diagrams showing the time dependence of the cosmic expansion, the temperature of the radiation, and the respective mass densities of matter and radiation under different assumptions for the initial constituents of the Universe – single figures encapsulating the essential history of the Cosmos since early epochs.

Finally, Alpher and Herman returned to the question of galaxy formation setting in once matter density exceeded radiation density. Following Gamow's application of the instability criterion of James Jeans, they then found condensations whose masses and sizes appeared to be extremely sensitive to the initial matter densities assumed at the time of element production. They considered this rather unsatisfactory, but pointed out that the Jeans criterion for condensation formation had been intended to hold in a non-expanding space, and that a satisfactory theory would have to take "universal expansion, radiation, relativity and low matter density" into account. They noted that the Soviet theoretical physicist Evgeny M. Lifshitz, several years earlier, in 1946, had published a paper in which he showed that small fluctuations in density would not grow in a general-relativistic expanding universe. Such a cosmological model appeared to be stable, unable to form galaxies through the collapse of massive extended regions.[11]

Lifshitz's result seemed unshakable and was to haunt theorists for several further decades!

Edwin Salpeter, Ernst Öpik, and the Triple-Alpha Process

With hindsight, one can seriously question only one aspect of Bethe's paper on energy generation in stars. He had concluded that the energy generation problem was totally dissociated from the formation of heavy elements – and to a large extent it was. But the separation was not as absolute as Bethe had maintained even in the abstract of his 1939 paper, which read (with his parentheses and italics):

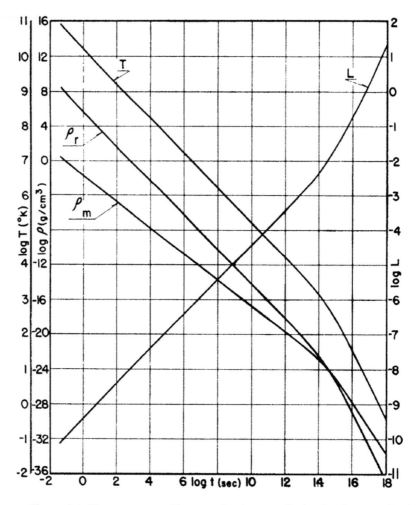

Figure 8.1. The temperature T, matter density ρ_m, radiation density ρ_r, and scale factor of the Universe L, in one of the cosmological models Alpher and Herman considered. In this model today's cosmic matter and radiation densities were, respectively, taken as $\rho_m'' \sim 10^{-30}$ and $\rho_r'' \sim 10^{-32}$ g/cm^3. For the epoch of deuterium and helium formation the corresponding densities were assumed to be $\rho_m' \sim 10^{-6}$ and $\rho_r' \sim 1$ g/cm^3. Current best estimates for today's mass densities are ρ_{r*} $\sim 4.6 \times 10^{-34}$ g/cm^3 for radiation and $\rho_{m*} \sim 0.4 \times 10^{-30}$ for baryonic matter. (*Reprinted figure with permission from Ralph A. Alpher and Robert C. Herman, Physical Review, 75, 1092, 1949, © 1949 by the American Physical Society*).

... *no elements heavier than He^4 can be built up in ordinary stars.* This is due to the fact ... that all elements up to boron are disintegrated by proton bombardment (α-emission!) rather than built up (by radiative capture). The instability of Be^8 reduces the formation of heavier elements still further. The production of neutrons in stars is likewise negligible. The heavier elements found in stars must therefore have existed already when the star was formed.

Figure 8.2. Edwin E. Salpeter as a young man. At age 27, he showed that the high temperatures at the center of giant stars lead to the fusion of helium nuclei to form carbon and thence the heavier elements oxygen and neon. (*American Institute of Physics Emilio Segre Visual Archives, Physics Today Collection*).

If Bethe meant main sequence stars, in referring to "ordinary stars," he was right. But, once those stars use up the hydrogen reservoir on which they had been surviving all along, they begin a new, if shorter life as giants, and there they depend on helium to sustain their energy generation, which also leads to the production of heavy elements.

The way that this comes about was shown by Edwin E. Salpeter, a 27-year-old physicist. He had arrived at Cornell University as a postdoc in 1949 to work with Hans Bethe. Within two years, he had become a member of the physics department and remained there for the remainder of his professional life.

In the summer of 1951, Salpeter accepted an invitation from William A. (Willy) Fowler, an experimental nuclear physicist at the California Institute of Technology, who was measuring cross sections of nuclear processes that might be of importance in astrophysical processes. Although academic in its aims, Fowler's group was largely sponsored by the Atomic Energy Commission (AEC) and the Office of Naval Research (ONR) as part of the envisioned amalgam of applied and basic science proposed in 'Science – The Endless Frontier.'

Fowler realized the value of having young theorists around, who could look into the astrophysical implications of the reaction rates he and his colleagues were studying, and Salpeter became a regular visitor during the summers, although he continued to stay at Cornell during the academic year.[12]

That first summer at Caltech, in 1951, Salpeter started work on a short paper that proved to be a crucial step toward solving both the energy generation problem in red giants and the production of heavy elements in stars. Willy Fowler and his research group had recently shown that although the beryllium isotope ^{8}Be is unstable, it does have a meta-stable ground-state that can be *resonantly excited* through the merger of two α-particles with energies as low as $\sim$95 keV, corresponding to temperatures of $T \sim 10^9$ K. This seemed a highly likely central temperature for stars that had exhausted all their hydrogen and had then gravitationally contracted.

Salpeter estimated that ^{8}Be nuclei at these temperatures would have an equilibrium abundance of one part in 10^{10} relative to the abundance of He^4, a sufficiently high concentration to permit a further collision with an ambient He^4 nucleus.[13] This would form a carbon ^{12}C nucleus with the liberation of a γ-photon at an energy of 7.4 MeV. Once this much energy was radiated away, there would be no way back. The carbon atom would be stable. It was a one-way street that provided the energy for a star to keep radiating after it had used up all its hydrogen supply and was sustaining itself on the helium the hydrogen had formed during the star's main-sequence phase.

Once carbon ^{12}C was formed, the high abundances of ^{4}He would also ensure the formation of the oxygen isotope ^{16}O, and then neon ^{20}Ne, through successive addition of ^{4}He nuclei. Each of these reactions released energies of a few MeV, ensuring that none of the successively formed nuclei could disintegrate. A progression of massive elements was sure to form, leading eventually to the isotopes magnesium ^{24}Mg, silicon ^{28}Si, sulfur ^{32}S, argon ^{36}Ar, and calcium ^{40}Ca, in a series of reactions that began as

$$^4\text{He} + {}^4\text{He} + 95\text{ keV} \rightarrow {}^8\text{Be} + \gamma, \tag{8.2}$$

$$^4\text{He} + {}^8\text{Be} \rightarrow {}^{12}\text{C} + \gamma + 7.4\text{ MeV}, \tag{8.3}$$

$$^4\text{He} + {}^{12}\text{C} \rightarrow {}^{16}\text{O} + \gamma + 7.1\text{ MeV}, \tag{8.4}$$

$$^4\text{He} + {}^{16}\text{O} \rightarrow {}^{20}\text{Ne} + \gamma + 4.7\text{ MeV}, \tag{8.5}$$

Owing to the increasing electrostatic repulsion of the heavier and more positively charged nuclei, Salpeter noted, the reaction rates would decrease for increasingly massive nuclei, although the energy generation rate, and thus also the production of heavy elements, would increase as the 18th power of temperature T – that is, as T^{18}.

The main conclusions were that (Salpeter's italics): *"A few percent of all visible stars of mass 5 M$_\odot$ or larger are converting helium into heavier nuclei, the central temperature being about ten times larger (and radius ten times smaller) than that of a main-sequence star of the same mass."* He hoped that some connection would "be found between stars undergoing this process, on the one hand, and carbon-rich and high-temperature stars (Wolf-Rayet, nuclei of planetary nebulae), on the other (and possibly even with some of the *variable stars* and novae)."

In a second paper that appeared the following year, 1952, Salpeter also conjectured about what would happen once the helium supply was exhausted.[14] A further contraction would again raise the temperature. Two carbon ^{12}C nuclei could then combine to form heavier elements, and at slightly higher temperatures ^{16}O and ^{20}Ne might undergo similar interactions. At even higher temperatures more complex processes could be conjectured, which eventually could lead to the formation of elements as massive as uranium before a further collapse followed by a supernova explosion might entirely eject these heavy elements back into a star's surrounding. In contrast to the α-particle adding processes, Salpeter was careful to point out, these subsequent phases in a star's history were "quite tentative and speculative."

In this second paper, which took up only slightly more than two pages, Salpeter laid the foundations of two main themes. The first concerned energy production in stars that had exhausted all their hydrogen, notably the most massive main-sequence stars and red giants; the second dealt with the synthesis of heavy elements which, despite their abundance on Earth, had previously remained unexplained.

Unbeknownst to Salpeter and to almost every one else, the Estonian astronomer Ernst Öpik, by then working at the Armagh Observatory in Ireland, had in 1951 independently suggested the same process for forming carbon ^{12}C. But Öpik was unaware of the beryllium resonance, without which the predicted carbon formation rate would have been considerably lower.[15]

Fred Hoyle, William Fowler, and the Abundance of Heavy Elements

The ^{12}C production rate that Salpeter provided, however, also was too low, although not by as much as Öpik's. Looking back half a century later, Salpeter ruefully remarked on this shortcoming. He recalled Hans Bethe's advice to him when he had first come to Cornell and decided to switch from high-energy particle theory to astrophysics. Bethe had advised a three-pronged course of action "(a) Be prepared to change fields; (b) Use only the minimum mathematical technique necessary; (c) In the face of uncertainty be prepared to use conjectures and shortcuts and take risks." But, Salpeter regretfully recalled, "I had learned lessons (a) and (b) but not yet (c). . . . I just did not have the guts to think of resonance levels which had not been found yet! A short while later Fred Hoyle demonstrated both chutzpah and insight by using known abundance ratios of ^{12}C, ^{16}O and ^{20}Ne to show that there JUST HAD to be an appropriate resonance level in ^{12}C and he was able to predict its energy" (Salpeter's capital lettering).[16]

* * *

Figure 8.3. Fred Hoyle around the time he was working on the production of heavy elements. (*By permission of the Master and Fellows of St John's College, Cambridge*).

Fred Hoyle, an immensely inventive theoretical astrophysicist at Cambridge, appreciated Salpeter's triple alpha process, but figured that its carbon production rate must be considerably higher to agree with the observed abundance ratios of carbon, oxygen, and neon isotopes, ^{12}C, ^{16}O, and ^{20}Ne, at levels of $1/3 : 1 : 1$. The neon abundance he cited is almost an order of magnitude too high by currently accepted values, but this did not change his central prediction that there had to be a resonance in the production rate of carbon that would lie at energies about 7.7 MeV above the ^{12}C ground state. This added resonance, on top of that of ^{8}Be, would make the addition of a third alpha particle more probable.[17]

As Willy Fowler later recalled,[18]

Hoyle came to Caltech in 1953 and walked into one of our Kellogg Staff meetings and announced that there had to be an excited state in ^{12}C at 7.70 MeV. He needed this state to serve as a resonance in ^{8}Be$(\alpha, \gamma)^{12}$C [impact of an α-particle on a beryllium nucleus to form carbon with the emission of an energetic photon] for which the threshold energy is 7.367 MeV. In turn he needed the resonance to obtain substantial conversion of

Figure 8.4. William (Willy) Alfred Fowler around the time he was conducting laboratory measurements on nuclear processes he expected to form heavy elements in stars. (*American Institute of Physics Emilio Segre Visual Archives, Physics Today Collection*).

helium into carbon in red giant stars under the conditions of density and temperature calculated by Martin Schwarzschild and Allan Sandage.

Willy Fowler and his associates at Caltech scoffed at Hoyle's audacity. Nobody had ever predicted a nuclear resonance on astrophysical grounds alone. But in an unpublished paper Fowler presented at an American Physical Society symposium I organized for the Society's meetings held in Washington in April 1987, Willy explained, and subsequently wrote[19]:

> 'The Lauritsens [Charlie and Tommy] and I were not impressed and told Hoyle to go away and stop bothering us. But Ward Whaling gathered his graduate students and research fellows together, went into the laboratory, used the $^{14}N(d,\alpha)^{12}C$ reaction to show that ^{12}C did indeed have an excited state at 7.653 ± 0.008 MeV, very close to Hoyle's prediction.

Charles Cook and the Lauritsens and I produced Hoyle's state in the decay of ^{12}B to ^{12}C and showed that it broke into 3 alpha-particles. Thus by time reversal invariance it could be formed from 3 alpha-particles. That made a believer out of me and in the fall of 1954 I went to Cambridge on a Fulbright to work with Hoyle.

Hoyle's paper, submitted in December of 1953, differs somewhat from Fowler's account probably because it may have taken Hoyle some time to write this long paper. By then the Caltech measurements had been published and Hoyle used the results to support his thesis. But the Caltech paper leaves no doubt about Hoyle's prescience. Their short paper begins with:

> Salpeter and [sic] Öpic have pointed out the importance of the ^{8}Be$(\alpha,\gamma)^{12}$C reaction in hot stars which have largely exhausted their central hydrogen. Hoyle explains the original formation of elements heavier than helium by this process and concludes from the observed cosmic abundance ratios of ^{16}O : ^{12}C : ^{4}He that this reaction should have a resonance at 0.31 MeV or at 7.68 MeV in ^{12}C.

They conclude, "We are indebted to Professor Hoyle for pointing out to us the astrophysical significance of this level."[20]

I have gone into this account in some detail because Hoyle's audacious prediction has become part of astrophysical folklore.

Following the events of the early 1950s just cited, Hoyle and Fowler joined forces, and working with the young British couple, Geoffrey and Margaret Burbidge, began collaboration on a major work published in 1957. Possibly the most celebrated astrophysical paper of the second half of the twentieth century, often referred to only by its authors' initials, B^{2}FH, the paper described in great detail the build-up of heavy elements in stars.[21]

I will return to this paper later, but first need to backtrack to work being pursued around 1950 to determine whether elements beyond hydrogen could have been produced early in Gamow's Universe.

Hayashi, Alpher, Follin, Herman, and the Formation of Primordial Elements

Chushiro Hayashi, a 29-year-old nuclear physicist working at Naniwa University in Japan, came across Alpher and Hermans's *Physical Review* paper shortly after its publication in 1949. He saw an important flaw that neither Alpher and Herman nor Gamow had noticed.[22,23] At the high temperatures of the early universe, neutrons n and protons p quickly reach thermal equilibrium with electrons e^-, positrons e^+, neutrinos ν, and antineutrinos $\bar{\nu}$ through processes like

$$n+e^+ \rightleftharpoons p+\bar{\nu} \quad ; \quad n+\nu \rightleftharpoons p+e^- \quad ; \quad n \rightleftharpoons p+e^-+\bar{\nu}. \tag{8.6}$$

These processes would have been rapid at temperatures T at which particle energies kT exceeded the electrons' rest mass energies, $m_e c^2$, that is, temperatures at which electrons, positrons, neutrinos, and antineutrinos were about as abundant as photons. Here k is the Boltzmann constant and m_e the electron mass. At temperatures and energies exceeding the rest mass of the μ meson, neutron–proton conversion rates would have been even higher. It therefore made little sense to postulate that the early Universe had been populated primarily by neutrons. Neutrons and protons must have been in thermal equilibrium all along at temperatures higher than $\sim 2 \times 10^{10}$ K, where the β-reactions (8.6) proceeded far more rapidly than the drop in temperature due to the cosmic expansion.

As the expansion proceeded, however, the distances between particles would increase, β-process would be less frequent, and temperatures would drop significantly. Hayashi's calculations indicated that the neutron/proton ratio would then freeze out at a proportion of about 1:4. This, he showed, was inevitable if the initial temperature of the Universe had ever been sufficiently high to establish thermal equilibrium. After freeze-out, the n/p ratio would further decline, but only through the relatively slow radioactive decay of neutrons into protons.

Assuming this n/p freeze-out ratio, and recognizing that nuclei other than helium ^{4}He were unlikely to form in significant numbers during primordial epochs, Hayashi calculated the ratio of hydrogen to helium nuclei produced at these early times. Helium production required the fusion of two neutrons and two protons. Hence, for every 2 initial neutrons, and the corresponding 8 initial protons, there would be 6 remaining protons for each ^{4}He nucleus produced. Hayashi remarked that this 6:1 ratio was close to the H:^{4}He ratios in stellar atmospheres and meteorites, respectively, 5:1 and 10:1, as judged at the time.[e]

The detailed calculations Hayashi produced, and the conclusions he reached, were a remarkable achievement: They provided a first quantitative indication that at least helium formation had taken place at primordial cosmological times and that the temperature in the Universe must at one time have been of the order of $\geq 10^{10}$K.[24]

Hayashi's calculations assumed that the primary contributor to the mass/energy density of the Universe was radiation. Although this was so for the epoch and temperature at which protons and neutrons combine to form deuterons, which can be transformed further into α-particles, that is, ^{4}He nuclei, the assumption fails at higher temperatures. At temperatures above 10^{10} K, collisions between photons produce electron–positron pairs whose further collisions give rise to neutrino-antineutrino pairs. Together, the mass densities of these four particle species then exceed the mass density of radiation.

The problem of including this interconversion of radiation and matter was next addressed by Alpher and Herman, now joined by James W. Follin, Jr., their

---

[e] Today's best estimate of the freeze-out ratio is $\sim 1:6$, and the the ratio of hydrogen to helium in the Solar System is $\sim$10:1 – still in good agreement with Hayashi's conclusions.

colleague at the Johns Hopkins Applied Physics Laboratories.[25] Their detailed calculations, based in part on Hayashi's work, led to small but significant changes in the n/p ratio at freeze-out, placing it roughly in the range of 1:4.5 to 1:6.0, which led to better agreement with observed abundance ratios of hydrogen and helium. Although their paper did not repeat the prediction of the photon temperature expected in the Universe today, they did predict that the ratio of neutrino temperature T_ν to photon temperature T_γ should be $T_\gamma/T_\nu \sim 1.401$ today. Given that the photon temperature is 2.73 K, the neutrino temperature should be $\sim$1.95 K. Their prediction assumed zero rest mass for the neutrinos.[f]

The paper by Alpher, Follin, and Herman, like that of Hayashi, confined itself to the era during which light element formation was just setting in. It no longer speculated about the formation of heavy elements. Perhaps this was because the prospects for heavy element production in the early Universe were beginning to look dim.

In a review article that Alpher and Herman published that same year, 1953, they wrote "the principal difficulties still extant [in the production of heavy elements] are those associated with the gaps at A = 5 and 8 ... While the theory is in a quite approximate state at the present time, it is hoped that the inclusion of the growing knowledge of the properties of nuclei and nuclear reactions will satisfactorily explain detailed features of the abundance data, provided, of course, that the extremely difficult detailed calculations can be carried out."[26]

Gamow, Alpher, and Herman were fully focused on an evolving model of the Universe that had a well defined beginning. But an alternative to this model was also being considered in the 1950s. It envisioned a universe that had always existed and would continue to exist for all time.

The Steady State Theory of Bondi, Gold, and Hoyle

To understand why the possibility of an infinitely old universe was even considered, it is necessary to look back at the age attributed to Lemaître's relativistically expanding universe in the late 1940s. As judged by the best available expansion rates of the Universe at the time, that is, by measurements of the Hubble constant, the Universe appeared to be no older than $\sim 2 \times 10^9$ – two billion – years. This was puzzling because radioactive decay of elements in terrestrial rocks seemed to imply a terrestrial age exceeding 3×10^9 years, and the ages of massive stars in globular clusters seemed to be of order 5×10^9 years. The ages of these stars could be estimated, by the 1940s, knowing that stars are composed primarily of hydrogen and that massive stars are more luminous than less massive stars.

---

[f] The neutrino background remains observationally inaccessible, even today. The minute neutrino rest mass can currently be estimated only coarsely, but the calculations of Alpher, Follin, and Herman may also be expressed in equivalent number densities, and these may some day become measurable and provide additional information on the era of light-element formation.

Luminous stars rapidly convert hydrogen into helium and, having depleted their store of central hydrogen, leave the main sequence of the Hertzsprung–Russell diagram to move onto the red giant branch.

The most massive stars remaining on the stellar main sequence of globular clusters, all of whose stars were assumed coeval, thus were the stars about to move off the main sequence. This permitted determination of the duration over which they had sustained the rate at which they radiated energy before consuming all the hydrogen in their cores.

That the stars might be older than the Universe raised a conundrum. Surely, Earth and the stars could be no older than the Universe? How could the relative ages indicated by the cosmic expansion and the age of Earth be reconciled?

A new insight was brought to the discussion in 1948 in two separate versions of the *Steady State Theory of the Universe*. One was presented by two young researchers, Hermann Bondi and Thomas Gold, both Austrian-born, who had become close friends when conscripted into World War II service on British radar.[27] The other was published by their more senior colleague, the Cambridge astrophysicist Fred Hoyle, whom they had met when he too had been conscripted into wartime service.[28]

Both versions of the theory postulated, albeit along somewhat different lines, that the Universe continually created new matter – out of nothing – at just such rates that the mass density, that is, the number of atoms, stars, and galaxies per unit volume, forever remained constant. This meant that the Cosmos would have appeared the same throughout all time. It would never have been more compact than it is today. The ratio of *young stars* and old would forever remain the same. Even though individual stars evolved, their relative numbers would remain constant. Bondi, Gold, and Hoyle called this the *Steady State Theory of the Universe*, because everything remained steady for all time. As galaxies receded from each other in a constant expansion, new galaxies would form to fill the additional space the expansion had created. It was an imaginative new model of the Universe!

Bondi and Gold invoked a new principle they called the *perfect cosmological principle*. This was an extension of a principle widely accepted by cosmologists of the time – the *cosmological principle*. That principle postulated that the Universe displays an essentially identical appearance to all astronomers, not just to astronomers located in our Galaxy, but also to others observing from anywhere else in the Universe.

The perfect cosmological principle was even more stringent in requiring the appearance of the Universe to remain essentially identical for all observers for all time. Although varying in detail from place to place and epoch to epoch, the Universe as we view it today should *statistically* have appeared identical to observers billions of years ago, and should maintain this appearance forever. Only in this way, Bondi and Gold argued, could one expect to understand cosmic evolution.

They referred to *Mach's principle*, which maintained that the inertia of a body is determined in some as-yet-unspecified way by the large-scale structure of the

Universe.[29] They suggested that the constants of Nature, that is, the speed of light, the mass and charge of an electron, or the value of Planck's constant, might similarly be determined by the disposition of the universe at large. They wrote:

> Present observations indicate that the universe is expanding. This suggests that the mean density in the past has been greater than it is now. If we are now to make any statement regarding the behavior of such a denser universe, and the way in which it is supposed to have reached its present condition, then we have to know the physical laws and constants applicable in denser universes. But we have no determination for those. An assumption which is commonly made is to say that . . . there would be no change in the physical laws which an observer would deduce in the laboratory. . . . [But if] we adopt Mach's principle . . . we cannot have any logical basis for choosing physical laws and constants and assigning to them an existence independent from the structure of the universe.

To these arguments, Hoyle added that the idea of a Steady State Universe "seemed attractive, especially in conjunction with aesthetic objections to the creation of the universe in the remote past. For it is against the spirit of scientific enquiry to regard observable effects as arising from 'causes unknown to science', and this in principle is what creation-in-the-past implies."[30]

Hoyle modified Einstein's field equations to accommodate the creation of matter. The most essential step was the incorporation of a creation tensor, $C_{\mu\nu}$, to specify the rate at which matter was continually being created in the universe. This neatly replaced the cosmological term that Einstein had originally incorporated in his equations of general relativity to produce his static cosmology of 1917.

Hoyle's creation tensor yielded a cosmic expansion rate similar to that of the de Sitter model, three decades earlier, except that the atomic matter density of the universe no longer was zero but corresponded to the observed density if the value of $C_{\mu\nu}$ was appropriately chosen.

Hoyle, Fowler, the Burbidges, and the Formation of the Heavy Elements in Stars

We now need to return to the previously mentioned paper of Eleanor Margaret Burbidge, her husband Geoffrey Ronald Burbidge, William A. Fowler, and Fred Hoyle.[31]

In 1957, as Hoyle, Fowler, and the Burbidges were writing their paper, it was not clear whether the Universe had ever had a beginning or whether it might be infinitely old. If it had a beginning, as Gamow, Alpher, and Herman had assumed, then at least some of the elements could have been formed at earliest times when the Universe was very hot. If the Universe was infinitely old, the only place known to be sufficiently hot to account successfully for the synthesis of elements heavier than hydrogen was the interior of stars.

Burbidge, Burbidge, Fowler, and Hoyle (B^2FH) did not commit themselves on "whether all of the atomic species heavier than hydrogen have been produced in stars without the necessity of element synthesis in a primordial explosive stage of the universe." Instead, they intended their paper to restrict itself to element synthesis in stars and "to lay the groundwork for future experimental, observational, and theoretical work" that might ultimately provide conclusive evidence for the origin of elements in stars.[32] They pointed out that from the standpoint of the nuclear physics alone it was clear that their conclusions would be equally valid for a primordial synthesis in which the initial and later evolving conditions of temperature and density were similar to those found in the interior of stars.

The paper considered eight different types of nuclear reactions, covering all the major factors at play in stars and able to account for the existence of the more abundant observed isotopes. This paper, more than a hundred pages in length, presented a theory whose outlines still reflect our understanding today, although many details have since changed.

Of course, much has happened since the paper appeared in 1957. By now we know that the Steady State Theory is no longer in contention. We see the Universe appearing quite different at great distances, and therefore early times, than it does in our immediate vicinity today. It is definitely evolving.

The consensus that has emerged is that most of the helium observed in the Universe today was primordially produced in the first few minutes of the existence of the Universe, as indicated by Hayashi in the late 1940s. At least some of the observed deuterium, lithium, and beryllium also appears to date back to those epochs. The heavier elements all have been formed in more recent times in the interior of stars, more or less as predicted by B^2FH.

* * *

Even this esoteric astrophysical-landmark paper, B^2FH, bears the acknowledgment "Supported in part by the joint program of the Office of Naval Research and the US Atomic Energy Commission." It indicates, again, how readily defense-related and basic scientific information was being shared among scientists shortly after the war. Many had worked on classified problems during the conflict. After cessation of hostilities, they found ways of continuing to work on interesting problems without posing a threat to national security.

Notes

1. An Attempt to Interpret the Relative Abundances of the Elements and their Isotopes, S. Chandrasekhar & Louis R. Henrich, *Astrophysical Journal*, *95*, 288–98, 1942.
2. Expanding Universe and the Origin of the Elements, G. Gamow, *Physical Review*, 70, 572–73, 1946; erratum: *71*, 273, 1947.
3. The Origin of the Chemical Elements, R. A. Alpher, H. Bethe, & G. Gamow, *Physical Review*, *73*, 803–04, 1948.

4. *Geochemische Verteilungsgesetze der Elemente. IX. Die Mengenverhältnisse der Elemente und der Atom-Arten*, V. M. Goldschmidt, *Shrifter utgitt av Det Norske Videnskaps-Akademi i Oslo I. Mat.-Naturv. Klasse.*, 1937. No. 4, Oslo, i Kommisjon Hos Jacob Dybwad, 1938.

5. The Origin of Elements and the Separation of Galaxies, G. Gamow, *Physical Review*, 74, 505–6, 1948.

6. The Evolution of the Universe, G. Gamow, *Nature*, 162, 680–82, 1948.

7. Evolution of the Universe, Ralph A. Alpher & Robert Herman, *Nature*, 162, 774–75, 1948.

8. On Relativistic Cosmology, G. Gamow, *Reviews of Modern Physics*, 21, 367–73, 1949.

9. Ralph Alpher and Robert Herman, transcript of an interview taken on a tape recorder, by Martin Harwit on August 12, 1983, archived in the American Institute of Physics Center for History of Physics, pp. 13-15.

10. Remarks on the Evolution of the Expanding Universe, Ralph A. Alpher & Robert C. Herman, *Physical Review*, 75, 1089–95, 1949.

11. On the Gravitational Stability of the Expanding Universe, E. M. Lifshitz, *Journal of Physics of the USSR*, 10, 116–29, 1946.

12. A Generalist Looks Back, Edwin E. Salpeter, *Annual Reviews of Astronomy and Astrophysics*, 40, 1-25, 2002.

13. Nuclear Reactions in Stars without Hydrogen, E. E. Salpeter, *Astrophysical Journal*, 115, 326–28, 1952.

14. Energy Production in Stars, E. E. Salpeter, *Annual Review of Nuclear Science*, 2, 41–62, 1953.

15. Stellar Models with Variable Composition. II. Sequences of Models with Energy Generation Proportional to the Fifteenth Power of Temperature, E. J. Öpik, *Proceedings of the Royal Irish Academy*, 54, Section A, 49-77, 1951.

16. Ibid., A Generalist Looks Back, Salpeter pp. 8–9.

17. On Nuclear Reactions Occurring in Very Hot Stars I. The Synthesis of Elements from Carbon to Nickel, F. Hoyle, *Astrophysical Journal Supplement Series*, 1, 121–46, 1954.

18. Recollections of Early Work on the Synthesis of the Chemical Elements in Stars, William A. Fowler, from an unpublished talk at an American Physical Society meeting, Crystal City, VA, on April 22, 1987. Fowler sent a copy of this talk to Martin Harwit, the session's organizer, on April 27, 1987, adding, "If the question of publication arises I would like to be consulted first." Regrettably, this is no longer possible.

19. Ibid., Recollections, W. Fowler, 1987.

20. The 7.68-MeV State in C^{12}, D. N. F. Dunbar, et al., *Physical Review*, 92, 649–50, 1953.

21. Synthesis of the Elements in Stars, E. Margaret Burbidge, G. R. Burbidge, William A. Fowler, & F. Hoyle, *Reviews of Modern Physics*, 29, 547–650, plus 4 plates, 1957.

22. Ibid., Remarks on the Evolution, Alpher & Herman.

23. Proton-Neutron Concentration Ratio in the Expanding Universe at the Stages preceding the Formation of the Elements, Chushiro Hayashi, *Progress of Theoretical Physics*, 5, 224–35, 1950.

24. Ibid., Proton-Neutron Concentration, Hayashi.

25. Physical Conditions in the Initial Stages of the Expanding Universe, Ralph A. Alpher, James W. Follin, Jr., & Robert C. Herman, *Physical Review*, 92, 1347–61, 1953.

26. The Origin and Abundance Distribution of the Elements, Ralph A. Alpher & Robert C. Herman, *Annual Review of Nuclear Science*, 2, 1–40, 1953.

27. The Steady State Theory of the Expanding Universe, H. Bondi & T. Gold, *Monthly Notices of the Royal Astronomical Society*, 108, 252–70, 1948.

28. A New Model for the Expanding Universe, F. Hoyle, *Monthly Notices of the Royal Astronomical Society*, 108, 372–82, 1948.

29. *Die Mechanik in ihrer Entwickelung – Historisch-Kritisch dargestgellt*, Ernst Mach, F. S. Brockhaus, Leipzig, 1883, p. 207ff.
30. Ibid., A new Model, F. Hoyle, p. 372.
31. Ibid., Synthesis of the Elements, Burbidge, Burbidge, Fowler & Hoyle.
32. Ibid., Synthesis of the Elements, Burbidge, Burbidge, Fowler & Hoyle, p. 550.

9

Landscapes

Scientific structures consist of layered domains – landscapes of activities – most of which remain invisible to the working astronomer focused on solving problems through a rational approach. The influence these external activities exert on astronomy, nevertheless, is profound. The heightened priority that the United States and many other countries were assigning to science after World War II led to expansion. Astronomers now organized themselves into groups whose members collaborated to compete more effectively for national support of projects of common interest. Increasingly linked networks of astronomers were springing up and interacting with networks permeating other scientific fields. Links to governmental networks were also arising.

Landscapes of Activity

The astrophysicist's landscape, Figure 9.1, represents his view of the Universe. It mirrors the interaction of stars with their surroundings, the interplay of galaxies within their clusters, and our heritage from an earlier phase in the evolution of the Universe in which some of the lighter chemical elements we see around us today were first forged at temperatures of a billion degrees Kelvin.

The everyday task of the astrophysicist is to gather additional information and make the new pieces of observational and theoretical evidence fit neatly into

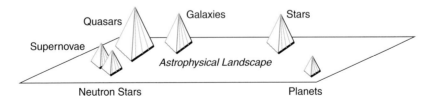

Figure 9.1. A schematic astrophysical landscape.

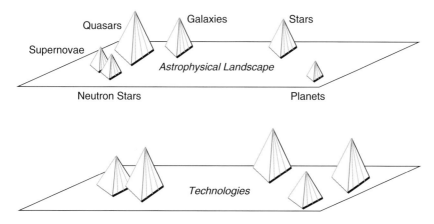

Figure 9.2. The astrophysical landscape is often thought of as depicting innate features of the Universe, largely independent from day-to-day developments in a technological landscape evolving to meet industrial or governmental needs.

this landscape, making sure that nothing is missing and, to the extent possible, everything is tidy and explained. This can be an arduous process. Each newly gathered piece of evidence clamors for a place in the landscape but often will not fit. To accommodate it, other pieces of the landscape have to be moved around, and so the landscape continually changes.

But although the astrophysicist often thinks it is, his landscape is not totally isolated. Evolution in apparently unrelated activities normally does not intrude on the landscape, Figure 9.2. But when those activities grow beyond a critical point, as shown in Figure 9.3, they may abruptly force their way into that landscape, affecting its appearance and subsequent evolution. As sketched in the lower half of Figure 9.3, astrophysics is then in temporary disarray. Unrecognizably transfigured, the landscape needs urgent reconstruction.

In the post–World War II era, as we found in Chapter 7, discoveries made by observing in newly accessible wavelength ranges, time after time, shook up the field and reconfigured the landscape. Often the astronomical community could not even catch its breath to pick up the pieces and reassemble them before the next discovery disrupted the efforts. Each major discovery came as a surprise because the landscape that had previously been assembled had meticulously attempted to exhibit continuity and coherence, leaving little room for unexpected external influences. Until World War II, our astrophysical theories had been too narrowly defined by observations gathered almost exclusively at visible wavelengths. Now, new instrumentation developed elsewhere began to intrude.

The war introduced new technologies that led to the discovery of phenomena that had until then left no visible imprint of themselves. Viewed in isolation, these technologies constituted a totally different, previously largely ignored domain – a landscape of their own – from which time and again disturbances emanated to alter the astrophysical landscape. The tools these technologies could provide

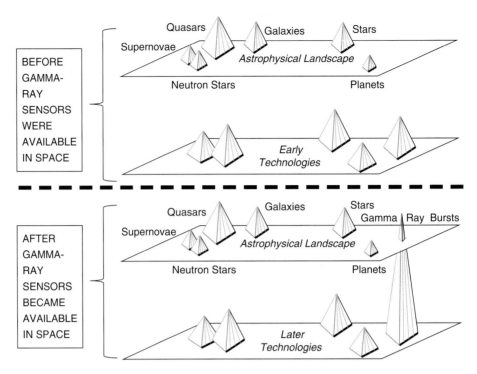

Figure 9.3. Transformation of the astrophysical landscape through the launch of γ-ray sensors into space. The United States' Vela Nuclear Test Detection Satellite (NTDS) system was established in space through a series of launches between 1963 and 1970 to detect potential Soviet nuclear bomb tests in outer space. This introduced γ-ray sensing instruments into space, as shown in the strip marked *Later Technologies* in the bottom half of the figure. Before the establishment of the NTDS system, as pictured in the *Astrophysical Landscape* in the top half of the figure, nothing was known about the possible existence of astronomical γ-ray sources. Once the NTDS system had been established, it became clear, as represented in the lower *Astrophysical Landscape*, that γ-ray bursts sporadically arrived from unpredictable directions far out in space. They could not be surreptitious bomb tests. But what they were, nobody knew.

had – and even today have – no assigned place in the astrophysical landscape, that is, in our views about the world we live in. They are considered a mere means for gathering data, whereas the astronomical landscape is the end product of those data pieced together to represent the world we inhabit. Once the new data are fitted into the landscape, all traces of how they were obtained are erased. It no longer seems to matter.

But, if you are cognizant of the technological landscape, a landscape just as real and perhaps just as complex as the astrophysical landscape, you possess information about all the conceivable ways in which the Cosmos might ultimately be studied and have a map that tells you where observational astronomy has thus far failed to take a foothold. This is where new observations will most likely produce novel insights. The Universe is so complex that theoretical predictions have usually

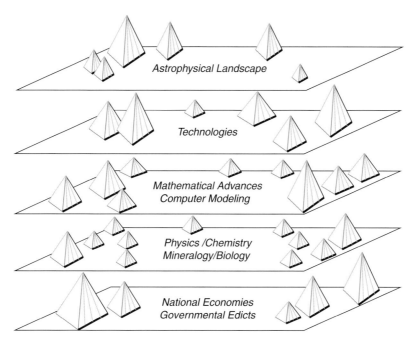

Astrophysical Landscape

Technologies

Mathematical Advances
Computer Modeling

Physics /Chemistry
Mineralogy/Biology

National Economies
Governmental Edicts

Figure 9.4. Astrophysics evolves not only in response to changes internal to its own landscape or to newly introduced instrumentation. It also is affected by intrusions from other layers of activity, respectively reflecting the development of novel tools in theoretical physics, new ways of visualizing and assimilating information, developments in other sciences, budgetary constraints, or political decisions.

failed to extrapolate from known observations into areas where observations are still lacking.[1]

*   *   *

As indicated in Figure 9.4, novel technologies are not the sole external influences that shape the astrophysical landscape. Mathematical models of the underlying cosmic topology and geometry, and the ability to compute complex interactions or analyze vast accumulations of data, can also lead to new world views. Discoveries of novel physical or chemical processes, or advances in mineralogy, can similarly cast astronomical processes in a new light. And, certainly, governmental decrees, often based on a nation's current economy or societal priorities, can command new directions of research.

The direction of our research is at least partially influenced by societal forces. Control of the purse channels scientific projects. Investigations that looked promising are dropped for lack of funding; new projects are started because they can be financed. Rarely do scientists swim upstream against the budgetary flow. Most astronomers follow the money.

Any of these external influences can overnight determine that work on tidying one domain of the astrophysical landscape will stop, while some other, more remote, less developed domain that had been too difficult to investigate under previous external influences becomes accessible.

In the first half of the twentieth century, astronomy was affected by such external factors mainly through the importation of theoretical tools devised by mathematicians and theoretical physicists, and observational tools often developed by physicists, chemists, or engineers. But the extent of these influences was far less intrusive than we saw emerging in the second half of the century. There, a heightened degree of governmental support and a conscious effort to amalgamate basic and applied research more closely, led to far closer linkage. The example in Figure 9.3 exhibiting the effect of national priorities on astronomical findings, as exemplified by the discovery of γ-ray bursts, is just an indication of how the conduct of astrophysics changed after World War II and the onset of the Cold War.

Communal Networks

To a large extent, the different landscapes I have described mirror the work of different professional communities, some of which may specialize in distinct areas of astronomy, astrophysics, or cosmology, while others are engaged in disparate fields of science, industry, or government. Mark Newman, Steven Strogatz, and Duncan Watts, three pioneers studying social networks, have shed light on the links crisscrossing such communities.

Two types of activity tend to provide astronomers an opportunity to establish respect, trust, and thus, mutual influence. The first is through collegially conducted research, as indicated by jointly published papers. The second is through joint service on professional boards or committees, or other close contacts.[2]

These links can be represented in two ways. One indicates the specific activities involved. The upper half of Figure 9.5 shows five activities, say, the work of five scientific oversight or advisory boards numbered 1 to 5, on which some eleven professionals A through K served in groups. Professional D sat on a board with A, B, C, E, F, and G, and thus had direct links to them. H is directly linked only to I and F, and via F through a second step to D. The lower half of the figure drops all reference to the linking activity and merely shows which scientists are linked to each other, making the linkage between them more explicit.

Within any one professional group, be they astrophysicists, instrument builders, theoretical physicists, or government officials, the interaction between individuals can be close because they have worked together on some form of activity. Links across different disciplines or professions tend to be relatively infrequent but can be among the most influential, Figure 9.6.

Although only a few astrophysicists have close ties to government officials and perhaps rather more of them to instrument builders or theoretical chemists, these few links between members of different professions suffice to bring all the different communities into effective working contact. Across the entire community of

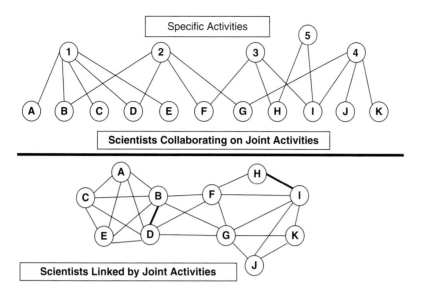

Figure 9.5. Scientists linked through joint activities. The circled areas in the diagrams marked by letters or numbers are referred to as the *nodes* in a network. The links between them are referred to as *edges*. In the upper panel, scientists A through K are loosely connected through activities 1–5, in which some of them are jointly engaged. The bottom panel shows how directly or indirectly these same scientists are linked, but disregards the linking activities. Note the stronger bonds between B and D, and also between H and I. In contrast to others linked by only a single joint activity, both of these pairs of scientists are linked through two shared activities. (*Based on a drawing by M. E. J. Newman, S. H. Strogatz, & D. J. Watts, Physical Review E, 64, 026118–2, 2001*).

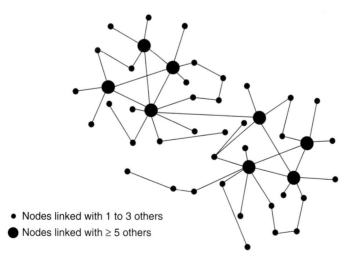

Figure 9.6. Scientists as nodes in a network. Note that this network has two heavily populated regions joined by only a single link. This suggests that astronomers at home in their respective regions have little in common and work in different sub-disciplines.

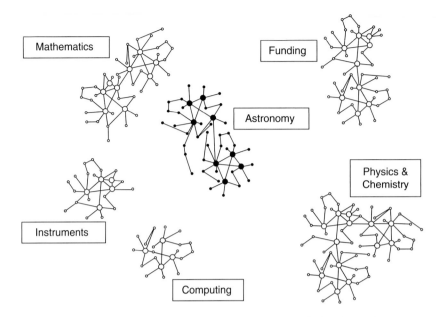

Figure 9.7. Astronomers are part of a larger network of scientists and mathematicians and computer programmers with whom they exchange information; engineers with whom they devise, construct, and maintain instrumentation and observatories; and agencies or benefactors through whom they are funded. In this figure the links between these different communities are not explicitly shown, partly because those links can arise through different varieties of interactions. We return to one form of cross-linking in Figure 9.12.

scientists, the network of astronomers is merely part of a much larger complex, Figure 9.7.

Mark Newman of the University of Michigan at Ann Arbor has analyzed some of these networks by tracing the collaboration between astronomers on research papers listed on the widely accessible preprint website *arXiv astro-ph* during a five-year period from 1995 to 1999.[3,4] He argues convincingly that two astronomers who have completed a co-authored paper must know each other well if the two were the sole authors or the number of their co-authors was small. Studying co-authorship then becomes a way of assessing the extent of cross-linkage across the profession.

Three interesting characteristics emerge from Newman's work.

First: Shown on the right of Figure 9.8, the formation of a *giant component* of the community consisting of roughly 90% of all astronomers who are so closely linked through joint publications that, on average, it takes only four or five steps, or three to four intermediaries, for any one member of the community to contact any other whom he or she might not personally know through joint authorship, but who could be reliably contacted through just a short chain of intermediaries who have co-authored papers.

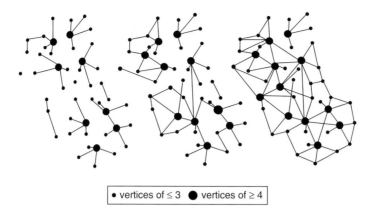

| vertices of ≤ 3 | vertices of ≥ 4 |

Figure 9.8. A phase transition forming a giant linked component. At left, individuals or small groups pass by each other without hindrance. At center, some begin to bind into larger, semi-permanent groupings until, at right, they gel into a giant component, which few fail to join. This phase transition is analogous to that of a gas of molecules first cooling to form a liquid, still permitting aggregates of molecules to pass by each other with ease until, at right, the liquid freezes to form a solid in equilibrium with a few gaseous molecules. Note the small group at upper right remaining detached throughout.

Total number of papers	22,029	papers
Total number of authors	16,706	authors
Mean number of papers per author	4.8	papers
Mean number of authors per paper	3.35	authors
Mean number of collaborators per author	15.1	colleagues
Size of the giant component, g.c.,	14,849	authors
Second largest component	19	authors
Mean separation between g.c. authors	4.66	steps
Maximum separation between g.c. authors	14	steps

Figure 9.9. The network of astronomers who co-authored scientific publications in 1995–1999. The number of papers, the mean number of authors per paper, the extent to which the astronomical community was interconnected through joint publications, and the characteristics of the giant component, (g.c.), are all reflected in this sample derived from the *arXiv astro-ph* astronomical preprint archives for the five-year interval spanning 1995–1999. (*Excerpted from M. E. J. Newman, Proceedings of the National Academy of Sciences of the USA, 98, 404, 2001*).

In Newman's investigation, the maximum distance between the two most remote astronomers, both of whom were members of the giant component linked through joint publications, was 14 of these links, or so-called *geodesic steps* across the network, as shown in the last row of Figure 9.9. For other scientific disciplines,

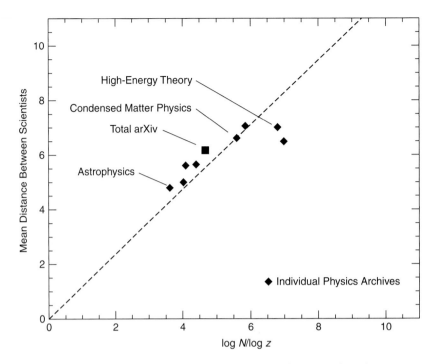

Figure 9.10. The mean distances separating pairs of scientists in various sub-disciplines of physics. These are deduced from data for the years 1995–1999 derived from the *arXiv* preprint website. *N* is the total number of authors who published papers listed in the archives, and *z* is the mean number of collaborators per paper. The *arXiv* physics archives compile data for separate subdisciplines, some of which are identified here, along with data summed over all physics archives and designated *Total arXiv*. The dashed line gives the expected slope on this plot for random networks – networks in which *N* points are connected so each point is randomly connected to *z* other points; the plot is constrained to pass through the origin. M. E. J. Newman, who compiled the data, notes the small number of steps that separate members of a scientific discipline through co-authorships, and conjectures that this is a crucial characteristic of a functioning scientific community. (*After M. E. J. Newman, Proceedings of the National Academy of Sciences, 98, 408, 2001*).

Newman found similar characteristics. In Figure 9.10 the scale on the left gives the mean distance between scientists, which ranges from about 5 to 7 for different disciplines.

Second: Newman found that the giant component linked roughly 14,850 astronomers among a total of about 16,700.[5] Only about 10% of all astronomers published work that had no cross-linkage to this giant component or group. The maximum size of the co-author linkage within this second largest component numbered only 19 linked authors – a grouping nearly one thousand times smaller than the giant component. There were thus hundreds of publishing astronomers who worked by themselves or in small groups, and whose work may have been largely disconnected from the interests of, and perhaps totally ignored by, the vast majority of astronomers, many of whom may learn about interesting new results by word of mouth from trusted, closely linked colleagues.

Newman's third finding is that a rapidly declining number of astronomers having a large cadre of co-authors places a very few individuals with copious co-authors as key intermediaries along the shortest path joining other members of the field, making these highly published authors the most closely connected, best informed, and thus most influential members of the field.

This is no small effect. Newman finds that the most highly linked individual in the field by virtue of co-authorship is substantially more highly linked than the second highest member who, in turn, is substantially more highly linked than the third most highly linked member.

But it is not just the number of co-authors to whom these individuals are linked that counts. Experimental high-energy physics collaborations often have many hundreds of collaborators, most of whom hardly know each other. In the last few years of the twentieth century, the size of collaborations of astronomers, particularly those carrying out observations on some of the largest ground-based or space telescopes, began to approach comparable sizes. Newman has established a criterion he calls the *collaboration strength*; it ranks individuals highly if they publish many papers with relatively few co-authors. The strength of the link to each co-author is weighted in inverse proportion to the number of co-authors on a paper.

The status of a highly linked intermediary has the potential of making an individual influential in sifting and transmitting new ideas. Newman suggests that these leading individuals are well positioned to fashion opinions, decide what makes sense and what does not, which ideas are to be accepted and which are not, which areas of astronomy deserve further pursuit and funding and which do not. Knowledge is power. At the other extreme, the subset of around 10% of the community that has no links to the giant component has virtually no influence on developments in the field.

Joint publications, however, are not the sole contacts that forge firm links between astronomers. Joint service on advisory panels, visiting committees, and planning boards offers similar opportunities to fashion respect, trust, and consensus on directions that astronomy should be taking. And those who publish and co-author widely also find themselves asked to serve on influential committees.

Newman has studied the boards of Fortune 1000 companies, the thousand largest U.S. corporations, and finds that directors who sit on only one of these boards are statistically likely to find themselves in the company of other directors who also sit only on one board, whereas directors who sit on many boards tend to have fellow directors who similarly sit on numerous boards, Figure 9.11. These are the most influential individuals. They spread their communal influence through the interlocking directorates on which they sit.

Astronomical boards and committees are similarly constituted. Most of the Survey Committee members conducting the Astronomy and Astrophysics Decadal

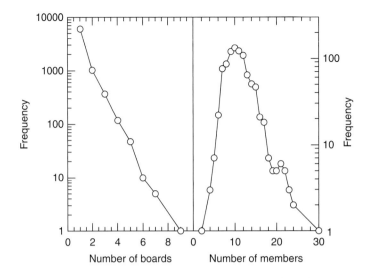

Figure 9.11. Board membership of Fortune 1000 Companies: Shown on the left are the numbers of boards on which different directors sit. Shown on the right are the numbers of board members different companies engage. (*Reproduced with permission from M. E. J. Newman, S. H. Strogatz, & D. J. Watts, Physical Review E 64 026118–13, 2001, ©2001 by the American Physical Society*).

Survey of 2010 – the review that prioritized projects that the U.S. astronomical community should pursue in the decade 2010–2020 – sat on many other advisory boards as well. Through these connections they could influence decisions on new directions of research, on new facilities to be constructed, on the allocation of budgets and allotment of observing time, in short on how astronomy and astrophysics are viewed, pursued, and evolve.

These are the people also who recommend promising young colleagues for promotion; nominate established scientists for prestigious prizes, editorships of leading journals, and directorships of observatories; direct the proceedings of academies; write defining review articles on topics of current interest; and generally see to it that the conduct of science meets the highest standards. All this, of course, is what public service and leadership in any community is all about.

But in science there is a down side to all this. It features the undesirable aspects of Fleck's thought collectives, which erect, interlock, and maintain a cross-cutting consensus, a set of thought conventions, a platform on which communal constructs find root despite sincere belief that scientific truths are based solely on dispassionate observation.

Random Networks

The structures of the networks just described show how the acceptance of new ideas and ways of thinking can come about.

In the mid-1990s Duncan J. Watts, then a graduate student at Cornell, and his thesis advisor Steven Strogatz began to depict the ways in which human contacts are established between specific individuals randomly selected from a large population, and how ideas subsequently propagate.[6] Interestingly, they established that ideas propagate very much like the spread of infectious diseases.

A case of influenza may originate in a village in some country and raise mild alarm as it at first is confined to a few adjacent neighborhoods. Then a single infected individual boards an airplane for another country. Aboard are people from several continents, some eventually traveling on to return to their home countries. A few of these are infected through close proximity on the flight. As the infection is disseminated by the dispersing fellow passengers, the disease may next be spotted and spread in East Asia, Europe, or North America. Soon, the disease is labeled a pandemic.

The spread of diseases and ideas, or the spread of gadgets and technologies, may be modeled as links randomly established between arbitrarily placed dots on a sheet of paper. The links may spring up more readily between neighbors separated by short distances than between points far apart. We then speak of a probability distribution, the probability being high through contact between immediate neighbors and low between widely separated individuals. But once these probabilities are roughly determined, we can show how, over time, the links between different individuals arise and increase until, eventually, a continuous link can be traced from almost any individual to any other, along a path of intermediaries who all are linked to each other across a giant component without an intervening break.

Because this way of establishing links specifies only a probability that any two individuals separated by a given distance might become linked, without any reference to who these individuals are, we speak of it as a *random process*. The network thus established is called a *random network*.

One of the findings of Watts and Strogatz was that, in a random network the size of the population of, say, the whole Earth, contact between just about any two randomly selected individuals can be established along as few as a mere half dozen links. This result was a mathematical explanation for a previously established observation that any two people in a population as large as that of the United States or even the entire Earth were separated by only about six such links – a finding colloquially referred to as *six degrees of separation*.[7]

Watts and Strogatz referred to their theory as the *small-world theory*. Among many other predictions, the theory appears to provide a prescription of how old ideas can be toppled and give way to the acceptance of new directions to follow in a scientific field such as astronomy. It does this not through a depiction of the logical factors at play, but rather as a schema involving linkages between workers in the field, motivated purely by reliance on trusted colleagues who influence the acceptance of new ideas, theories, or thought conventions.[8]

This is where the network that bonds astronomers and astrophysicists through jointly conducted and jointly published research plays its most significant role. Of special importance is Newman's collaboration strength between individuals.[9]

The small-world theory also depicts how remote fields of research influence developments in astronomy. Just as in the spread of infections, a few rare links that range over large distances can be more effective than a much larger number of links on smaller scales. In astrophysics, a single new insight, tool, or method, contributed by a mathematician, chemist, mineralogist, or engineer, may have far greater impact than the more frequent interactions among largely like-minded astrophysicists.

External Influences

This network depiction coincides with our previous finding that, in addition to influences on the course of astronomy from within the field, we also witness powerful external influences, Figure 9.12. The landscapes I had earlier

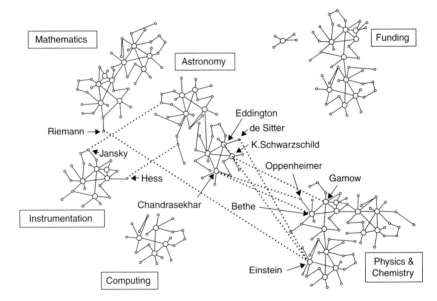

Figure 9.12. Scientists interact mainly with colleagues in their own field, but occasionally also cross-link with those in others. Dotted links indicate personal contributions made to other fields. Scientists within a field are connected by a network of solid lines. Arrows point to the fields to which the named scientists made their most enduring contributions. The dotted lines joining different fields are meant to show the contributions to astrophysical understanding made, either directly or indirectly, by scientists and mathematicians who worked mainly in other fields. Note the indirect link from mathematics to physics and then to astronomy, joining the mathematician Bernhard Riemann to Einstein, who used Riemann's differential geometry to construct his general relativity, which in turn led to advances in astrophysics and cosmology.

drawn correspond to different realms of human activity, different scientific or technical fields or funding structures.

The influence exerted by workers from one field on developments in another often occurs through the transmission of new tools. Scientists and engineers such as Viktor Hess, who discovered cosmic rays; Karl Jansky, who first detected radio emission from the center of the Milky Way, or the U.S. Naval Research Laboratory scientists who first measured X-rays from the Sun, made important contributions through the new observing tools they provided astronomers. They became members of the astronomical community, if at all, only after they made their contribution to the field, rather than before.

The same holds true of many leading theoretical physicists who made fundamental contributions to astronomy and cosmology. Among these were people such as Albert Einstein who provided the first modern theories of gravitation and cosmology (Chapter 3); Niels Bohr and Meghnad Saha, whose atomic and ionic models led to a clear understanding of the temperature and elemental abundances in the atmospheres of stars (Chapter 4); J. Robert Oppenheimer, who first worked out the pathways leading to neutron stars and stellar black holes (Chapter 5); and Hans Bethe, who derived the two main nuclear processes that make the Sun shine (Chapter 6). These men certainly thought of themselves primarily as theoretical physicists bringing useful new tools to bear on astronomical questions. The journals in which they chose to publish their findings all were issued by physical rather than astronomical societies.

The Craftsmen of Theoretical Tools

I want to halt here to look back at how our understanding of the Universe changed in the first half of the twentieth century, to note the role that new theoretical tools played in the development of astrophysical theory, and to show how those who fashioned the tools used them gradually to unveil the Cosmos we view today.

The new tools and the gains they brought about included relativity theory, which had its major impact on energy generation in stars, the limiting masses of white dwarfs, and the structure and evolution of the Universe; the old quantum theory, whose insights on the spectra of atoms and ions provided a first understanding of the chemical constitution of the atmospheres and thus presumably also the interior of stars; quantum mechanics and its influence on nuclear theory, which led to the understanding of what makes stars shine and how long they have existed; and quantum statistics, which elucidated processes deep in the interior of stars and the ultimate fate of massive stars.

Theoretical tools, like most others, come in different shapes and forms; and, just as with other tools, shape and form is rarely a measure of utility. A novel physical principle or insight, a mathematical formulation that sheds new light, a useful approximation that cuts through complexities – all can play central roles in advancing astronomy.

Two factors may be kept in mind. The first is readily recognized and the present chapter devotes itself to documenting it: The type of theoretical tool a physicist or mathematician crafts and contributes to the development of astrophysics reflects the toolmaker's personality and preferred style of engagement.

The second consideration may be more difficult to accept, and so needs emphasis:

The theoretical tools available to us shape the way we conceive and perceive the World!

Einstein was unable to find a way to incorporate gravity in his theory of relativity until Marcel Grossmann showed him that Riemannian geometry would provide the requisite mathematical tools. Similarly, Eddington was puzzled by the apparent paradox that white dwarfs might be unable to cool below a temperature at which electrons should combine with nuclei to form atoms. Ralph Fowler, who had been helping Eddington at the time, and who was familiar with quantum statistics, quickly cut through the paradox to a clear understanding of white dwarf structure.

Temperament and Style in Theoretical Astrophysics

Most of the new theoretical tools adopted by twentieth century astrophysicists were imported from physics. Often those who had honed the tools contributed to their application in astronomy but briefly – showing the way that they should be used, before hurrying back to pure physics to pursue other interests. Among these were Bethe, Einstein, Ralph Fowler, Alexander Friedmann, Oppenheimer, and Saha, who were diffident about staying on, preferring to leave the application of their theories to more traditional astronomers. Chandrasekhar, Gamow, Hoyle, and Salpeter were major exceptions. All four continued to exert a profound influence on further developments in astrophysical thought, although Gamow later moved on to theoretical work on genetics.

I have tried to show how the contributions of the different theorists, and the types of tools they provided astronomy, depended on personality. To nonscientists, this statement might seem strange, because they tend to take scientific findings to be products of Nature, rather than of the tools that were used in a discovery, or the personality of a scientist. The tools for which an astrophysicist reaches, and the vision he or she follows in sculpting a novel finding, out of an amorphous bulk of data, are often overlooked.

Earlier I showed how Einstein concentrated on discovering the architecture of Nature, the major principles governing all of physics. Chandrasekhar, along somewhat different avenues, wished to reveal Nature's inherent mathematical beauty. Einstein saw the Universe in terms of guiding principles, Chandrasekhar in terms of mathematical structures. Emmy Noether could have pointed out that her eponymous theorem showed the two approaches to be but two sides of the same coin.

Gamow's preoccupations were quite different. His favorite way of contributing was with a surprising new insight. His greatest contribution might have been his keen appreciation that nuclear particles could quantum-mechanically tunnel into or out of nuclei, through a process energetically forbidden by classical physics. Understanding this new feature was fundamental to unraveling the nuclear processes that sustain the energy output of stars and the buildup of heavy elements in Nature.

Gamow realized he was no good at messy calculations. Somehow he lacked the temperament. He may have sensed it would take an intellect that only Hans Bethe possessed to envisage all the different alternatives and select only those that Nature herself would choose for energy production in stars. Bethe was the encyclopedic master calculator. His quick insights helped him to surgically cut through a fog of nuclear processes to reveal the two major reaction chains that make main-sequence stars shine.

Edwin Salpeter, somewhat like Bethe, his early mentor, could handle wide varieties of different problems. In contrast to Einstein he had no major roadmap for his career. In an autobiographic sketch late in life, he wrote, "I belong to the minority of scientists who are 'ping-pong players', i.e., react to influences from others on a short time scale without much systematic planning of their own.... A 'ping-pong player' reacts to outside players, but has to decide which of several possible influences to react to, so decision making is important."

Fred Hoyle, like Salpeter, also worked on wide-ranging problems. Utterly fearless, he concentrated on whatever happened to intrigue him as a potential advance – the Steady State cosmology, the build-up of heavy elements in stars, the audacious prediction of a resonance in the carbon nucleus that laboratory physicists had not yet found. He did not worry as much as others whether his conclusions were right or wrong, as long as the questions he worked on fascinated him. Eventually, right or wrong would get sorted out; in the meantime it was good to at least get started. He wrote fantastic science fiction stories, which sometimes, as in his 'October the First is Too Late' or 'The Black Cloud', seemed based on entirely plausible concepts, which the stern referee of an astronomical journal might have rejected as too wild, but that less punctilious science fiction readers would welcome.

These great theorists walked a stage as wide if not wider than any great actor, infusing similar energies and passions into their work – yielding performances that later generations would follow with awe as they were replayed, perhaps not with the help of old film clips, but through perusal of journal articles written decades earlier.

Table 9.1. *Ages of Theoretical Astrophysicists at the time of their major contributions*

Proposer(s) of Theories	Subject of Theories	Date(s) of Birth	Date of Theory	Ages
Bethe	Stellar Energy Sources	1906	1938–39	31
Bohr	Atomic Structure	1885	1913	28
Stoner, Chandrasekhar	White Dwarf Mass Limit	1899,1910	1930,1935	31, 25
Eddington	Radiative Transfer in Stars	1882	1916	34
Einstein	Mass–Energy Equivalence	1879	1905	26
Einstein	Relativistic Cosmology	1879	1917	38
Fowler	White Dwarf Degeneracy	1889	1926	37
Friedmann	Relativistic Evolving Universes	1888	1922	34
Gamow	Nuclear Tunneling Cross Sections	1904	1928	24
Gamow	Cosmology	1904	1948	44
Lemaître	Cosmic Expansion	1894	1927	33
Payne, Russell	Hydrogen-Dominated Stars	1900,1877	1925,1929	25, 52
Oppenheimer	Neutron Stars and Black Holes	1904	1938–39	35
Saha	Spectra of Ionized Gases	1893	1920	27
Salpeter, Hoyle	Heavy Element Production	1924,1915	1951,1953	26, 38
Schwarzschild	Cosmic Singularities	1873	1916	43
Weizsäcker	Energy and Elements	1912	1938	26

The Ages of Theoretical Astrophysicists at Peak Productivity

Most of the major theoretical advances I have cited in preceding chapters were made relatively early in the lives of the contributors. We may best surmise this from Table 9.1. The first column of the table lists the scientist(s) who contributed a theoretical advance; the second names the subject of the theory; the third cites the date(s) of birth of those involved; the fourth marks the years during which the theory was developed; and the last provides the ages of the scientists at the time of their contributions. I have included the finding that hydrogen is the most abundant element in the atmospheres of stars because this invoked so much theoretical interpretation.

Only six of the theories were proposed by theorists as old as 37 to 52. The ages of the 14 others ranged from 24 to 35.

A shorter but perhaps equally indicative table might be constructed listing the ages at which some of the same theorists, as well as others, outspokenly resisted several of these innovations.

In 1925, when Henry Norris Russell strongly urged Cecilia Payne to excise her high derived hydrogen abundances from her thesis, he was 48 years old, and continued to deny the high abundance for another four years, until the evidence could no longer be resisted. He redeemed himself later, in 1939 at age 62, by becoming a strong advocate of Bethe's theory of stellar energy production.

Albert Einstein was finally willing to accept an expanding Universe at age 52, in 1931, by which time the evidence was overwhelming. He had been resisting the notion since age 43, when Friedmann's papers first appeared in 1922. On the other hand, despite congratulating Karl Schwarzschild on his initial general-relativistic paper on a point-mass, Einstein later felt that such masses could have no physical reality. In his paper of 1939, written at age 60, he explained why.

Eddington too was unwilling to accept a high hydrogen abundance in stars, at least in 1926, when he was writing his book, 'The internal constitution of the stars,' at age 43.[10] His most outspoken opposition to the collapse of massive white dwarfs arose in 1935, at age 52.

Perhaps these failings at an advancing age are remarkable only because those who proved themselves so wrong had made such important contributions earlier and because their opinions later in life carried such great weight.

Einstein used to joke that Fate was punishing his early disdain for authority by making an authority out of him late in life.[11]

Fleck's Thought Collective

Ludwik Fleck's thought collective is rooted in the respect assigned to aging astrophysicists whose creativity once led the field in new directions. Older astrophysicists may oppose advances for some time because the proposed innovations don't "feel right"; they are not based on the same approaches that led these creative individuals, who are in authority now, to their greatest successes when they were young.

Why not? It is because those approaches quickly caught on, became part of the community's tool kit, and have by now been drained of everything they could offer to further advance the field. Where major problems remain to be solved, it is most likely that all available tools have already been tried, have failed, and that the community will need to await a completely new approach, a new tool that nobody has thought of yet. This may not necessitate uncovering a new phenomenon, but rather realizing the existence of an unwarranted assumption, a prejudice that nobody recognized but that now has to be rejected.

This is the kind of rejection that led Tsung-Dao Lee and Chen-Ning Yang to suggest, in 1956, against all previous expectations, that Nature might be able to distinguish between left- and right-handed systems, at least where interactions involving weak nuclear forces were involved. To this end, they proposed how the violation of these expectations might be experimentally observed.[12] Experiments then showed that neutrinos always have what is now called left-handed spin, whereas antineutrinos have right-handed spin.

Astoundingly, the Universe distinguishes between left- and right-handed particle spins! How might Mach's principle ever account for this? Perhaps the principle simply does not conform to Nature.

Often the older creative individuals may even recognize the requirement for a whole new approach to effect the next major advance and are working on one themselves. But the attack on the problem they are pursuing may not be working. Einstein was concentrating toward the end of his life on unifying gravity and electromagnetism to create a new overarching theory. This was probably hopeless because the weak and strong nuclear forces were not yet known. Eddington was working on his 'Fundamental Theory,' which also led nowhere. In the meantime, neither fully appreciated what relativistic quantum mechanics and quantum statistics was doing to advance physics and astrophysics.

Astrophysics advanced rapidly in the first half of the 20th century, and then continued to accelerate, possibly because a larger number of scientists were active in the field. Certainly, science moves faster now than it did in Fleck's times. The increasingly rapid pace often forces those in authority and reluctant to accept change eventually to agree that a new advance is genuine. These belated acknowledgments may be teaching the community to take authority somewhat less seriously – an attitude that can only be good for science.

Tools versus New Ideas

Historians have often emphasized the influence that new ideas have had on the evolution of science. My emphasis, here, has been on theoretical tools and their impact.

In my earlier book, 'Cosmic Discovery,' I tried to convincingly show that the most significant observational astronomical discoveries, particularly in the twentieth century, swiftly followed the introduction of new observational tools, often by strangers to astronomy, mainly physicists and engineers, most of whom had little familiarity with the problems that were at the forefront of astronomical thought at the time.[13]

The same circumstances seem to have led to new directions in theoretical astrophysics. The invention of relativity, both special and general, had an enormous impact on astrophysical thought, as did the inventions of quantum mechanics, quantum statistics, and atomic and nuclear theory. The application of these theoretical tools to cosmology, the theory of stellar structure and evolution, and eventually also to the origin of the chemical elements, totally changed our conceptions of the world around us.

New ideas seem to have played but a minor role. Ideas often are "in the air"; they may be discussed by colleagues over a glass of wine but go no further. They often are no more than speculations that would need to be sifted out from among many other equally intriguing speculations, none of which have any greater or lesser merit than others. Priority disputes over who may have expressed an idea earliest have always been around. They suggest that ideas often arise in the minds of many individuals and are "cheap," in the sense that many of them require little insight but a great deal of further effort to bring them to fruition – effort that few are prepared to invest because the right tools are not at hand to readily pursue them.

In science, one requires tools that will sift out the good ideas applicable to describing Nature and reject others that fail. The scientist who can develop these tools and divine a new truth is the one who provides the greatest service. Thus Newton's development of the mathematical tools, which permitted him to show that an inverse square law of gravitational attraction could explain Galileo's measurements on falling bodies, Kepler's laws of planetary motions about the Sun, the trajectories followed by *comets*, and the tides induced on Earth by the Moon, was the great achievement that launched innumerable further investigations of Nature.

Without the observational and theoretical tools of astrophysics to discriminate between applicable and inapplicable ideas, the ideas by themselves have little impact.

We saw how John Michell, in 1783, conceived that massive stars might exist with gravitational fields so strong that even light could never escape. But despite the passage of nearly two centuries, the pursuit of this idea, even by as great a scientist as Laplace, led nowhere until general relativity provided the tools with which this idea could be pursued. Schwarzschild's idea was no newer than that of Michell or Laplace, but the tools Einstein had provided him made his work useful.

When Alexander Friedmann in St. Petersburg saw in Einstein's general relativity the mathematical tools for describing cosmological models that might expand, contract, or oscillate, nobody thought that the Universe we inhabit would display this kind of behavior. Certainly, Einstein did not. He was still firmly wedded to the idea of a static universe.

Once a tool is found useful in one application, it is likely to be used in approaching other problems as well. Saha's equation, after proving useful in dealing with the ionization of atoms in the atmospheres and interior of stars, came to be used also to deal with the equilibria of neutrons with atomic nuclei in Weizsäcker's and in Chandrasekhar and Henrich's studies of the build-up of heavy elements.

One may also compare and contrast the application of new tools to sift through different ideas. Stoner, Landau, Chandrasekhar, and Oppenheimer and his students all used an identical set of quantum theoretical and relativistic tools, but Stoner, Chandrasekhar, and Oppenheimer applied them to ideas that proved to be applicable to astrophysics, whereas Landau's ideas have, at least to date, led nowhere.

Those who introduced new tools into astronomy and astrophysics over the years served the field well. Their tools often could be applied to the most diverse problems.

Acceptance of New Approaches, Theories, and Fashions

New approaches to research in any subfield of astrophysics tend to be spearheaded by groups that mutually agree on pursuing new lines of attack. This collaborative approach is common to most scientific disciplines, as Fleck first emphasized. The group that embarks on a given approach then tends to vigorously fend off any alternative lines of attack on the investigation they are conducting, often to the bewilderment and consternation of others who may have equally valid or possibly even more interesting methods for investigating the problem, and who cannot comprehend the narrow focus, bordering on mania, of members of the leading group.

Although one may consider communally imposed scientific fashions harmful, a sign that most scientists merely follow each others' views – forming a community that acts in accord and, therefore, portrays the Cosmos to suit mutually favored tastes – there do exist distinct advantages to fashion in science. By adopting a current fashion, scientists agree that there is a problem worthy of attack. If one member of the group finds a useful way to make progress, there are many colleagues who have also thought about the problem and who are able to critique the new approach, improve on it, and debate its advantages and shortcomings with equally well-informed colleagues to gain further insights. The problem under attack is then solved more quickly, or else more rapidly encounters insurmountable obstacles and reaches a dead end.

If the approach does not pay off, the group can turn to more fruitful-appearing approaches and start all over again. Soon an idea for a new direction surfaces, a new lead-group emerges, sometimes with the same membership or some slightly re-arranged version of it, less often with an entirely new membership altogether – and a headlong, single-minded chase starts again. The main advantage of adherence to fashions is thus an ever-available, well-informed community familiar with the details of a problem.

What determines the acceptance of these new fashions? How solidly are the criteria for the changed directions in hand? How quickly does the new group gather its membership? How long does each chase persist before running its course?

These are questions that appear to be elucidated by *social cascades*, first modeled by Duncan Watts in 2002.[14]

Before describing how such cascades arise, however, I want to turn to two other topics. The first, the topic of Chapter 10, deals with further theoretical advances marking the second half of the twentieth century. The second, discussed in Chapter 11, involves ways in which the astronomical community was beginning to plan its future. I will then return to a discussion of social cascades in Chapter 12.

Notes

1. *Cosmic Discovery – The Search, Scope and Heritage of Astronomy*, Martin Harwit. New York: Basic Books, 1981.
2. Scientists Linked Through Joint Activities, M. E. J. Newman, S. H. Strogatz, & D. J. Watts, *Physical Review E*, *64*, 026118, 2001.
3. Entries for these papers can be found distributed over the range http://arxiv.org/list/astro-ph/9501 to http://arxiv.org/list/astro-ph/9912
4. The structure of scientific collaboration networks, M. E. J. Newman, *Proceedings of the National Academy of Sciences of the USA*, *98*, 404–09, 2001.
5. Scientific Collaboration Networks – Network Construction and Fundamental Results, M. J. E. Newman, *Physical Review E*, *64*, 016131, 2001.
6. *Six Degrees – The Science of a Connected Age*, Duncan J. Watts. New York: W. W. Norton, 2003.
7. Ibid., *Six Degrees*, Watts.
8. *Small Worlds – The Dynamics of Networks between Order and Randomness*, Duncan J. Watts, Princeton University Press, 1999.
9. Ibid., The structure of scientific collaboration networks, Newman.
10. *The internal constitution of the stars*, Arthur S. Eddington, Cambridge University Press, 1926.
11. *Albert Einstein: Einstein sagt – Zitate, Einfälle, Gedanken*, Alice Calaprice. Munich: Piper Verlag 1997, p. 51.
12. Question of Parity Conservation in Weak Interactions, T. D. Lee & C. N. Yang, *Physical Review*, *104*, 254–58, 1956.
13. Ibid., *Cosmic Discovery*, Harwit.
14. A simple model of global cascades on random networks, Duncan J. Watts, *Proceedings of the National Academy of Sciences of the USA*, *99*, 5766–71, 2002.

10

The Evolution of Astrophysical Theory after 1960

Two major innovations occupied theorists in the postwar era: The first began in the 1960s and involved the study of black holes. The second started around 1980 and renewed efforts to understand better the origins of the Universe. Not that these were the only theoretical advances. There was plenty of other work keeping theorists busy. The discovery of X-ray, infrared, and radio galaxies, quasars, pulsars, cosmic masers, and γ-ray bursts, to name just the most striking, begged for quantitative models that could explain these new phenomena. But most of the theoretical models that satisfactorily matched observations involved known conceptual approaches, albeit applied in new settings. In the vocabulary Thomas Kuhn established in the early1960s, in 'The Structure of Scientific Revolutions,' they constituted *problem solving*. They did not call for new *paradigms*, entirely new ways of conceiving Nature.[1]

At some level, the theoretical thrusts on black holes and investigations of the earliest moments in cosmic evolution overlapped. Both sought to improve our understanding of space and time through clearer insight into general relativity and – to the extent one might guess at it – quantum gravity. Knowing that highly compact masses might collapse, one had to ask why the early Universe didn't immediately collapse under its gravitational self-attraction to form a giant black hole rather than expand, as now observed.

* * *

During an interview conducted in January 1988, Robert H. Dicke, one of the most gifted physicists of the postwar era, equally familiar with sophisticated instrumentation and imaginative theoretical approaches, recalled the views of most physicists before World War II. "Except for a few places … relativity and cosmology were not regarded as decent parts of physics at all. … I asked Victor Weisskopf one time – he was [a professor] at Rochester when I was a graduate student – shouldn't a graduate student pay some attention to relativity? And he explained to me that it really had nothing to do with physics. Relativity was kind of a mathematical discipline."[2] After the war, Weisskopf, a well-known theoretical physicist, joined the physics faculty at MIT, which, like most other U.S. physics departments at the time, still offered no course in general relativity, even in the late 1950s.

All this was about to change, notably in the years between 1957 and 1963. At Princeton, Robert Dicke and John Archibald Wheeler – both veterans of projects the Office of Science Research and Development (OSRD) had set up under Vannevar Bush in World War II – turned to general relativity and cosmology with the same energy and self-confidence with which they and their fellow physicists had attacked the wartime problems of constructing radar systems and building atomic bombs. In Moscow, Yakov Borisovich Zel'dovich, an expert on explosions and leader of the Soviet nuclear bomb effort of the Cold War era, similarly turned to general relativity and cosmological work. As Bernard Lovell had noted, these alumni of large-scale wartime efforts knew how to assemble and direct teams of co-workers – by the post-war years mostly students and postdocs – to attack major theoretical problems.

In Cambridge, a rather younger Dennis Sciama with a strong appreciation for talent and promise began attracting gifted students, among them Roger Penrose, Stephen Hawking, and Martin Rees, to relativistic astrophysics and cosmology. And, in 1963, a group of mathematicians and theoretical physicists at the University of Texas in Austin initiated a series of influential 'Texas Symposia on Relativistic Astrophysics,' subsequently held at different sites across the globe in alternating years.

Princeton, Moscow, Cambridge, and Austin were to remain the four dominant hubs in the network of modern general relativity and cosmology for the next two decades, with Kip Thorne, a former PhD student and later collaborator of Wheeler, soon adding a fifth, increasingly influential node at Caltech.

This chapter and several of those that follow portray the advances in relativistic astrophysics and cosmology that began to flourish in the 1960s but then abruptly gave way, in the early 1980s, to a cosmology preoccupied with particle physics. Other institutions and research centers, often associated with major particle accelerators, then came to the fore as the cosmology network's dominant hubs.

Yet, throughout these decades, a theory of quantum gravity has remained tantalizingly beyond reach. General relativity and particle physics have not properly meshed. A major ingredient still eludes us!

Investigations of Black Holes

Although nuclear physics had occupied most theorists in the era immediately preceding and following World War II, one area of general relativity that began to strike a few quantum physicists of the late 1950s as worthy of pursuit was the investigation of black holes – even before that name had been coined.

Uncertainty persisted about whether these collapsed aggregates of matter could actually exist. Although Oppenheimer and Snyder had shown in 1939 that catastrophic collapse at the center of dense stars should be inevitable, doubts persisted about the *equation of state* of highly compressed nuclear matter, that is, the extent to which nuclear matter at the highest densities might successfully resist further compression. Research on atomic bombs had, by the late 1950s, led to better understanding of nuclear processes at high pressures.

Few physicists were as well acquainted with this Cold War research as John Archibald Wheeler, professor of physics at Princeton. Wheeler was interested in black holes, in part because they might yield theoretical insight on the interplay of quantum physics and general relativity. Deep in the interior of a collapsed star, both theories should be relevant and might provide clues to the structure of a unified theory of gravitation, electromagnetism, and nuclear matter.

Kip Thorne, a graduate student of Wheeler around that time, recalls Wheeler's belief in 1958 that black holes could not occur in Nature. Wheeler considered Oppenheimer and Snyder's theoretical paper of 1939 to have assumed such idealized initial conditions – among them perfect spherical symmetry in a nonrotating collapsing star – that the inevitable collapse they predicted could well be a chimera.[3]

Nature, Wheeler felt, was far too messy for those assumed conditions to ever occur. He expressed these views in a presentation at the Solvay Conference dedicated to 'The Structure and Evolution of the Universe,' held in Brussels in June 1958. At the end of Wheeler's talk, Oppenheimer rose and reiterated his confidence that the approach he and Snyder had taken two decades earlier was still correct.[4]

Wheeler might have found support from Fred Hoyle, who had expressed similar views in a 1946 paper on the formation of heavy elements in the collapse of a massive star. There, Hoyle had concluded that if the mass of a star exceeded Chandrasekhar's limit, it would collapse until conservation of angular momentum spun it up to the point of tearing apart. Hoyle thought that this would eject the star's outer layers in a supernova explosion and reduce the mass of the remaining stellar nucleus to less than Chandrasekhar's limit, where it would cool down to form a cold, stable white dwarf.[5]

As late as 1958, the potential existence of black holes still appeared unresolved.

The Asymmetries of Black Holes

That same year, 1958, David Finkelstein, at the time a 29-year-old physics postdoc at the Stevens Institute of Technology in Hoboken, New Jersey, published a short but important paper in the *Physical Review*. He examined the various time and spatial symmetry properties of empty space surrounding a massive point source as defined by general relativity, noting that Schwarzschild's equation (5.1) was obviously time-symmetric. If a particle could fall into a black hole, it should also be able to retrace its steps, reemerging from the black hole at infall speeds. But this made no sense. As seen from an observer at a large distance, a space-farer planning to fall into a black hole, say, right at noon on a given day, would never seem to disappear, even though appearing to approach the surface at the Schwarzschild radius ever more slowly. The spacefarer falling in, however, would glance at his watch, while crossing through the surface, and see it was right at noon.

Just as a distant observer would have to wait an infinite time to see the infalling spacefarer actually crossing into the black hole, it would take an infinitely long time for the observer to see a spacefarer ever escaping the hole. As a result, a distant observer would see any amount of matter falling inward toward the Schwarzschild surface, but nothing ever emerging through it.

This constituted a paradox. It also raised a second question. If the rate at which infalling matter approaching the Schwarzschild surface were sufficiently high, the radius of the surface should be found to grow, as seen by a distant observer. The infalling spacefarers should then appear to approach a halt at ever increasing radial distances from the central mass. But how could a sufficient amount of mass ever aggregate in a finite time to even create a Schwarzschild surface, if none ever could be seen crossing it?

Finkelstein put it succinctly[6]:

> How is it possible that causes which are symmetric can have effects that are not?...There must be a theoretical process in which...for example, a slow process of accretion or compression...makes an initially weak grav-itational source approach and finally pass through the critical situation in which the Schwarzschild radius of the source equals the extension of the source.

He suggested that space–time then might *buckle*, so that radiation and matter no longer propagated along time-symmetrical trajectories.

In important ways, Finkelstein's proposal transcended the views that Riemann had advocated a century earlier in his 1853 'Habilitationsvortrag,' where he posited that the chief characteristic of a space is its topology. Once this was known, one could seek the metric that defined distances between points and thus determine the curvature of the space.[7] Now, it appeared, even the topology could be altered when the curvature of space exceeded some maximum value. The differential geometry of space and its topology must somehow be inseparably intertwined and potentially time dependent in our Universe; topology no longer could be considered immutable.

Previous work on the gravitational field around a central mass had portrayed the field as static at any given point and time. Finkelstein showed that alterna-tive ways of thinking of the field could provide new insights. One of these is particularly easy to envision. Right from the start, Einstein had proposed that a person freely falling toward a central mass feels no forces. Special relativity should hold for this voyager, who would observe light propagating at its ordained speed, *c*.

We may imagine a spherical surface in empty space surrounding a massive star. Distributed over this surface are spacefarers all freely falling in. As the surface contracts, all the while remaining spherical, none of them feel a gravitational force as they all fall inward in unison. We may further think of many such concentric surfaces, each with its own spacefarers freely falling into the star.

Even if there were no spacefarers sitting on them, these concentric spherical surfaces would still collapse toward the central mass. In this view, empty space around a compact massive body is filled with spatial surfaces centered on the body and continually collapsing inward. At infinite distances from the star the infall velocity is infinitesimal. Closer in, the speed monotonically increases.

If the central mass is too small or insufficiently compact to have formed a Schwarzschild surface, the spherical spatial surfaces – and any spacefarers happening to sit on them – will collapse through the point centered on the mass and then re-expand out to infinity. But for a central body sufficiently massive and compact that a Schwarzschild surface exists, the infalling surfaces will be trapped on collapse through the Schwarzschild surface. They cannot reemerge. A fundamental asymmetry then exists.

Finkelstein showed that a second mathematical solution satisfying general relativity exists. Outside the Schwarzschild surface, the original solution Schwarzschild had devised still is correct. But the second solution shows that once inside the Schwarzschild surface, escape outward through the surface is impossible. Matter and radiation can only converge on the central mass point.

What Finkelstein had found was that Schwarzschild's original solution told only half the story. There was a further solution. Together they defined a *space–time manifold*, of which only a limited *completion* had been known to Schwarzschild.

I have dwelled on this insight of Finkelstein at such length because it removed a conceptual barrier that had kept many experts uncertain about whether or not black holes could ever exist in Nature, or whether they were even worthy of investigation. Kip Thorne of the California Institute of Technology – Caltech – recalls that Finkelstein's paper had a strong impact on Lev Landau and Landau's close collaborator Evgeny Lifshitz in Moscow, as well as on the 27-year-old Roger Penrose who was just finishing a PhD thesis in mathematics at Cambridge University and attended a talk by Finkelstein in London.

The mental block that was lifted by Finkelstein's display of the breakdown of time-reversal symmetry at the Schwarzschild surface was key to a new confidence that black holes had to be taken seriously. Eventually John Wheeler also was convinced and became a leading proponent for the existence of black holes.[8]

Rotating Black Holes and Quasars

By the early 1960s, the rejuvenation of general relativity was in full swing. Talented young mathematicians and physicists formed a new, enthusiastic community: Physics departments at American universities opened their doors to relativists. The Texas Symposia on Relativistic Astrophysics, initially organized by the University of Texas at Austin but by now held in different parts of the world every second year, provided a forum for astronomers, physicists, cosmologists, and mathematicians to exchange ideas.

At the first of these symposia held in December 1963, Roy Patrick Kerr, a 29-year-old New Zealander working at the University of Texas at Austin, presented a first paper shedding mathematical insight on rotating black holes. It came at a particularly opportune time. A paper published by Maarten Schmidt in *Nature* the previous March had shown the quasar 3C 273 to have a high redshift.[9] If the redshift indicated a great distance, the quasar had to be highly luminous and was likely to be very massive. Because of its point-like appearance, it also had to be highly compact. Another possibility was that the quasar was not as distant and that part of its redshift was gravitational. Either way, it looked as though a massive black hole might be involved.

Nature is unlikely to produce nonrotating black holes. Whatever matter falls into a gravitationally attracting object is likely to strike it off-axis, inducing angular momentum and causing the black hole to rotate. Kerr had just derived the metric that defines rotating black holes – now often referred to as *Kerr black holes*. 1963 was a good year for relativistic astrophysics; the first Texas Symposium became a hit.

In late July that year, Kerr had submitted a short summary of his results to the *Physical Review Letters*, which published his paper on September 1.[10] The paper is less than two pages long, giving only Kerr's final results.[a]

Less than a year after Kerr's discovery, the 34-year-old Ezra T. Newman and five of his co-authors at the University of Pittsburgh showed that rotating black holes could also carry electrical charge. Their paper too was quite short, less than a page and a half in length.[12] The authors initially thought that their mathematical expressions were defining the properties of a ring of mass and charge rotating about its axis of symmetry; but in a footnote their paper cited Roy Kerr as pointing out a difficulty with that interpretation. Instead, the metric they had proposed provided the most general description of a charged, rotating, black hole. Such black holes are now often referred to as *Kerr-Newman black holes*.

It took almost another decade for the realization to jell that a black hole is fully defined once one specifies its mass M, angular momentum J, and electric charge Q.

---

[a] For a long time the derivation of this metric and the rationale that drove Kerr to it were not well known. It took 45 years until 2008 for Kerr to publish his recollections of these early events as he saw them nearly half a century later and to show the reasoning that led him to the metric.[11]

All other information is erased during the formation of the black hole. Two black holes with identical parameters M, J, Q have indistinguishable physical properties. The metric given by Newman and his co-authors included all three parameters and thus provided a basis for further study of any black hole, no matter what the value of the three parameters might be.

A Topological Approach to Black Holes

Although a considerable amount was known about various possible types of black holes by 1964, it still wasn't clear that black holes actually can form. The work of Finkelstein, a few years earlier, had made this more likely; but the possibility persisted that some process could intervene to deny black hole formation.

* * *

On an autumn day in 1964, Roger Penrose, by now at Birkbeck College in London, was walking back to his office, deep in conversation with his colleague, Ivor Robinson, who was just then talking. At the back of his mind Penrose was worrying about a problem he had long been trying to sort out.

The previous two years had witnessed the discoveries of quasars. Astrophysicists were speculating about what these extremely luminous bodies might be. Could they be generating their vast output of energy through collapse into black holes? If so, how had the black holes formed?

Penrose wondered how real stars, with all their deviations from perfect spherical symmetry, could collapse to form black holes. The general relativistic depiction that Oppenheimer and Snyder had provided in their paper of 1939 had assumed perfect spherical symmetry. This was an idealization, of course. What would happen if the collapsing body was not completely spherical? Or, if it was spinning, would it fracture into smaller bodies on collapse? One could imagine the fragments then spiraling out again, partway through their infall, rather than continuing their collapse into a singularity.

Penrose had been thinking for some time about whether there might be some mathematical proof that would specify how close to perfectly spherical a collapse had to be, or how far it would have to proceed before a point of no return was reached where further collapse into a singularity became inevitable. How might such a proof be found?

As Penrose and Robinson were crossing a street, their conversation stopped for a moment, only to resume on the other side. But during the crossing Penrose had an idea that seemed to escape him as their conversation continued.

Only later in the day, after Robinson had left and Penrose had returned to his office, did he feel a sense of elation for which he could not account. Racking his mind of what might have happened during the day that could account for this, he finally recalled that, as he was crossing the street, it occurred to him he might attack the problem of imperfect bodies collapsing to form perfectly spherical black holes by topological means.

Topology deals with the study of objects that can be deformed, squeezed, or extended in arbitrary ways but still remain members of a recognizable set. Topologically, a cube and a sphere both belong to the same set, but a doughnut does not. Squeezed and otherwise deformed without tearing, a doughnut retains its central hole and cannot be deformed into a sphere. It belongs to a different set.

Penrose began by considering a perfectly symmetric spherical distribution of matter. After this has collapsed into a Schwarzschild black hole a spherical sheet devoid of matter will exist right at the Schwarzschild radius, $r = 2m$, in Figure 10.1.

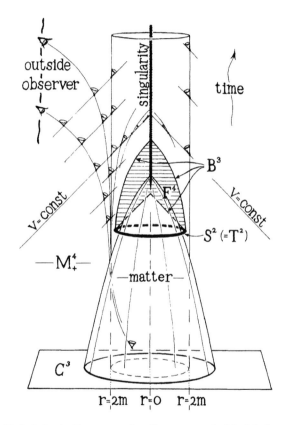

Figure 10.1. Spherically symmetric collapse around a black hole as envisioned by Roger Penrose. Four light rays emanating from a collapsing spherical surface at successive epochs are shown. Two emitted sufficiently early in the surface's collapse can escape to reach an observer at infinity. This observer will always see radiation emitted from matter on the surface while it is outside the Schwarzschild radius but not after it has collapsed through this radius and into the singularity. Radiation emitted in the outward direction from a spherical surface right at the Schwarzschild radius will remain there forever. Radiation emitted from anywhere inside the radius, no matter what its direction, will eventually fall into the central mass. (*Reprinted figure with permission from Roger Penrose, Physical Review Letters, 14, 58, 1965, ©1965 by the American Physical Society*).

At this surface two sets of null-geodesics – that is, paths followed by light beams – find their future directions converging. Light from outside the black hole will fall inward, and light from inside the black hole will be prevented from exiting and will fall inward as well. Penrose called such a surface a *trapped surface*. Independent of shape, sufficiently massive compact bodies would collapse into a geometric point!

Penrose argued that these trapped surfaces will exist whether or not the matter that had fallen into the black hole had a sharp boundary or whether or not it had been precisely spherical. Similar situations would hold also for Kerr black holes. The only significant feature was that the manifold $\mathcal{M}$ had to comprise both of the *completions*, both of the solutions Finkelstein had specified.[13]

Conceiving such topological structures and arguing their significance constituted a new approach to relativistic problems. Topology became a novel mathematical tool for probing further.

An Imaginative Idea

A few years after his topological insight, Penrose came up with another intriguing idea. In 1971, he and R. M. Floyd, a fellow-mathematician at Birkbeck College in London, published a radical new suggestion for extracting energy from a rotating black hole.[14] Recognizing the audacity of their proposal they wrote,

> The coupling we shall now describe between the angular momentum of the black hole and that of the external world depends on a somewhat contrived process, but once it has been established that such a process is theoretically possible, it becomes reasonable to ask whether processes of this general kind might occasionally even occur naturally. . . .
>
> Our process depends on the existence of a region between the two surfaces known as the stationary limit and the (absolute) event horizon. The stationary limit is a surface at which a particle would have to travel with the local light velocity in order to appear stationary to an observer at infinity; and just inside which no [such] particle . . . can remain stationary as viewed from infinity. The event horizon is the effective boundary of the black hole inside which no information can escape to the outside world. The stationary limit lies outside the event horizon. . . .

Briefly stated, a rotating black hole forces the immediate surrounding space to rotate along with it. The geometric coordinates of the space and thus the locations of particles and radiation within the space just above the event horizon of the rotating black hole are dragged along as the hole rotates. This effect is called *frame dragging*; the entire coordinate frame near the rotating black hole is dragged along by the black hole's rotation. To appear to be standing still, a particle needs to counteract this forced motion by flowing backward against the direction of the induced motion. But there is a region, whose outer radius is defined by the *stationary limit*, where no particles, not even light, can appear to stand still. Everything, including light, is dragged along.

Penrose and Floyd envisaged dropping a particle into the region within the stationary limit, but still outside the event horizon from within which nothing can ever return. Once there, they imagined the particle splitting apart, conserving energy in a quite normal way. Viewed from infinity, one of the fragments could then have negative mass–energy while the other would necessarily have positive mass–energy, enabling it to escape to infinity. The fragment with negative energy would fall into the black hole, reducing the hole's mass–energy, while the fragment with positive energy would carry away some of the rotational energy of the black hole.

This seemed like a neat conceptual way to extract energy from a rotating black hole. But was there any practical consequence?

Five years later it seemed that there was. But to see how this came about we have to first look at what, if anything, thermodynamics might tell us about black holes.

In his early years, when he was first pursuing some of the most puzzling phenomena that physical investigations were revealing, Einstein often turned to thermodynamic fundamentals to guide his thinking. Seven decades later, this approach was about to prove useful again.

The Thermodynamics of Black Holes

In 1972, the 25-year-old Jacob Bekenstein, born in Mexico City but with an undergraduate education in the United States, was grappling with a PhD problem that almost every other relativist considered wrongheaded. Bekenstein's thesis advisor at Princeton was John Wheeler, who considered Bekenstein's ideas just crazy enough to be worth pursuing. He encouraged his student to dig further to see whether he could make his ideas more convincing.

Several relativists had noticed that no matter what transformations a black hole underwent, its surface area usually increased; if it didn't increase, the area might stay constant, but it never diminished. For a nonrotating, electrically neutral black hole this is easy to see; the Schwarzschild radius is proportional to the black hole's mass; and, with the impossibility of any escape, the black hole's mass and the area of a spherical surface at the Schwarzschild radius can only increase.

Bekenstein was convinced that this property of a black hole's area was remarkably similar to that of the entropy of a system. *Entropy* is the thermodynamic property of a system that measures its degree of randomness. If we thoroughly mix the contents of a salt shaker and a pepper shaker, we end up with some speckled mixture. Separating the mix of brown and white grains back into its pure salt and pepper constituents requires work. The entropy of the mix can be thought of as a measure of the work required to separate the mix into its original components. The entropy of the mixed system is larger than the sum of the entropies of the individual amounts of salt and pepper in their separate shakers.

Thermodynamically, any time one mixes two substances, randomizing the locations of the two constituents, entropy increases. More generally, the second law of thermodynamics tells us that the entropy S of a closed system is never observed to decrease, no matter what changes the system undergoes. Because matter entering a black hole remains trapped there, and because any addition of mass–energy always increases the surface area A, Bekenstein proposed that A is a true measure of the black hole's entropy.

The skepticism Bekenstein encountered was understandable. By 1972, most relativists were convinced by a property of black holes that John Wheeler had jokingly summarized in six words, "A black hole has no hair." The "no hair theorem," as it came to be called, postulated that a black hole could be fully characterized by specifying its mass M, its electric charge, Q, and is angular momentum, J. Nothing more could be known about it. No matter how one assembled two black holes, whether one was produced by tossing Steinway grand pianos down the hole until its mass became M, or whether you opened a garden hose and filled it with water up to mass M, the two black holes would be completely indistinguishable as long as their electric charges and angular momenta were identical. The past history of their constituents would be totally erased. A black hole could exhibit no hair – no legs of pianos sticking out to hint at how it had been assembled.

Most relativists thus argued that black holes were the simplest of all imaginable bodies. Only three parameters, M, Q, and J, were required to describe them. So there was no randomness to them that an entropy could describe. Bekenstein's preoccupation with entropy, they felt, was nonsense. Nonetheless, Bekenstein persisted, and toward the end of 1972 he submitted a highly original paper to back up his arguments.[15]

By then, information theory was a well-known subject widely used by communications engineers. Originally invented by Claude E. Shannon, while working for the telephone company, his work 'On the Mathematical Theory of Communication' had appeared in the *Bell System Technical Journal* in two seminal papers, published in 1948.[16]

In the introduction to his papers, the 32-year-old Shannon clarified two matters: *Information*, as he defined it, had no relation to *meaning*. But it could be quantified, and the measure of information would need to be *logarithmic*. He wrote,

> The fundamental problem of communication is that of reproducing at one point either exactly or approximately a message selected at another point. Frequently the messages have *meaning*; that is they refer to or are correlated according to some system with certain physical or conceptual entities. These semantic aspects of communication are irrelevant to the engineering problem. The significant aspect is that the actual message is one *selected from a set of possible messages*. ... If the number of messages in the

set is finite then this number or any monotonic function of this number can be regarded as a measure of the information produced when one message is chosen from the set, all choices being equally likely.... [We] will in all cases use an essentially logarithmic measure.... It is practically more useful.... It is nearer to our intuitive feeling as to the proper measure.... It is mathematically more suitable.

Shannon defined the uncertainty of whether or not a message had been transmitted correctly through a noisy transmission channel as the *entropy* of the system. Like the formal similarity between thermodynamic entropy and the surface area of a black hole, information theoretical entropy provided a formal similarity to thermodynamic entropy.

Bekenstein noted that Shannon's entropy S differed somewhat from its thermodynamic counterpart. In thermodynamics, the product of entropy S and temperature T constitutes a quantity that has the units of energy. The information theoretical entropy, however, is a pure number. Because Bekenstein wanted to identify the black hole entropy with Shannon's entropy, he needed to replace the black hole temperature T by the product kT, where k is Boltzmann's constant. kT then became the temperature of the black hole measured in units of energy, and the black-hole entropy S became a dimensionless number, just like Shannon's entropy.

But did the analogy between Shannon's entropy and the area of a black hole go any further? Bekenstein asserted that it did. Writing about the three defining characteristics of a black hole M, J, and Q, he noted that different black holes having identical mass, angular momentum, and electric charge might nevertheless have quite different internal configurations depending, for example, on whether they had formed through the collapse of a normal star or a neutron star. Any of a large number of such configurations could give rise to identical values of M, J, and Q. To emphasize such differences Bekenstein introduced the concept of black-hole entropy – a measure of the information inaccessible to an external observer about the internal configuration of any particular black hole.

Noting the monotonic increase in black hole surface area A in response to the many different processes that had been investigated by others, Bekenstein sought some function $f(A)$ that could correspond to the loss of information and gain in entropy S that an increase in mass required. If he assumed that the entropy $S \equiv f(A)$ was directly proportional to $A^{1/2}$, which would also make it proportional to the total mass M, he found that this would not work. The merger of two nonrotating black holes of masses M_1 and M_2, with respective surface areas A_1 and A_2 could only yield a nonrotating black hole mass $M_3 \leq (M_1 + M_2)$ because some mass might be radiated away during the merger. But because information would be lost during the merger of the two black holes, meaning that entropy would have to increase, while the total mass could only decrease, an entropy proportional to mass $M \propto A^{1/2}$ would not work.

This objection could not be raised against an entropy directly proportional to the black hole area $S = \gamma A$, where the constant γ needed to have dimensions inversely proportional to area, because Shannon's entropy is dimensionless. The only "truly universal constant" occurring in nature, which meets these demands, Bekenstein argued, is the ratio $(c^3/G\hbar)$, where c is the speed of light, G is the gravitational constant, and $\hbar$ is Planck's constant divided by 2π. This suggested that the proportionality constant γ should be some multiple of order unity of $(c^3/G\hbar)$.

Bekenstein sought to estimate what this multiple might be. The procedure he selected was to consider a spherical particle of mass $\mu \ll M$ and radius b slowly lowered from afar and brought to rest at a small height b above the black hole's surface, that is, above its radius R. In somewhat simplified terms, his procedure allowed μ to fall from there into the black hole. He showed that this should increase the black hole's surface area by a minimum amount $2\mu b$. The minimum permissible product μb, however, is found to be that of a mass μ whose radius is defined quantum mechanically by the *Heisenberg uncertainty principle* and classically by the black hole radius requirements. Bekenstein thus found the minimum increase in area to be $2(\hbar G/c^3)$. He then equated this to the minimum increase in entropy permitted by Shannon's theory, $\ln 2$ – the natural logarithm of 2 – amounting to the loss of one *bit* of information. This gave him the value of the proportionality constant, $\gamma = (1/2)(\ln 2)(c^3/\hbar G)$, where $\ln 2 \sim 0.693$.

For a black hole of area A, this finally allowed him to write the Shannon entropy as

$$S = \frac{(\ln 2)}{8\pi} \frac{c^3}{\hbar G} A . \tag{10.1}$$

If Bekenstein multiplied this by the Boltzmann constant k, to regain the thermodynamic entropy for comparison to the entropy of ordinary stars, he found that the entropy of a solar mass black hole had to be of order 10^{60} erg K^{-1}. In contrast, the entropy of the Sun is only of order 10^{42} erg K^{-1}. This difference, amounting to a factor of $\sim 10^{18}$, starkly dramatized the highly irreversible character of black hole formation – the enormous amount of information lost as a black hole forms.

In analogy to ordinary thermodynamics, Bekenstein also defined a black hole temperature $T = [\delta S/\delta(Mc^2)]^{-1}$, where the partial derivative referred to conditions of constant charge and angular momentum.

Three years after Bekenstein's thermodynamic approach was first published, the Cambridge University theorist Stephen Hawking found that the numerical coefficient $(\ln 2)/8\pi$ in Bekenstein's entropy should be replaced by $1/4$, making

$$S = c^3 A/(4\hbar G) , \tag{10.2}$$

which was roughly nine times greater; otherwise, Bekenstein's insight, imported and adapted from information theory, an engineering discipline, had been remarkably correct.[17] With this, the black hole temperature became $T = \hbar c^3/8\pi kMG$. How Hawking had reached this result is remarkable.

Radiating Black Holes

A year after the publication of Bekenstein's article, most relativists were still highly skeptical. Stephen Hawking was particularly irritated by Bekenstein's approach, which he felt had misused "my discovery of the increase of the area of the event horizon."[18] But Hawking's views were about to change.

In September 1973 he visited Moscow, where Yakov Borisovich Zel'dovich and his graduate student Aleksei Starobinsky told Hawking they thought quantum mechanical fluctuations in the vacuum surrounding a rotating black hole should gradually radiate away the hole's rotational energy. On returning home, Hawking tried to see whether these ideas made quantitative sense. Like Penrose and Floyd, several years earlier, he started with entities that split into two fragments. Only, the entities he considered were vacuum fluctuations rather than conventional particles.

Heisenberg's *uncertainty principle* tells us that over brief time intervals, the precise energy of a system can be determined only approximately. During such a short interval even the energy in empty space is uncertain. Vacuum is capable of briefly creating particle–antiparticle pairs, which soon annihilate to disappear without violating either the Heisenberg uncertainty principle or long-term energy conservation. Although general relativity precluded any radiation from leaving a black hole, quantum mechanics, it seemed, might provide the means for a black hole to radiate energy.

Consider a vacuum fluctuation creating a particle–antiparticle pair just outside a black hole's event horizon. If, like in the Penrose/Floyd process, the particle with negative energy falls into the black hole, while the particle with positive energy propagates out to infinity, a distant observer would conclude that the arriving energy had been lost by the black hole. This much would have been in accord with the ideas of Zel'dovich and Starobinsky. But, to Hawking's consternation, he found that energy might be extracted in an almost identical way from a stationary black hole. He was worried about this mainly, as he wrote, because "I was afraid that if Bekenstein found out about it, he would use it as a further argument to support his ideas about the entropy of black holes, which I still did not like."[19]

As Hawking repeatedly checked his calculations, everything appeared in order. The emitted energy did not arise in the black hole but rather above, albeit close to, its surface. Particle–antiparticle or photon pairs fleetingly formed there may be created with one member of the pair having positive, the other having negative energy, the pair's net energy being zero. In the strong gravitational field near the black hole's surface the fragment having negative energy would fall into the black hole, and the fragment with positive energy could escape to infinity. The black hole then would lose mass, its surface area would shrink, and its temperature rise. What struck Hawking most was not only that the radiation spectrum of the emitted particles was just that expected from a radiating hot body, but that the number of emitted particles also was in accord with thermodynamics. "They all confirm[ed] that a black hole ought to emit particles and radiation as if it were a hot body with a temperature that depends only on the black hole mass: the higher the mass, the lower the temperature."[20]

In a letter he submitted to the journal *Nature* Hawking proposed,[21]

> Quantum gravitational effects are usually ignored in calculations of the formation and evolution of black holes. ... [They] may still, however, add up to produce a significant effect ... it seems that any black hole will create and emit particles such as neutrinos and photons at just the rate that one would expect if the black hole was a body with a temperature of ... $\sim 10^{-6}(M_\odot/M)$ K. As a black hole emits this thermal radiation one would expect it to lose mass. This in turn would increase the surface gravity and so increase the rate of emission. The black hole would therefore have a finite life of the order of $10^{71}(M_\odot/M)^{-3}$ sec. For a black hole of solar mass this is much longer than the age of the Universe. There might, however, be much smaller black holes which were formed by fluctuations in the early Universe. Any such black hole of mass less than 10^{15} g would have evaporated by now. Near the end of its life the rate of emission would be very high and about 10^{30} erg would be released in the last 0.1 sec. This is a fairly small explosion by astronomical standards but it is equivalent to about 1 million [25 Megaton] hydrogen bombs.

Hawking's 1974 finding was astounding! Because no complete quantum theory of gravity existed, he had to elucidate how at least the relevant parts of accepted quantum theory could be melded with general relativity in strongly curved spaces to yield credible thermodynamic results. In the two years that followed, 1975 and 1976, Hawking presented more detailed quantum theoretical calculations in two further papers to support his initial findings.[22,23]

With Hawking's results in front of us, we can see now why Einstein had concluded in 1939 that black holes could not possibly be formed. Einstein was convinced that, if black holes could form, it should be possible to assemble them gradually from a swarm of mutually attracting particles radiating away energy. But, as Hawking showed, the formation of a black hole had to proceed very quickly. Because a small black hole of mass $M \sim 10^3$ tons must radiate away 10^{30} ergs – constituting its entire mass – in 0.1 seconds any long-lived black hole would have to accrue mass at an initial rate orders of magnitude more rapidly than 10^4 tons per second as it first formed. Thus, long-lived black holes must start their lives through a catastrophically rapid collapse to reach a threshold mass at which they radiate less rapidly and mass may be added more gradually to further lower the radiation rate.

I have dwelled on these fundamental questions concerning black holes not because the extraction of energy from black holes by these means has been found to be significant. To date, it has not. But their elucidation made the formation of black holes through catastrophic processes more apparent. How black holes actually generate highly energetic outflows is quite another matter.

The Blandford–Znajek Effect

Black hole radiation is not the dominant way for a black hole to emit energy.

Some first attempts to explain how quasars could be so luminous, sometimes orders of magnitude more luminous than ordinary galaxies, were independently and almost simultaneously proposed by Edwin E. Salpeter in the United States and Yakov Borisovich Zel'dovich in the Soviet Union.[24],[25]

They took the quasar to be a massive black hole. An extended cloud of gas falling toward the hole would be gravitationally focused in its wake, producing a shock where the converging portions of the gas stream collided. If the velocity of the approaching gas was low, the region of maximum collision would be close to the massive hole. Much of the gas would ultimately spiral in toward the hole, radiating thermal energy both at the initial shock and as the gas gained temperature during accretion onto the hole.

Five years after these preliminary ideas were proposed, Donald Lynden-Bell, a 34-year-old astrophysicist at the Royal Greenwich Observatory in England, considered the effects of magnetic fields embedded in the inward-spiraling gas forming a disk around the black hole. In 1969, he proposed the production of cosmic rays in the magnetized *plasma* and their emission of radio waves as they traversed the magnetic fields.[26]

All three of these efforts, however, concerned the emission of energy by gas falling into the black hole.

Black holes are expected to have high angular momenta, because matter falling into the hole is unlikely to travel precisely along radial trajectories and any small deviation from radial infall will transfer angular momentum to the black hole.

The first significant attempt to examine natural means of extracting energy from a black hole was launched in 1976 by the 27-year-old Roger D. Blandford, who had just finished his doctoral work at the Institute of Astronomy at Cambridge University, and his fellow graduate student Roman L. Znajek, still putting the finishing touches to his thesis. Their work on 'Electromagnetic extraction of energy from Kerr black holes' applied magnetohydrodynamic methods to the study of rotating black holes and was strikingly complete and persuasive – although also quite complex.[27]

The basic thrust was that a rotating magnetic field will be threaded by magnetic field lines induced by currents flowing in the encircling equatorial disk. This configuration will set up electric fields sufficiently strong to break down the

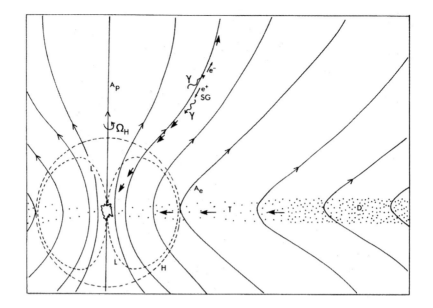

Figure 10.2. Schematic cross-section of a rotating black hole and its magnetosphere, postulated by Roger D. Blandford & Roman D. Znajek. The axis of rotational symmetry and central axis of the poloidal magnetic field is marked A_p, while more equatorial field surfaces are labeled A_e. The sense of rotation is indicated by Ω_H. The Kerr black hole's event horizon is shown as a dashed circle marked H. To the upper right of the horizon, a magnetospherical spark gap, SG, is producing electron–positron pairs, e^+, e^- from the vacuum. These are accelerated and interact with ambient photons to produce γ photons, whose mutual collisions can produce further electron–positron pairs and thus a cascade. Positrons are shown flowing into the hole and electrons outward. Particles flow along the indicated surfaces $A =$ constant only as long as these are normal to space-like surfaces. At loci L this condition breaks down. A disk D lies in the symmetry plane of the Kerr hole and is denoted by heavy stippling. The disk does not extend all the way into the black hole but breaks up in a transition region T within which matter falls from the disk into the hole. This is indicated by lighter stippling. (*From R. D. Blandford and R. L. Znajek, 'Electromagnetic extraction of energy from Kerr black holes,' Monthly Notices of the Royal Astronomical Society, 179, 1977, Figure 1, p. 445*).

vacuum and produce cascades of electron–positron pairs and a surrounding stationary axi-symmetric magnetosphere. The production of the cascade is promoted by electrons accelerated to high energies. These, in turn, will scatter ambient photons to produce gamma rays. The collision of two gamma rays can then produce further electron–positron pairs, thus maintaining the cascade.

Blandford and Znajek then pictured an active galactic nucleus as a massive rotating black hole surrounded by a magnetized accretion disk. If they assumed that the magnetic field lines through the black hole took on an approximately paraboloidal configuration, the extracted energy would be beamed into two counter-directed jets, emanating from galactic nuclei, as observed in many radio sources, Figure 10.2.

The rotational energy of a black hole consists of two parts, an irreducible part associated with the hole's entropy and a reducible one that may be extracted, at least in principle. Blandford and Znajek estimated that the efficiency, that is, the actual energy extracted divided by the maximum extractable energy, might be as high as $\epsilon \sim 38\%$ in one of the reasonably credible magnetic field configurations they considered. This is a very high efficiency compared to that with which mass can be converted into useful energy in nuclear processes, the most efficient of which, the conversion of hydrogen into helium, has an efficiency of only $\epsilon \sim 0.7\%$.

For a galactic nucleus, the complexity of the interactions involved made it difficult to predict how much of this energy would emerge as *nonthermal* rather than thermal emission. Nonthermal emission would be more likely to produce powerful outflows, cosmic ray particles, and other forms of energy otherwise difficult to account for in a galactic setting. On balance, Blandford and Znajek concluded their paper with this summary:

> [We] have discussed a mechanism for extracting rotational energy from a Kerr hole. The presence of the three principal ingredients: angular momentum, a magnetic field and a massive black hole, seems difficult to avoid within a galactic nucleus. It appears that this could be an effective agency for transforming gravitational energy into nonthermal radiation.

* * *

Stellar Black Holes

By the late 1970s, the field of black hole studies was becoming active. Not only were massive black holes forming the nuclei of galaxies, stellar black holes also were being pursued and identified with X-ray binaries, pairs of stars generally consisting of a white dwarf, neutron star, or black hole, orbiting close to a companion *evolved star* with a large radius and correspondingly low surface gravity.

In this setting, the intense gravitational pull of the compact star tidally strips matter from its larger companion. If this material has too much angular momentum, it cannot directly fall onto the compact partner but, instead, accretes onto an *accretion disk* orbiting the star. Viscous friction transfers angular momentum from the faster rotating inner parts of this disk to its periphery. Deprived of angular momentum the inner portions move closer to the surface of the compact star, eventually reaching a last stable circular orbit and then spiraling onto the star.

The distinction between X-ray binary stars in which the massive companion is a neutron star or white dwarf, rather than a black hole, is generally possible only if the mass of the compact component can be established. This involves determination of the binary stars' orbits, a somewhat complex procedure because the compact component may be detectable solely in X-rays, whereas the evolved star is detected solely at visible wavelengths. If this observational difficulty is resolved and one then finds that the X-ray emitting companion has a mass well in excess of $3M_\odot$, one may expect that this is a black hole because the most massive neutron stars known have masses $\lesssim 2M_\odot$.[28] But even this is not necessarily so. One may still need to rule out the possibility that what appears to be a single massive star is not something else – such as a close neutron star binary.

With the emphasis on black holes that had sprung up by the late 1970s, their role in the earliest phases of a highly compact expanding universe was also under consideration. Complex questions regarding quantum effects in strong gravitational fields and their impact on cosmology also were being debated. A further concern requiring deeper insight was the nature of vacuum, empty – or apparently empty – space, and its role in cosmic evolution.

Sidney Coleman's True and False Vacuum States

Sidney Richard Coleman, born in Chicago in 1937, came to Harvard in 1962, right after completing doctoral work in theoretical physics at Caltech. He soon became a professor of physics, staying at Harvard until his death in 2007. In 1972, he and Erick Weinberg, then a graduate student at Harvard, published a paper on radiative spontaneous symmetry breaking.

Symmetry breaking generally depends on small effects that at first appear unimportant in describing a physical system. If we stand a pencil on its sharpened point on a perfectly horizontal table top, symmetry arguments suggest that the pencil shall remain standing upright. Gravity alone does not preferentially make it fall in one direction rather than another. But we know that pencils don't remain upright; they soon fall in some unpredictable direction. Arbitrarily small disturbances cause a pencil to tip over and come to rest lying on the table, a state of lower potential energy.

We may stop the pencil from falling over, by slightly grinding off its point, making it flat. The pencil will then stand upright on its flattened point. That vertical position is now stable, being a position of minimum potential energy compared to any immediately neighboring attitudes – meaning that if the pencil is very gently tipped ever so slightly to one side, it will recover and come to rest again on its flattened point, an attitude of minimum local potential energy. Tipping the pencil more than this very slight amount, however, causes it to fall, because there is an even lower state of potential energy in which the pencil lies flat on the table.

We can thus speak of two states of minimum potential energy, a shallow minimum and a deep minimum. A very slight disturbance can break the symmetry, that is, overcome the potential barrier that keeps a pencil standing on its flattened point and make it fall into the deeper potential energy minimum where it lies flat on the table.

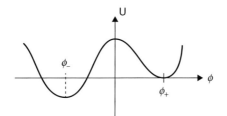

Figure 10.3. Coleman's sketch of the energy U of a system with two relative energy minima, located at positions ϕ_+ and ϕ_-, along some characterizing parameter ϕ, where only the lower minimum located at ϕ_- is an absolute minimum. In Coleman's theory the location ϕ_- corresponds to a true vacuum state, whereas the location ϕ_+, which could initially be stable, would soon be perturbed by quantum effects and rendered unstable, causing the system to gravitate to state ϕ_-. Coleman called the state at ϕ_+ a *false vacuum* and the state at ϕ_- a *true vacuum*. (*Reproduced from Sidney Coleman, Physical Review D, 15, 2929, 1977, with permission. ©1977 American Physical Society*).

Coleman and Weinberg's paper predicted a similar pairing of potential energy minima, one shallow, the other deep, that would produce transitions among quite different configurations of particles and fields in high-energy physics. Here the disturbances breaking the original symmetries were small radiative effects.[29]

Four years after this initial venture, Sidney Coleman applied a similar viewpoint to show that even vacuum – empty space – could have similar pairs of energy minima, as in Figure 10.3, one minimum having a lower energy than the other. During this shift from one energy minimum to the other, the vacuum would undergo a phase transition, analogous to the phase transition that turns liquid water into steam at high temperatures.[30]

Coleman explained this transition in terms that anyone who has watched water boiling in a pot will recognize. If the walls of the pot are smooth, the water may become superheated, that is, heated above its normal boiling temperature before a large bubble explosively bursts forth. Small bubbles that might nucleate will not have sufficient energy to overcome the surface tension of the water and will collapse almost as soon as they form. Just as superheated water could equally well be either in a liquid or a vapor phase, depending on relatively minor controlling factors, Coleman considered that the vacuum might similarly transit from some particular phase to another. He called vacuum in a higher energy state a "false vacuum" and called a lower energy state into which the false vacuum could transit the "true vacuum."

Coleman proposed that an infinitely old universe must culminate in a true vacuum state, no matter how slowly the false vacuum decayed. However, in the earliest instants of the Cosmos, the energy per unit volume was likely to have been very high, so that the state of the Universe could have greatly differed from any vacuum state, true or false. As the Universe then expanded and cooled, it might first have settled into a false vacuum before transiting to the true vacuum state. If so, he hoped this had already happened.

The Inflationary Universe

Born in New Brunswick, New Jersey in 1947, Alan Guth had been an undergraduate at MIT and had stayed on to obtain a PhD in theoretical physics in 1971. He then spent three years as an instructor in physics at Princeton before going on to postdoctoral appointments successively at Columbia, Cornell, and at the Stanford Linear Accelerator (SLAC). By the late 1970s, now in his early thirties, he was beginning to worry that he might never find a suitable permanent position.[31]

About the time Guth came to Cornell, grand unified theories had begun to greatly interest particle theorists. Writing in late 1979, Guth and his fellow postdoc at Cornell, Henry Tye, had begun working on one aspect of these theories. As they explained in a jointly authored article in *Physical Review Letters*,[32]

> [G]rand unified theories ... of the electromagnetic, weak, and strong interactions ... combined with classical gravity, attempt to describe all physics which occurs at energy scales well below the Planck mass, ... 1.2×10^{19} GeV, at which point gravitational interactions will become strong. They, in principle, allow one to extrapolate the history of the universe back to a temperature of say $T = 10^{17}$ GeV [at cosmic age] ($t \sim 10^{-41}$ sec). With reasonable success, such extrapolations have been used to obtain crude theoretical estimates of the net baryon-number density of the universe. These models contain stable magnetic monopoles with mass M_m which is typically (although not necessarily) of order 10^{16} GeV. ...
>
> Recently, Zel'dovich and Khlopov and also Preskill have attempted to estimate the abundance of magnetic monopoles which exist today as a result of production in the very early universe. Both studies ... concluded that the number of monopoles would be unacceptably large.

Where did these monopoles come from? None had ever been observed! Why were Guth and Tye even concerned about them?

Magnetic Monopoles

The quantum history of magnetic monopoles begins in 1931, when Paul Dirac noted the well-known symmetry between electric and magnetic fields in Maxwell's electromagnetic theory but went on to emphasize its one asymmetry, the apparent absence of isolated magnetic charges analogous to the isolated electric charges of electrons and protons. Experiments had consistently shown that magnetic fields invariably involve magnetic dipoles, rather than monopoles, that is, isolated magnetic charges. Dirac now proposed that theoretical considerations did not prevent the existence of magnetic monopoles.[33] Quantum considerations led him to believe that magnetic monopoles should exist and that the magnetic charge, μ, of a monopole should be quantized in units of $\mu = \hbar c/2e$. That the monopoles had never been seen, Dirac thought, was because "the attractive force between two one-quantum poles of opposite sign is $(137/2)^2 = 4692.25$ times that between electron and proton. This very large force may perhaps account for why poles of opposite sign have never yet been separated."

In a paper written in 1981 Helge Kragh recalled the 50-year history of these particles.[34] Dirac's ideas were largely ignored in the immediate years after 1931 but a gradually increasing interest beginning around 1960 led to several hundred publications citing Dirac's paper, cresting at nearly 50 citations a year around 1977 before slightly declining between 1978 and 1980.

An accelerating rise in publications began shortly after the appearance of two influential papers in 1974. The first was an article on *Magnetic Monopoles in Unified Gauge Theories* written by the 28-year-old Gerardus 't Hooft working at CERN at the time.[35] The second, independently developed by A. M. Polyakov at the Landau Institute of Theoretical Physics in Moscow and submitted for publication only five weeks after 't Hooft's, was titled 'Particle spectrum in quantum field theory.'[36] Both papers concluded that magnetic monopoles should exist under certain circumstances. The masses of these could not yet be determined but were likely to far exceed those envisaged by Dirac. In some theories they might be of the order of the Planck mass, $\sim 2 \times 10^{-5}$ g.

Others soon became interested in the magnetic monopoles as well, among them Yakov Borisovich Zel'dovich, the giant of Soviet cosmology, relativistic astrophysics, and virtually every other part of modern physics.

Zel'dovich was born in 1914, in Minsk, the capital of Belarus but, at the time, part of the Russian Empire. His family moved to Saint Petersburg a few months later, and this is where he grew up. At age 17 he became a laboratory assistant at the Institute of Chemical Physics of the Soviet Academy of Sciences, with which he stayed associated until late in life. When Germany invaded the Soviet Union in 1941 the Institute first moved from Saint Petersburg, by then renamed Leningrad, to Kazan and then to Moscow. The young scientist's doctoral work dealt with chemical processes, and he spent roughly a decade working on problems of combustion, detonation, and chain reactions. In 1946 he began work on nuclear weapons, taking the lead on theoretical work on both fission and fusion bombs. He continued working on nuclear weapons until 1963. Meanwhile, in the early 1950s, he had also begun studying elementary particle interactions and transformations. He did not begin working in astrophysics and cosmology until 1963, but soon became one of the leaders in the field.

On becoming aware of the work of Polyakov and 't Hooft, Zel'dovich and the 27-year-old Maxim Yurievich Khlopov, who had just joined Zel'dovich's research group at the Institute of Applied Mathematics of the Academy of Sciences of the USSR in Moscow, took a quick look at the expected destruction rate of cosmic magnetic monopoles. They used an elementary thermodynamic and diffusion theoretic approach to determine the number density that might have persisted down to current times, and estimated the contemporary number density to be $\sim 10^{-19}$ cm^{-3}, far above observational upper limits.[37] They concluded "[T]he hypothesis of free monopole existence seems to contradict the modern picture of the hot [Friedmann] universe.... The discovery of free monopoles would be of greatest importance not only for particle physics, but for ... cosmology as well, because it would cause us to abandon our main views on ... big bang theory."

John P. Preskill at Harvard reached similar conclusions the following year.[38] He too argued that an unacceptably large number of monopoles and anti-monopoles might have been produced in the early universe, suggesting a potential discrepancy between grand unified models and big-bang cosmology.

Guth's Theory of Inflation

At Cornell, Guth and Tye were working on vacuum phase transitions likely to occur at the high temperatures expected at early cosmic times. Their calculations, like those of Preskill, were based on grand unified theories coming into play at particle and photon energies of order $\sim 10^{16}$ GeV, that is, at temperatures of order $\sim 10^{29}$ K. They took these temperatures to have prevailed at the earliest epochs.

As the Universe subsequently expanded and cooled, its early vacuum phase could have undergone a series of transitions along the lines that Coleman had described in his paper three years earlier.[39] The Universe would have transited through one or more false vacuum phases before reaching the *true vacuum* we may be experiencing today. A predominant consequence of these early phase changes would be the separation of the single interaction among subnuclear particles and photons into what today appear to be three quite distinct interactions, the strong, the weak, and the electromagnetic interactions.[40]

In trying to find ways of avoiding the glut of magnetic monopoles that Zel'dovich and Khlopov and Preskill had found, Guth and Tye concentrated on the production of monopoles during the stage of false vacuum expansion. They examined production resulting from topological defects occurring wherever bubbles in the rapidly expanding false vacuum collided. As topological entities analogous to knots, the monopoles were expected to be stable and would not decay.

Why then had they not been observed?

Guth kept worrying about this question when he moved on to Stanford for another year's postdoctoral stay. He recalled that Henry Tye had previously noted that they were somewhat arbitrarily assuming the expansion of the Universe to remain adiabatic, that is, at constant entropy. Now, as he checked on this, he found to the contrary, that the vacuum transition would produce an exponentially accelerating expansion in which a different set of conservation principles would apply. The expansion would be governed by a quantity akin to, or even precisely the same as induced by Einstein's original cosmological constant Λ. The behavior during the exponential expansion would be identical to that of de Sitter's expanding model – devoid of matter, but expanding. In some ways, Guth was returning full circle to a cosmology of 1917, though applied in a completely new setting.

Coleman's paper had suggested bubble expansion at the speed of light, the fastest possible expansion within a non-expanding space. But when space itself expands, the speed of expansion can be incomparably higher. This is what made the rapid inflationary expansion possible.

At this point Guth also recalled two cosmological puzzles that the Princeton physicist Robert Dicke had raised in a colloquium talk held at Cornell in the fall of 1978.[41] The generally accepted model of the universe, as Dicke pointed out, was that of a space in adiabatic expansion. The first difficulty with this was: How could two portions of the Universe viewed along opposite directions seem to appear largely identical when, according to prevailing cosmological models they had always been expanding away from each other at speeds exceeding that of light – meaning that they could never have been in physical interaction?

The other problem Dicke had raised was more subtle. He noted that the Universe today is roughly 10^{10} years old, with an apparent mass density that would keep it expanding with what, at the time, appeared to be only a quite gradual deceleration. If the Universe had any significant positive initial curvature it should have expanded and then collapsed within a tiny fraction of a second. If its initial curvature was negative, it would have expanded with a rapidly dwindling mass density. Taken together, this meant that the initial curvature had to be fine-tuned to zero curvature to an accuracy of about one part in 10^{55}. Surely no physical system could be that finely tuned without some disturbance throwing it out of balance.

Guth now recognized that an exponentially expanding false vacuum phase in the early universe would solve all three problems if it rapidly expanded by a factor z exceeding 3×10^{27}. The monopole problem would go away because a continuing expansion that increased the linear dimensions of the Universe by this large a factor would so dilute the number of monopoles per unit volume that the observable portion of the Universe viewed today might contain a total of only a very few monopoles if any. No wonder then that none had ever been found!

The enormous early expansion also reduced the curvature of the Universe, just as the curvature of any greatly expanded body is monotonically reduced. And, finally, as Guth envisioned it, the homogeneity problem would also disappear if he dropped the assumption that at early times some conventional equilibrium state might have existed.

Guth was not sure he had answered all questions that might be raised. He pointed to some weaknesses in his theory, but felt it had enough merit to submit it for publication.[42]

The weaknesses were quickly addressed by a variant of inflation successively introduced by Andrei Linde in the Soviet Union and by Andreas C. Albrecht and

Paul J. Steinhardt in the United States, and the theory of inflation came to be rapidly accepted.[43,44]

With this, Guth's years as an itinerant postdoc also came to a happy end with an invitation to join the Department of Physics at MIT.

Initial Fluctuations

We might still wonder where the initial fluctuations to seed Guth's exponentially expanding regions might have originated.

In the early 1970s, only a few years after the cosmic microwave background radiation had been discovered, Edward Robert Harrison in the United States and, two years later, Yakov Borisovich Zel'dovich in the Soviet Union, proposed that the density of the early Universe would have had to be quite inhomogeneous.[45,46] Density inhomogeneities dating back to the dawn of time, they thought, should be scale-invariant and scalar, meaning that they should be observable on all scales and be devoid of spatial anisotropies. They proposed that these inhomogeneities should eventually reveal themselves as fluctuations in the brightness distribution of the microwave background radiation across the sky and in the size distributions of clusters of galaxies.

When the inflationary cosmology was proposed a decade later by Guth, Linde, and Albrecht and Steinhardt, the arguments of Harrison and Zel'dovich became particularly compelling. In principle, they might provide an almost magical way of checking, today, whether the inflationary universe made any sense at all. In the inflationary model, the early exponential expansion of the Universe would so rapidly isolate primordial fluctuations, as their mutual expansion exceeded the speed of light, that the fluctuations would have no opportunity to interact and dissipate each others' energies until the cosmic expansion rate eventually declined. The larger-scale fluctuations would thus remain frozen in time, revealing themselves only late in the cosmic expansion through their imprint on the microwave radiation mapped across the sky and the distribution of galaxy clusters.

That the preservation of these primordial fluctuations might reveal to us even now, billions of years later, the primordial fluctuations that had marked our Universe in the very first instants of cosmic time, was heady. The prospect that the Universe could still reveal its origins to us today – if we just looked for them with sufficient care – was breathtaking!

* * *

Within a year or two, however, the particle physicists pursuing the predictions of inflationary models discovered the need for a correction. The primordial fluctuations initially postulated by Harrison and Zel'dovich were too large, and would have left a stronger imprint on the microwave background radiation than actually observed. With radio receivers available in the early 1980s the background radiation appeared to be almost featureless. Two decades later, far more powerful instrumentation would reveal the long-sought features in exquisite detail; but in the interim, by 1983, Andrei Linde had introduced a variant of the theory called *chaotic inflation*, which showed a potential way out of the dilemma.[47]

The mass of a black hole is proportional to its radius. Its density, that is, its mass divided by its volume thus increases with diminishing mass. Chaotic inflation assumes that the Universe initially was extremely dense and hot. At the highest densities it could have been filled with the smallest, least massive, but highest density black holes consistent with general relativity and quantum theory. Their mass, named after the father of quantum theory Max Planck is called the Planck mass, $m_P = (\hbar c/G)^{1/2} \sim 2.2 \times 10^{-5}$ g.

The energy density of this extremely hot matter corresponds to a temperature of order $T_P \sim 10^{32}$ K $\sim 10^{19}$ GeV. This is the temperature range in which the Grand Unified Theories (GUTs) of particle physics postulate that the strong and weak nuclear forces and the electromagnetic force should all be equally strong.

Before a cosmic time $t_P \sim 10^{-43}$ sec, different segments of a region even as small as that of a Planck mass would have been unaware of each others' existence, unable to mutually communicate to arrive at any kind of equilibrium. A state of chaos would have reigned. Distinctions between space and time might not even have been clear; a state referred to as a fluctuating *space–time foam* might have prevailed.

The chaotic inflation scenario proposed by Linde assumed that the evolution of the vacuum in this chaotic state was determined by a *scalar force field* holding sway under GUTs conditions. The field, mediated by a set of postulated particles of mass m far smaller than the Planck mass, $m << m_P$, and called the *inflatons*, is hypothesized to have been responsible for transmitting the forces generated by the scalar fields – more or less as photons or virtual photons mediate electromagnetic fields.

Linde postulated that, at these early times, most of the mass–energy within a region encompassing a Planck mass would have resided in the vacuum, rather than in the inflatons. By the time the Universe had even slightly aged, our current laws of physics could have begun to exert a measure of control across regions somewhat larger than the radius of the Planck mass and with energy densities somewhat lower than its density. The internal force fields should have become more homogeneous, although scale free fluctuations imprinted by the initial chaotic conditions would continue to pervade. Because of the high prevailing densities rapid general-relativistic expansion should simultaneously have set in.

For a uniformly expanding vacuum of constant energy density, the density acts like a cosmological constant Λ, and leads to exponential expansion, just as in the cosmological model postulated by de Sitter in 1917. But the inflationary expansion is orders of magnitude faster because of the far higher energy density. The Universe can expand by factors of order $z \sim 3 \times 10^{27}$ in less than 10^{-33} sec, and potentially can expand by far more.

Because of the chaotic initial conditions, Linde suggested that the vacuum within the confines of a Planck mass might have originated in a high-energy state and could gradually have decayed down to its lowest possible state. This would ultimately have reduced the exponential expansion rate, without changing the end result of a highly inflated Universe.

If the scalar field was a so-called *Higgs field*, named after Peter Higgs of the University of Edinburgh, who first postulated its existence, it would break the symmetry between the different forces, endowing the particles mediating the different interactions with different masses, and introducing the asymmetries distinguishing the weak, strong, and electromagnetic interactions. Energy conservation then would heat all these particles to very high temperatures as the vacuum energy plunged to near zero. Thereafter the Universe would expand and cool, along lines that Gamow had earlier foreseen.

One tantalizing question still needed to be answered. If inflation had carefully preserved and amplified initial fluctuations, could these perhaps have both left their imprint on the microwave background and also seeded the formation of galaxies at later times? Linde argued that this question could be answered positively: If the vacuum exhibited energy fluctuations generated by, and proportional to, the inflaton mass m, a value of $m \sim 10^{-4} m_P$ could be shown to give rise to fluctuations sufficiently large eventually to seed galaxy formation once the Universe had sufficiently cooled. Below a mass m in this range, galaxies would not form. Above it, the microwave background radiation would exhibit fluctuations larger than those observed.[48]

It all sounded promising.

How galaxies might have formed had troubled theorists for half a century. In 1946, Evegeny M. Lifshitz in the Soviet Union had found that an expanding universe in thermal equilibrium would quickly dampen its original fluctuations, making them so weak that gravitational contraction into galaxies could never occur.[49] But if the Universe had at one time been chaotic, then a state of true thermal equilibrium might never have arisen. The inflationary expansion could have rapidly increased the distances between individual fluctuations, isolating them from each other and preventing their mutual dissipation to establish thermal equilibrium. The extreme uniformity of the Universe envisaged by Lifshitz might then never have existed.

* * *

In the chaotic inflationary scenario, different regions isolated at the dawn of time can conceivably undergo vastly different evolutionary histories. Some regions might be so dense that they quickly collapse. Others might continue to expand exponentially through eternity. Still other regions, like our's, may have undergone similar phases of inflation followed by a transition to a radiation-and-particle-dominated phase. However, not all of these regions might have had fluctuations of just the right magnitude to form galaxies, stars, and planets hospitable to life.

The assumption of the inflationary scenario is that our's may be a very special realm; but that just such a realm of the Universe was needed to give rise to life and to people even able to study the Universe. We do not know anything about the potential existence of other, quite different realms the Universe may equally well encompass because, if they exist, they lie far beyond our own realm's horizon, the only portion of the Universe we have ever been privileged to view. Some of these remote portions may, for all we know, have mutated and spawned daughter realms that go on to generate yet others. Through continuous regeneration along such lines, there might conceivably never have been a start of time, only an endless history in which existing cosmic realms continue to spawn others – an almost unrecognizably different variant of a steady state universe, steady but chaotic.

The inflationary universe has given rise to the idea that humans are here solely because our realm of the Universe evolved as it did. Some astronomers have elevated this idea to a principle – the *Anthropic principle*, which states that the Universe had to be the way it is in order to give rise to humans. As we saw in Chapter 3, however, some principles, such as the principle of relativity, have had important consequences, while others, like Mach's principle, have not. The anthropic principle has, to date, yielded no apparent new insights on the nature of the Universe. This does not mean that it never will, but it probably does mean that some significant added ingredient will be needed to enable the principle to predict new and testable consequences.

* * *

In the decades following Guth's and Linde's pioneering papers, a veritable army of young high-energy particle theorists modeled wide varieties of inflationary scenarios, as well as mechanisms that could have brought the inflationary phase to an end, to produce the particles and radiation observed in the Universe today. Potentially each of these models might leave a subtly differing imprint on the microwave background and on the distribution of matter in galaxies today. Once the necessary but also exquisitely difficult observations can be undertaken, we might know which of these many different models is in best accord with the early history of the Universe.

Answers to such questions are beginning to emerge today, as a progression of increasingly sophisticated, powerful observatories are deciphering the evidence deeply buried in the cosmic microwave background radiation.

The only problem is that, through three decades of theoretical modeling, we have by now inherited such a large number of variants of the inflationary scenario that we may lack sufficiently powerful tools to discern a single scenario, among these many, that uniquely fits the Universe we now observe.

Will we ever succeed in convincingly tracing the history of the Universe and discovering its true primordial origins?

Notes

1. *The Structure of Scientific Revolutions*, Thomas S. Kuhn, University of Chicago Press, 1962.
2. *Origins – The Lives and Worlds of Modern Cosmologists*, Alan Lightman and Roberta Brawer. Cambridge MA: Harvard University Press, 1990, p. 204.
3. On Continued Gravitational Collapse, J. R. Oppenheimer & H. Snyder, *Physical Review*, 56, 455–59, 1039.
4. *Black Holes and Time Warps – Einstein's Outrageous Legacy*, Kip S. Thorne. New York: W. W. Norton Company, 1994, pp. 209–54.
5. The Synthesis of the Elements from Hydrogen, F. Hoyle, *Monthly Notices of the Royal Astronomical Society*, 106, 343–83, 1946.
6. Past-Future Asymmetry of the Gravitational Field of a Point Particle, David Finkelstein, *Physical Review*, 110, 965–67, 1958.
7. Über die Hypothesen, welche der Geometrie zu Grunde Liegen, Bernhard Riemann, *Abhandlungen der Königlichen Gesellschaft der Wissenschaften zu Göttingen*, Vol. 13, where the 'Habilitationsvortrag' delivered on June 10, 1854 is reproduced.
8. Ibid., *Black Holes and Time Warps*, Thorne, p. 244.
9. 3C 273: A Star-like Object with Large Red-Shift, M. Schmidt, *Nature*, 197, 1040, 1963.
10. Gravitational Field of a Spinning Mass as an Example of Algebraic Special Metrics, Roy P. Kerr, *Physical Review Letters*, 11, 237–38, 1963.
11. Discovering the Kerr and Kerr-Schild metrics, Roy Patrick Kerr, http://arxiv.org/pdf/0706.1109v2.pdf (dated 14 January 2008).
12. Metric of a Rotating, Charged Mass, E. T. Newman, et al., *Journal of Mathematical Physics*, 6, 918–19, 1965.
13. Gravitational Collapse and Space-Time Singularities, Roger Penrose, *Physical Review Letters*, 14, 57–59, 1965.
14. Extraction of Rotational Energy from a Black Hole, R. Penrose & R. M. Floyd, *Nature Physical Science*, 229, 177–79, 1971.
15. Black Holes and Entropy, Jacob D. Bekenstein, *Physical Review D*, 7, 2333–46, 1973.
16. *The Mathematical Theory of Communication*, Claude E. Shannon, *Bell System Technical Journal*, July and October 1948. The papers are reproduced in their entirety in *The Mathematical Theory of Communication*, Claude E. Shannon & Warren Weaver. Urbana: University of Illinois Press, 1949.
17. Black holes and thermodynamics, S. W. Hawking, *Physical Review D*, 13, 191–97, 1976.
18. *A Brief History of Time*, Stephen W. Hawking, Bantam Press, 1988, p. 110.
19. *Ibid. A Brief History*, Hawking, p. 111.
20. Ibid., *A Brief History*, Hawking, p. 111.
21. Black Hole Explosions?, S. W. Hawking, *Nature*, 248, 30–31, 1974.
22. Particle Creation by Black Holes, S. W. Hawking, *Communications in Mathematical Physics*, 43, 199–220, 1975.
23. Ibid. Black holes and thermodynamics, Hawking, 1976.
24. Accretion of Interstellar Matter by Massive Objects, E. E. Salpeter, *Astrophysical Journal*, 140, 796–800, 1964.
25. The Fate of a Star and the Evolution of Gravitational Energy upon Accretion, Ya. B. Zel'dovich, *Doklady Akademii Nauk, 155, 67*; translated and published in *Soviet Physics – Doklady*, 9, 195–97, 1964.
26. Galactic Nuclei as Collapsed Old Quasars, D. Lynden-Bell, *Nature*, 223, 690–94, 1969.

27. Electromagnetic extraction of energy from Kerr black holes, R. D. Blandford & R. L. Znajek, *Monthly Notices of the Royal Astronomical Society*, 179, 433–56, 1977.

28. A two-solar-mass neutron star measured using Shapiro delay, P. B. Demorest, et al., *Nature*, 467, 1081–83, 2010.

29. Radiative Corrections as the Origin of Spontaneous Symmetry Breaking, Sidney Coleman and Erick Weinberg, *Physical Review D*, 7, 1888–1910, 1973.

30. Fate of the false vacuum: Semiclassical theory, Sidney Coleman, *Physical Review D*, 15, 2929–36, 1977.

31. Ibid. Origins, Alan Lightman and Roberta Brawer, p. 476.

32. Phase Transitions and Magnetic Monopole Production in the Very Early Universe, Alan H. Guth and S.-H. H. Tye, *Physical Review Letters*, 44, 631–35, 1980; 44, 963, 1980.

33. Quantised Singularities in the Electromagnetic Field, P. A. M. Dirac, *Proceedings of the Royal Society of London A*, 133, 60–72, 1931.

34. The Concept of the Monopole. A Historical and Analytical Case-Study, Helge Kragh, *Historical Studies in the Physical Sciences*, 12, 141–72, 1981.

35. Magnetic Monopoles in Unified Gauge Theories, G. 't Hooft, *Nuclear Physics B*, 79, 276–84, 1974.

36. Particle spectrum in quantum field theory, A. M. Polyakov, *ZhETF Pis. Red.*, 20, No. 6, 430–33, September 20, 1974; translated and published in *JETP Letters, 20*, 194–95, 1974.

37. On the Concentration of Relic Magnetic Monopoles in the Universe, Ya. B. Zel'dovich & M. Yu. Khlopov, *Physics Letters B*, 79, 239–41, 1978.

38. Cosmological Production of Superheavy Magnetic Monopoles, John P. Preskill, *Physical Review Letters*, 43, 1365–68, 1979.

39. Ibid., Fate of the false vacuum, Coleman.

40. Unity of All Elementary-Particle Forces, Howard Georgi & S. L. Glashow, *Physical Review Letters*, 32, 438–41, 1977.

41. Inflationary Cosmology and the Horizon and Flatness Problems: The Mutual Constitution of Explanation and Questions, Roberta Brawer, a Master of Science in Physics thesis, Massachusetts Institute of Technology, 1996, p. 69. The thesis is publicly available at http://dspace.mit.edu/bitstream/handle/1721.1/38370/34591655.pdf?sequence=1

42. Inflationary universe: A possible solution to the horizon and flatness problems, Alan H. Guth, *Physical Review D*, 23, 347–56, 1981.

43. A New Inflationary Universe Scenario: A Possible Solution of the Horizon, Flatness, Homogeneity, Isotropy and Primordial Monopole Problem, A. D. Linde, *Physics Letters B*, 108, 389–92, 1982,

44. Cosmology for Grand Unified Theories with Radiatively Induced Symmetry Breaking, Andreas Albrecht & Paul J. Steinhardt, *Physical Review Letters*, 48, 1220–23, 1982.

45. Fluctuations at the threshold of Classical Cosmology, E. R. Harrison, *Physical Review D*, 1, 2726–30, 1070.

46. A Hypothesis, Unifying the Structure and the Entropy of the Universe, Ya. B. Zel'dovich. *Monthly Notices of the Royal Astronomical Society*, 1P–3P, 1972.

47. Chaotic Inflation, A. D. Linde, *Physics Letters B*, 129, 177–81 1983.

48. Particle physics and inflationary cosmology, Andrei Linde, *Physics Today*, 40(9), 61–68, September 1987.

49. On the Gravitational Stability of the Expanding Universe, E. Lifshitz, *Journal of Physics of the USSR*, 10, 116–29, 1946.

11

Turmoils of Leadership

In Chapters 7 and 9 we saw how strongly the conclusions we reach about the structure of the Universe can be shaped by intrusions that have no formal role in the philosophy of science. Technological innovations, governmental priorities, economic factors, as well as scientific advisory and oversight boards all play a role in the shaping of astronomy. Astrophysics is continually under pressure from forces pulling in different directions, destabilizing ongoing efforts, and keeping the field off balance.

How astrophysics nevertheless manages to advance is difficult to grasp in the abstract. The present chapter attempts to sketch realistically how modern astronomy is conducted and the vital role a well-coordinated communal approach can play as long as it remains sufficiently flexible to recover from inevitable setbacks. As we shall see, steadfast leadership is the essential ingredient of ultimate success.[a]

A Logical Way to Proceed

By the mid-1970s, it was easy to see how the influx of new detection techniques introduced by the military in World War II and the Cold War had led to undreamed-of discoveries. Earlier theories had given us no inkling to expect the existence of quasars, pulsars, or the detection of cosmic masers, all initially revealed by radio astronomy. Similarly significant discoveries were being made in the gamma-ray, X-ray, and infrared regimes.

Techniques initially devised for industrial purposes had meanwhile been developed further by gifted instrumentalists, raising expectations of equally astonishing future discoveries. Suggestions for new ways of searching the skies and mission designs for realizing these ventures were inundating NASA.

---

[a] Readers interested in a more detailed history of the efforts to design and construct the Hubble Space Telescope will find the book by Robert W. Smith, 'The Space Telescope: A Study of NASA, Science, Technology, and Politics' particularly informative.[1]

Franklin D. (Frank) Martin, in charge of advanced astrophysics projects at NASA Headquarters, was well aware of these. In March 1978 NASA's Astrophysics Division issued a set of 24 brochures, each summarizing the findings of a technical study of a different space mission recommended by the Space Science Board of the National Research Council, or by a report to the NASA Administrator by an *Outlook for Space Study Group* that had been assembled to identify directions the United States should take "in the civilian use of exploration of space for the remainder of the century," but highlighting the period from 1980 to 1990.[2] Each of the brochures was brief and explicit, comprising barely 10 pages of text, drawings, dimensions, and cost estimates. Wrapped in colorful covers ranging from blue for the proposed gamma- and X-ray probes, to red for the far-infrared and submillimeter missions, and brown for relativistic and gravitational investigations, the stack of slender brochure bindings arrayed on one's bookshelf attractively displayed the colors of the rainbow.

The number and diversity of proposed missions, however, posed two problems, as Martin quickly realized the following year, when he was appointed Director of NASA's Astrophysics Division in 1979. An endless stream of astrophysicists was dropping by his office to urgently advocate the priority of any one of these projects above all others. And Congress was finding it difficult to understand why NASA was asking support for so many different missions.

In an interview with the historian and writer Renee M. Rottner, 30 years later in 2009, Martin recalled the problem he had to face of selling these projects to the politicians, one at a time. "All those things were really expensive. Yet the science of these missions was necessary. But what I thought, as a representative of the taxpayer's money, was that there had to be a strong discovery element in all these missions, given…that we really hadn't looked at the universe across the whole electromagnetic spectrum."[3]

Around 1976, Martin came across an article titled 'The Number of Class A Phenomena Characterizing the Universe' which I had published in 1975.[4] It emphasized the promise that the new technologies of the postwar era were offering astronomy and highlighted how much more could be learned about the Universe through joint observations across the entire electromagnetic spectrum. Observing a celestial source solely in the X-ray wavelength regime would teach us little, unless we also determined how observations of the source in visible light, the infrared, radio, and gamma-ray regimes were related to each other. That combination of data would spell out the physical processes at play. Anything less than full wavelength coverage could easily lead to misapprehensions about conditions prevailing in the source. It was the combination of observations across the entire spectrum that promised to yield greater understanding of the Universe. Moreover, as the article proposed, "Phenomena that have not yet been discovered are most likely to be found by searching with observational techniques in the currently inaccessible domains…"

As Frank Martin recalled in his interview with Rottner, "Somebody walked down the hall and handed it to me....I read it and I knew exactly what to do...Harwit's article really spoke to me – that's what we should be doing....All those things were pieces of this huge puzzle....He made it very clear about the criteria for determining what kinds of mission should be pursued.... [they should involve] every part of the electromagnetic spectrum." Frank added, "We were able to package all this stuff together, in such a way that folks in Washington DC and at NASA could understand...[and] open up this huge discovery space for science and the country."

Not until 30 years later, when I read Martin's interview, was I aware that the article I had written had been useful to him. This may have been because, as Frank mentioned to Rottner, "I told Harwit he wrote this [article] for two people and probably only one – it was me – the director of Astrophysics at NASA, because many of these things he talked about couldn't be done from the ground."

In 1981 I published a book 'Cosmic Discovery, The Search, Scope and Heritage of Astronomy' that elaborated on the article written six years earlier.[5] It provided a more detailed rationale for a comprehensive set of surveys conducted across the entire electromagnetic energy range. Without that full inventory, a thorough understanding of the Universe would elude us.

* * *

Charles J. (Charlie) Pellerin, who at age 38 succeeded his friend and former boss Frank Martin as director of NASA's Astrophysics Division in 1983, understood these factors equally well. Martin and Pellerin both had recognized early that NASA had an incomparable opportunity to advance our understanding of the Universe in ways which, at the time, only the United States and NASA could propose. It would involve launching into space, well beyond Earth's interfering atmosphere, four large telescopes, respectively operating at γ-ray, X-ray, optical, and infrared wavelengths. Radio observations, which could be made with ground-based telescopes, would have to complement this thrust, but the National Science Foundation was primarily responsible for those. NASA would have to build the observatories to be launched into space.

The practical problem Pellerin faced in late 1984 was to build these observatories to cover all four energy ranges, preferably in synchrony with each other to enable simultaneous observations across the entire electromagnetic spectrum. This would be a challenging task under any conditions, but a new obstacle had just arisen. Surmounting it would take a concerted effort!

Governmental Priorities

In his January 25, 1984 State of the Union message, President Ronald Reagan had announced, "Tonight, I am directing NASA to develop a permanently manned space station and to do it within a decade. A space station" Reagan assured his audience "will permit quantum leaps in our research in science, communications, in metals, and in lifesaving medicines which could be manufactured only in space…"[6] Pellerin, like everyone within NASA, knew that this directive would urgently refocus the agency's efforts. NASA had promised Reagan that the station would cost no more than eight billion dollars at 1984 rates. Every knowledgeable space engineer in the country knew that this was unrealistic.

A more immediate problem for astrophysics, in view of the President's new directives, was that Congress would not understand why yet another set of astronomical observatories was needed in space. The President's plan envisioned the Space Station to be launched in 1991–92 as including two unmanned free-flying, co-orbiting platforms, respectively dedicated to astronomy and Earth observations. An Orbital Maneuvering Vehicle, serving as an unmanned space tug for moving payloads around in the vicinity of the Space Station, would permit astronauts from the Station to service these payloads.

In view of this sudden change in priorities, Pellerin now needed to revise his programs to fit in with the directives of the President. If they expected to realize their aims, astronomers would have to articulate and advocate their needs much more effectively than in the past. Pellerin understood he would have to sell Congress and the administration on a long-term program encompassing an entire family of telescopes to conduct observations in all accessible wavelength ranges, and align this with the President's plans for the Space Station. The program would be expensive, and Pellerin knew it would require the full support of the leading U.S. astrophysicists – theorists as well as instrument builders.

Scientific Plans

Every 10 years, the U.S. National Academy of Sciences conducts a survey to set astronomical priorities for the decade ahead. The 1980s decadal review on astronomy and astrophysics had been carried out under the auspices of the National Research Council of the National Academy of Sciences in the United States. It was chaired by the theoretical astrophysicist George Brooks Field, founding director of the Harvard–Smithsonian Center for Astrophysics (CfA) in Cambridge, Massachusetts. The recommendations of this survey were meant to serve as a guide to the U.S. government in its implementation of a national astronomy program.

The review for the upcoming decade of the 1980s had recommended a long-planned mission, the Advanced X-ray Astrophysics Facility (AXAF) as the community's highest upcoming priority in space astrophysics. The Shuttle Infrared Telescope Facility (SIRTF) which in 1984 was renamed the Space Infrared Telescope Facility – no longer to be attached to the Space Shuttle – had also garnered high marks. Launching not just one but both of these missions would be possible only if Congress was willing to assign NASA the substantial funding required.

This led to an acute problem: By late 1984 Congress and the White House had already given approval to work that would lead to the launch of two other expensive NASA missions, the Gamma Ray Observatory (GRO) and the Hubble Space Telescope (HST) both of which had received the astronomical community's backing during the 1970s.[7] Now, before these earlier requests had even been fully met, competing segments of the astronomical community were already clamoring for yet another two missions, SIRTF and AXAF.

In practice, the job of the chair of the decadal review does not end when the review's recommendations are released. For the remainder of the decade the chair is expected to cruise the halls of Congress tirelessly, reminding members of the House of Representatives and the Senate, as well as the staffs of their legislative committees and staffers at the White House, that the astronomical community had spoken with one voice in its review, and that a level of continuity and discipline is required if government funding for science is to be wisely spent.

On March 8, 1984, Field wrote Senator Jake Garn, Chairman of the Senate Committee in charge of appropriations for NASA, pointing out that the 1980 decadal survey had "placed high priority on both the Advanced X-ray Astrophysics Facility, AXAF, and the Shuttle Infrared Telescope Facility, SIRTF." NASA's plans for timely flights of both these observatories would require that each underwent a Phase B study, for which the President's recently proposed FY85 (fiscal year 1985) budget was inadequate.[b]

Field wrote, "I strongly urge that your committee augment the NASA budget by approximately $10M to make these studies possible."[8]

The two missions also faced other threats. Nobody understood this better than Pellerin. The astronomical community was badly split. Every so often an infrared astronomer would drop by Pellerin's office to urge priority for SIRTF. Then an X-ray astronomer would call to argue for AXAF.

Pellerin recalled, "The argument might have been worth having if one [of these] supplanted the need for the other. However, either mission increased the necessity for the other ... I needed to create a fresh, compelling strategy that would mobilize everybody behind both missions, including the two [the Hubble Space Telescope, HST, and the Gamma Ray Observatory, GRO] already underway."[9]

As the end of 1984 approached, Pellerin instructed his deputy, George Newton, to invite a number of astrophysicists to a one-day meeting at the start of the new year.

[b] A NASA mission progresses through a series of phases before launch, beginning with a *Conceptual Study* followed by a *Phase A* preliminary analysis. It may then progress to *Phase B*, which entails a clearcut definition of system requirements that a successful mission must meet, a review of the system's design, and an independent *non-advocate* review. If these reviews are satisfactory, the mission may be promoted to *Phase C/D*, involving more detailed design and development, and a series of stringent reviews before launch. Finally, once in space the actual start of operations signals the beginning of *Phase E* – the onset of work the launch was designed to accomplish.

I had received an initial call and told Newton I'd gladly participate. He called again just before Christmas and asked whether I'd be willing to chair the meeting. I said "Yes." But after we hung up, I was puzzled. Normally, a NASA division director appoints someone he knows well and can trust to chair his advisory committee without embarrassing him.

This was different. Pellerin and I had never met.[c]

To those who had agreed to attend, a formal letter of invitation went out on December 21. It specified that the meeting should "stimulate discussion of the contributions of [each of NASA's proposed astrophysics missions] in context with those of others, and help illustrate the strength of the total program and its scientific justification … Output from the meeting is expected to be [a] unified understanding and its description in a set of graphic materials suitable for briefing informed lay people."[12]

An Initial Meeting

In the late afternoon of January 2, 1985, I arrived in Charlie Pellerin's office at NASA Headquarters in Washington. We spent the evening planning the next day's meeting and found ourselves getting along well.

I had prepared a number of sketches in advance to meet Charlie's request for "graphic materials." They were simple doodles of the kind that astrophysicists draw on a blackboard to explain their thoughts to colleagues: no equations, no formulae – just line drawings with a few phrases. Charlie agreed that these might make for a good start.

As the meeting got underway, the next morning, I looked around the group Charlie had assembled. In addition to five members of his own staff and three subcontractors handling logistics, Charlie had called on some of the leading scientists shaping astrophysics theory and the future of γ-ray, X-ray, optical, infrared, and radio astronomy in the United States.[13] Most of us had flown into Washington the previous evening.[d] The day began with Charlie describing what he saw as the major issue, the presentation of a coherent program that had a sound scientific

---

[c] A quarter century later I came across an interview Pellerin had also provided Renee Rottner, in 2009.[10] He told her he had read 'Cosmic Discovery,' which argued for exploring the entire electromagnetic energy range.[11] Pellerin remarked, "I thought, Martin Harwit wrote this book – why don't I get [him] to run this meeting? I'd never met him, so I called Dave Gilman [who said] 'He's a great guy'." Years earlier, Gilman had been a graduate student at Cornell, where he sat in on one of my classes. Later, Dave and I had often chatted informally. But back in 1984, I wasn't aware that Pellerin had checked all this out.

[d] Bob Brown from the Marshall Space Flight Center; Carl Fichtel and George Pieper from the Goddard Space Flight Center; George Field, Josh Grindlay, Robert Noyes, Irwin Shapiro and Harvey Tananbaum, all from the Harvard–Smithsonian Center for Astrophysics; Riccardo Giacconi from the Space Telescope Science Institute; Bill Hoffmann and George Rieke from Arizona; Ken Kellermann from the National Radio Astronomical Observatory, NRAO; Jerry Ostriker from Princeton; Ed Salpeter from Cornell; and Rainer Weiss from MIT.

rationale and could be explained to policy makers and to a wider community in simple and appealing terms. With GRO and HST well along in construction, it was time to focus on AXAF and SIRTF, particularly in the context of ongoing solar work and ground-based radio-astronomical efforts.

Shortly before lunch, we addressed the primary task Charlie had assigned us – the drafting of visual materials to help him explain the astrophysics program. I showed the sample viewgraphs, the doodles that I had discussed with Charlie the previous evening. The group then split into smaller teams and got to work on constructing similar viewgraph sketches to illustrate how the four observatories would help us gain insight on key astrophysical problems, including:

- the origin of the universe and structure of the cosmos;
- the formation of galaxies and clusters;
- large-scale motion and dark matter;
- quasars and active galactic nuclei;
- black holes;
- gas dynamics;
- star formation; and
- planetary systems.[14]

By day's end, these viewgraphs had been projected on a screen and critiqued by the whole team. The sketches were given to Valerie Neal of the Essex Corporation of Huntsville, Alabama, who was there to support our work and whose doctoral studies fortunately had included the history of science. The group agreed to meet again in two months. In the meanwhile Valerie and I were to come up with a first draft of the booklet we had been asked to produce to explain better the astrophysics community's priorities to the government officials who would have to agree on their merit and the need to fund them.

A More Permanent Group

Although the group had not been expected to convene again until March 26, Charlie called another meeting for February 22. By then, Valerie and I, assisted by Brien O'Brien, an illustrator also of Huntsville, had assembled a first draft of drawings. We circulated these among the group – which for lack of a formal name we began to call the *Astrophysics Council* – and then discussed potential revisions.[15]

Charlie's original purpose in convening the Council had been to assemble a group that could solve his immediate problem of explaining the new thrust of the Astrophysics Division to his superiors at NASA, to Congressional staffers, to the astronomical community, and to the public; but the energy flowing out of the January 3 meeting had indicated that a more permanent council could prove useful to him in other ways as well.[16] The informal structure of the Council provided him a rotating, rather than a more formally selected, membership of leading astrophysicists who could debate and resolve issues so that decisions he reached would have solid community support. It also helped him to market his ideas in Congress.

Keeping the Astrophysics Council going, however, required broadening its membership to make it more representative of the entire astronomical community. It would also require official consent.

Congress is specific about who may or may not advise government agencies. Thus NASA was permitted to set up *Management Operations Working Groups* but could not arbitrarily institute its own advisory panels. A ruling on this issue was eventually handed down a few months later, in May 1985, by NASA's Logistics Management and Information Programs Division, which determined "that the Astrophysics Management Operations Working Group is not an advisory committee within the meaning of the Federal Advisory Committee Act (P.L.92–463) and hence not subject to the provisions of that statute." With this double negative, the Astrophysics Council was ruled legitimate but under a more official-sounding name, the *Astrophysics Management Operations Working Group, (AMOWG)*. By then, however, habit proved hard to break and most members continued to call it the *Astrophysics Council*, a practice I will continue to follow here.[17]

The Great Observatories Comic Book

That winter of 1985, George Field happened to drop by Charlie's office. Charlie recalls, "I said, 'George, my only dilemma is what to call this program of observatories, it's so great.' "George offered, "Why don't you call it *the Great Observatories?*" And the family of four observatories now had a name![18]

By late June 1985, the brochure of text and doodles in colorful comic-book style was ready and Charlie ordered a 15,000 print run. It proved immediately effective. Through the brochure the Council attempted to show the timelessness of astronomy and astronomy's role in public education. It alluded to some of the technical advances astronomy had contributed. Yet, many fundamental questions remained about how the Universe had started, how stars and galaxies had formed, and whether black holes exist. The processes we needed to understand spanned huge temperature ranges, requiring telescopes covering comparably large spectral regimes. Such telescopes could work only outside Earth's atmosphere, which is opaque to the radiation the telescopes needed to detect. An inability to cover all these wavelength ranges could lead to our missing important discoveries.[19]

Marketing the Great Observatories

Meanwhile, Riccardo Giacconi, a Council member, Director of the Space Telescope Science Institute on the campus of Johns Hopkins University, but also one of the world's leading X-ray astronomers, had been invited to a small luncheon gathering at the White House. Among those meeting with President Reagan were Edward Teller, generally considered to have fathered the hydrogen bomb; Jay

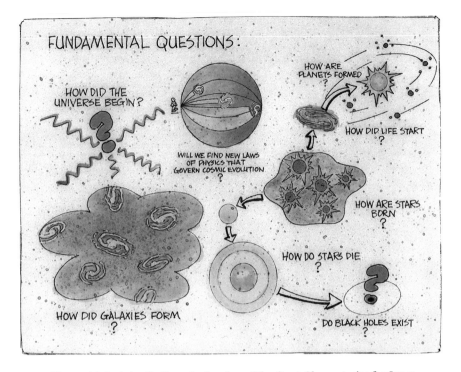

Figure 11.1. A doodle from the brochure 'The Great Observatories for Space Astrophysics.' The original drawing was in vibrant colors. The reproduction here is in gray tone. The caption for this image proclaimed: "Our thoughts about the long-term future of the human race involve fundamental questions about the nature of the cosmos – its past and its future, its governing laws, its harsh explosions, and its potential for hospitable planetary systems." The main cosmological questions it raised were, "How did the Universe form and evolve in the first few seconds? Can we learn more about the basic laws of physics from the effects they have had on the structure of the Universe?" Both questions still remain unanswered. (*From* 'The Great Observatories for Space Astrophysics,' *prepared under the auspices of the NASA Astrophysics division, Charles J. Pellerin, Jr. Director, by Martin Harwit and Valerie Neal in consultation with the Astrophysics Management Operations Working Group. Graphic design and illustration by Brien O'Brien*).

Keyworth, the President's Science Advisor; Tom Paine, who chaired the National Commission on Space; Donald Regan, the President's Chief of Staff; and a small handful of others – five scientists and five administrators all told. Riccardo talked briefly about the astrophysics program. And Edward Teller, who in the 1930s had done leading work in astrophysics, spontaneously backed Riccardo's enthusiasm for the program by telling the President that astrophysics may be one of the most important scientific endeavors of our time. Subsequently, NASA Administrator James M. Beggs received a White House request for further "information on astrophysics." The meeting obviously had made a favorable impact.[20,21]

By law, NASA employees are prohibited from lobbying Congress. Charlie and his staff could merely *inform* the public about the promise of the Great Observatories.

Members of the Astrophysics Council were under no such restriction. Those who could, walked the halls of Congress to bring the Great Observatories to the attention of Congressional staffers and to speak out in their favor. Nobody was more active than Harvey Tananbaum, who at the time was the team leader for the AXAF mission study and mirror development efforts.

In the fall of 1985, David Gilman in Charlie's Astrophysics Division and Diana Hoyt, on the staff of Congressman Manuel Luján, ranking Republican on the House Committee on Science and Technology, arranged a series of lectures for Congressional staff. The format was a set of six one-hour talks on "Space Science for the Non-Scientist." The talks were presented roughly at weekly intervals late in the afternoon.

On October 30, I gave the first of these presentations, "Space Astronomy – An Historical Perspective." This was followed by Claude Canizares of MIT speaking on "The Violent Universe – High Energy Astrophysics;" Blair Savage from Wisconsin on "The Visible and Ultraviolet Universe – Planets to the Edge of the Universe;" Michael Hauser from the Goddard Space Flight Center on "The Infrared Universe – Cold Gas and Dust;" Eugene Parker from Chicago on "The Sun – The Rosetta Stone;" and Irwin Shapiro from Harvard on "The Future of Space Astronomy."[22]

The Associate Administrator

On March 26, two months after the Council's first meeting, and even before our brochure had been completed, the Council invited Charlie Pellerin's immediate boss, Dr. Burton I. Edelson, Associate Administrator for the Office of Space Science and Applications (OSSA) to meet with us.

A graduate of the U.S. Naval Academy, Edelson had become interested in satellite communications at the Naval Research Laboratory (NRL) in the 1950s. On retiring from the Navy in 1967, he had joined Comsat Laboratories, where he was active on a program to develop satellites for commercial communication. He became director of Comsat Laboratories in 1972. In 1982, Edelson was appointed NASA Associate Administrator for OSSA. Like James M. Beggs, whom President Reagan had appointed NASA Administrator the previous year, Edelson had graduated from the Naval Academy in 1947. Charlie explained that Beggs and Edelson had been roommates at the Academy and were close personal friends.

Where Pellerin was single-minded about the direction astrophysics should be heading, Edelson seemed always anxious to reach consensus among the various committees advising the National Academy of Sciences or the NASA Administrator, Edelson's own advisory committee, and his three OSSA division directors responsible, respectively, for Astrophysics, Earth Science and Applications, and Solar System Exploration. His search for consensus may have been a reasonable policy, given that Edelson was one of the very few officials ever to have served NASA as Associate Administrator for OSSA, whose background was largely in technology and industry rather than in science. He may have realized it would be a mistake to rely on his own judgment on matters of policy or priority. Such decisions might more safely emerge from a consensus among those who did have the required background.

Knowing of Edelson's preference for consensus, Charlie had set up the Astrophysics Council in early 1985 partly to obtain expert opinion, but also to have access to prominent scientists, especially Nobel Laureates, who might impress Edelson to heed their advice. Among the Laureates whom Charlie could list as members of the Council, and whom he on occasion asked to write letters to Edelson or to the NASA Administrator – initially James Beggs and later James Fletcher – were physicists Robert Hofstadter of Stanford, Carlo Rubbia from Harvard, and Leon Lederman of the Fermi National Accelerator Laboratory.

George Field was first to speak that morning. His task was to convey to Edelson the importance of the Great Observatories concept and the accomplishments we could expect from AXAF and SIRTF. He urged Edelson to pursue a FY 87 start for AXAF and a phase B study for SIRTF to commence at the same time. Edelson was clearly interested in the scientific aspects of the missions and asked many questions showing a personal curiosity about cosmology and life elsewhere in the Universe.[23]

Charlie's deputy, George Newton, made the next presentation. He described the role that the Space Station and its co-orbiting platforms could play in the program. Space Station astronauts would readily reach these platforms, service the telescopes we had mounted on them, and return to their Space Station habitat.

Throughout the Astrophysics Council's discussions that year, the Space Station was a given national priority. When the President of the United States instructs NASA to pursue construction of a new facility, the agency immediately shifts its priorities. Everything NASA does must then be adjusted to the President's new directives. Other projects must align their strategies to fit in. Otherwise they founder.

Most projects find ways to fit in however tortuous the new match becomes: At a meeting of the American Astronomical Society in January 1985, Charlie had spoken at length about the major missions the Astrophysics Division was hoping to launch by the mid-1990s. "What will it take to make this happen?" he asked rhetorically, and then answered,[24]

> First we need the Space Shuttle and the Space Station, which together will allow us to launch large, complex, scientific payloads, and to keep them operating for extended periods, 15–20 years in the case of HST, AXAF and SIRTF. When the Space Station is operational, in about 1992, we will have a servicing crew available in space to fix things on-orbit which before could not be fixed; to replace our instruments, as new technologies give us better resolution or higher sensitivity detectors; to replace batteries and tape recorders in our spacecraft if they fail; to deploy solar arrays or antennae if they stick; and, possibly, even to fix the science instruments

or to clean contaminated mirror surfaces, within the pressurized work areas of the Space Station. To use these new capabilities effectively we only need to make our systems simple to service, and to minimize our spares and support equipment by moving as far and as fast as possible towards common systems and common spares.

In the era of 1985, and with President Reagan's personal dedication to the Space Station, Charlie's vision seemed right on target. All the more remarkably, in our meeting with Burt Edelson, he advised caution. He listed the advantages but also disadvantages of a dependence on the Space Station, and asked that we reconsider all these factors and write him a formal letter explaining how AXAF and SIRTF might go forward in the then-current thrust to have NASA science closely tied to the Space Station.

I had not met Edelson before the March 26 meeting. He appeared affable, friendly, and urged us to explore all sides of different questions. It all sounded promising. We came away hoping that Burt would accept our recommendation for the family of four Great Observatories.

In a letter dated ten days after the meeting he had attended, I responded to Edelson's request for an explanation of our concepts in writing. The Council recommended "a promptly initiated cooperative venture with the Space Station as the most promising step for the Astrophysics Division to undertake."[25,26]

But as the year 1985 wore on, no tangible sign of support from Edelson was evident, despite several meetings and exchanges of letters with him. Other unsettling developments were also emerging. At a September 20 meeting of the Council with William Raney, Director of the Division of Utilization and Performance Requirements in the Office of Space Station, Raney said he wanted us to know that the cost of launches to service observatories would be high. Thus far OSSA never had to take these into account, but in future someone would have to pay for orbital operations and repairs, and these were likely to be charged to OSSA. We expressed surprise at this, as one of the advertised benefits of the Space Station for science projects had always been the potential for repairs, replacement and refurbishing.[27]

George Field later wrote me, "... the general tenor of the discussion of science on the Space Station was disquieting ... the notion that OSSA would have to pay for servicing and repairs. What is Space Station for, if not this? The evidence suggests that as on the Shuttle, science comes after engineering – and this was supposed to be different, as I recall according to written documents promulgated by Beggs and others ... The Council should pursue this until satisfactory answers are obtained from Edelson's office or higher."[28]

An Organization on Hold

But NASA's leadership was suddenly thrown into turmoil by the abrupt resignation of NASA Administrator Jim Beggs, on December 4, 1985, when he was accused of criminal intent to defraud the Government, a charge from which he was fully exonerated but only after years under indictment. In place of Beggs, William R. Graham, who had recently been appointed NASA Deputy Administrator, was named Acting Administrator the same day.[29]

Just before Christmas 1985, with no progress in sight, I turned to Herbert Friedman for advice. Friedman, one of the pioneers of X-ray astronomy, now was Chair of the Assembly of Mathematical and Physical Sciences at the National Academy of Sciences. I had known him for 20 years, ever since he had encouraged me and helped me get started on infrared astronomy using cryogenically cooled telescopes in space. Over the years, I had also come to respect his ever-thoughtful advice.

As it turned out, Herb Friedman, Burt Edelson, and their wives also were personal friends. Herb arranged for a luncheon meeting on January 3, 1986, at the Cosmos Club, a venerable Washington institution. Herb, Burt, Charlie Pellerin, Charlie's deputy George Newton, Harvey Tananbaum from Harvard sitting in for George Field, and I all attended.[30,31,32] Our aim was to come to some form of agreement with Burt on what the relation between the Space Station and the Great Observatories would be. The issue had been dragging on unresolved ever since our first meeting with Burt the previous March.

Over lunch, Burt declared himself willing to lend his support to the Great Observatories. AXAF and SIRTF would utilize the Space Station's co-orbiting platforms. But Edelson first wanted a white paper to show how this would work. He gave us a list of steps we would need to complete: He wanted us to secure the endorsement of the Astrophysics Council; Edelson's own Space and Earth Science Advisory Committee (SESAC); the AXAF and SIRTF Science Working Groups and their communities; the Space Science Board (SSB); its Committee on Space Astronomy and Astrophysics (CSAA); the Task Force on Science Uses of Space Station (TFSUSS), chaired by Peter Banks; as well as the Marshall, Goddard, and Johnson Space Flight Centers.

I said I thought we could probably get all of these endorsements, particularly because Friedman felt he could persuade the SSB which formally reported to him. He thought he could get the SSB to provide a two-page letter, whenever we let him know that we needed it. I undertook to make myself responsible for soliciting endorsements from the Council, SESAC, and CSAA. Harvey volunteered to handle the X-ray astronomy community, and George Newton was going to get in touch with Bill Hoffmann at the University of Arizona about the approval of the infrared astronomy community. We thought that Giovanni Fazio at the Harvard–Smithsonian Center for Astrophysics or Hugh Hudson at the University of California at San Diego could ask the Peter Banks Committee for their endorsement. This still left open how the NASA Centers were to be approached.[33,34]

At 4:15 pm, that same afternoon, of Friday, January 3, 1986, Charlie Pellerin, Burt Edelson, Harvey Tananbaum, and I also met with Acting NASA Administrator Bill Graham, to whom we presented the discussions that had taken place over lunch. During this session, Edelson took the lead. As I wrote George Field the next week, on January 7, I thought Graham looked terribly tired.[35]

A week thereafter, in great contrast, George Field, Charlie Pellerin, and I met with Bill Graham to make a more formal presentation. This time Burt Edelson was unable to join us and Graham was full of energy and encouraging. He pressed us to assure wide support for the Great Observatories at all levels of society and added that he thought President Reagan might enjoy a briefing on the program.[36,37] Two days later, Charlie wrote Graham, "If you are interested, we would be delighted to prepare [a] briefing package[38]

The *Challenger* Disaster and Its Wake

Within less than two weeks after our meeting with Graham all this fell moot. The *Challenger* disaster of January 28, 1986 changed everything. The break-up, 73 seconds after launch, of the Space Shuttle *Challenger* and the death of its seven crew members, including Christa McAuliffe, the first member of a *Teacher in Space Project* and mother of two children, aged six and nine, shocked the nation!

Two months after the disaster, at the April 17 meeting of the Council, Charlie discussed the impact on OSSA missions. We could expect a delay of two to three years for Shuttle payload space, particularly because military and commercial payloads would have priority.[39]

By the June Council meeting that year, there were questions about Shuttle launches from Vandenberg Air Force Base in California. A Vandenberg launch was necessary to take certain payloads into polar orbits. The Cosmic Background Explorer (COBE), a smaller but vital mission, required such a launch. The Air Force was looking into multistage rockets capable of launching payloads comparable in size and weight to those designated for a Shuttle launch. Other problems NASA was now facing were monthly costs in the $6 to 7M range just to keep the Hubble Space Telescope ready for launch. The payload could not simply be stored; the gyros needed to be exercised to keep them in good condition.[40]

Charlie began pointing out the great value of the considerably smaller, less expensive, Explorer Class missions. An Explorer Announcement of Opportunity had been sent out, and there was discussion of doubling the Explorer budget now, during the hiatus in Shuttle launches. However, Jeffrey Rosendhal on Burt Edelson's staff warned that AXAF could feel the competition for funds if the Explorer program was to receive a significant infusion of money.[41]

By the August 1986 Council meeting, Charlie talked about options open to the Astrophysics Program in view of budgetary developments and a likely decline in the number of Shuttle flights per year. Delta and Titan unmanned-rocket launches might have a major role to play; but few of the major payloads could be launched by Deltas. If the launch cost then had to be paid with Astrophysics Division funds, they would require approximately $60M per Delta and $250M per Titan launch.[42]

COBE, by then, was at an advanced stage of development. With modifications of a Delta support system, the mission could be launched into its polar orbit from Vandenberg Air Force Base. Payment for the launch vehicle, however, remained to be worked out. The Extreme Ultraviolet Explorer (EUVE) and the X-ray Timing Explorer (XTE) would be next in line.[43]

The final end to any association with the Space Station surfaced on April 4, 1987, when *The New York Times* reported, "President Reagan, who has enthusiastically supported a proposed space station, today scaled back the plans...to cut the rapidly rising estimated costs of the space station and thus to reduce political opposition in Congress."[44]

Harvey Tananbaum's ad hoc Planning Group

Over the summer of 1986, Harvey Tananbaum had begun setting up an ad hoc Planning Group for the Great Observatories, a broadly representative resource group that would write to members of Congress, potentially give talks on the Great Observatories, particularly AXAF and SIRTF, and generally publicize the effort both among colleagues and other supporters. A meeting held at the Center for Astrophysics in Cambridge, Massachusetts, on October 3 that fall discussed progress on AXAF and SIRTF, set objectives for the planning group, analyzed the activities of different Congressional committees, outlined plans to involve fellow scientists, and considered other ways to go forward.[45,46]

As an aid to those wishing to make presentations, Harvey, helped by the CfA staff, made available an immaculate Great Observatories 35-mm-slide set, comprising 175 slides dealing not just with the space missions but with astronomy and astrophysics as a whole, and how the Great Observatories fit into this grander scheme. These slides were useful to any member of the ad hoc Planning Group seeking to give a polished presentation. To accompany this slide set a book, 'New Windows on the Universe, The NASA Great Observatories,' published in 1987, devoted nearly a page of explanatory background text to each slide, whether it dealt with one of the Great Observatories and how it functioned or, for example, the significance of γ-ray bursts, eta Carinae, sunspots, dark matter, or radio or X-ray jets.[47]

The NASA Advisory Structure

The complexity of NASA's formal and informal advisory structure, Figure 11.2, now was becoming an increasing problem.

Two particularly powerful committees advised Burt Edelson. One was the National Academy of Sciences' Space Science Board (SSB) chaired by Thomas M. Donahue, past chairman of the University of Michigan's Atmospheric and Oceanic Sciences Department. The other was Edelson's own Space and Earth Science Advisory Committee, SESAC, headed by Louis (Lou) J. Lanzerotti, a leading

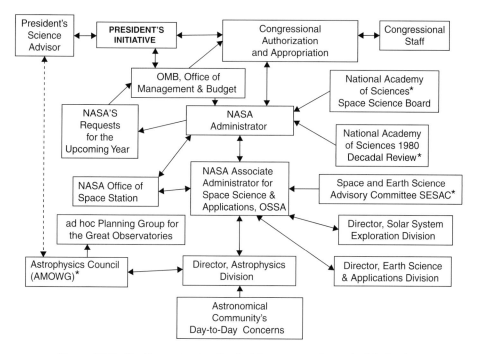

Figure 11.2. The forces shaping the decision to move forward with the Great Observatories: NASA's internal organizational structure, the pathways through which it communicated with the White House and the Congress, and the levels at which different advisory bodies exerted influence in the mid-1980s. Asterisks denote committees some of whose members may simultaneously have served also on other advisory committees shown. The dashed line from the Astrophysics Council to the President's Science Advisor was an informal link that provided the astrophysics community an opportunity to occasionally influence decisions at that level. A colloquium series the Astrophysics Council established to inform members of the Congressional staff about the aims of the Great Observatories provided a similar informal exchange between these two groups, though the link is not explicitly shown here.

Bell Laboratories specialist on geomagnetism, oceanography, space plasmas, and engineering problems raised by the impact of atmospheric and space processes on space and terrestrial technologies. Lou thus brought great expertise on missions being proposed by both the Solar System Exploration Division and the Earth Science and Applications Division within OSSA.

In many ways both Donahue and Lanzerotti reflected successful careers in areas that had gained special prominence in the 1960s and 1970s when great strides in planetary and geophysical research had taken place. Both were probably willing to agree to the increasing promise of space astrophysics in the 1980s; but astrophysical observatories in space still had to prove their full worth, and in the meantime, almost everyone agreed, the Astrophysics Division and its Hubble Space Telescope were demanding a large share of OSSA's budget.

The Space and Earth Science Advisory Committee

As early as March 6, 1985, shortly after he had decided to make the Astrophysics Council a permanent body to which he could turn for support, Pellerin circulated a list of potential members of the Council, who he hoped might agree to also serve in other roles. Included among these was a SESAC liaison to be designated by Lou Lanzerotti, SESAC chairman.[48] In due course, I received a letter from Edelson appointing me to SESAC.[49] But my participation on SESAC initially led to few gains. A forceful personality, Lou, it seemed to me, had definite priorities on where OSSA should be heading, and steered SESAC largely to pursue those directions. The level of frustration became palpable.

On November 27, that year, 1985, I wrote a note to my five fellow astrophysicists on SESAC:

> At last month's SESAC meeting in Washington, Claude Canizares and I were discussing the best ways in which the six of us, working as a group, might be able to exert our influence on behalf of the astronomical community we represent. At stake are progress that might be made toward launching AXAF, SIRTF, and possibly also a number of smaller astronomical missions. Six people on a committee of thirty is very little when one considers … (that) the Astrophysics Division is slated to receive around 40% of OSSA's budget. I suggest that we all make efforts to attend the SESAC meetings. Since this is not always possible, it would be helpful if those who cannot attend sent their views on particular issues to Lou Lanzerotti, with copies to the other astrophysicists, so we could speak up for these views at the meetings. Those of us who do attend might meet the night before the meetings either over a drink or for supper to see whether we share the same views on matters to be discussed at the meeting, so we might work out ways to best present those views.[50]

In a note to me dated December 12, Claude responded favorably to this suggestion about forming what he called "an astrophysics 'caucus' within SESAC."[51]

The Space Science Board

Blair Savage, an energetic member of the Astrophysics Council, was also chair of the Committee on Space Astronomy and Astrophysics (CSAA), which reported to the Space Science Board (SSB), headed by Tom Donahue. Surprisingly, the SSB seemed unsupportive of the Great Observatories. Our unhappiness grew with an announcement at the April 17, 1986 meeting of the Astrophysics Council that Tom Donahue's testimony to Congress, the previous week on April 10, 1986, had entirely omitted any mention of AXAF, thus giving the appearance of favoring geosciences and solar systems missions over astrophysics.[52]

I called Donahue to ask about this. He was noticeably taken aback. Fumbling through his papers while on the phone, he found that indeed he had omitted mentioning AXAF in his Congressional testimony. He said he'd try to correct this oversight if it was still possible. I also mentioned another concern about high praise he had given in his testimony to the Geopotential Research Mission – a mission that neither Edelson, nor his division directors, nor SESAC had considered recommending in the coming year's *new start* list.[e]

Tom's reply was disquieting. He thought that any mission should have a chance of obtaining a new start in any given year, no matter how long other missions had already been in the waiting queue. This statement flew in the face of the previous year's SESAC decision, and I found it difficult to understand how the chair of the SSB could act unilaterally in this fashion without consulting other advisory bodies. It also seemed fairly clear that Tom's testimony before Congress had not been shared ahead of time with other members of the SSB. Otherwise, one of his fellow board members would certainly have corrected Tom's oversight on AXAF and probably persuaded him to support some other mission rather than the Geopotential Research Mission, which was not about to go anywhere without NASA's explicit support.

Eventually an exchange of letters cleared the air in a reasonably productive way. It may not have had any profound consequences, but at least it was now clear that it might be better to coordinate our advice rather than express individual preferences without mutual consultation or regard for how specific advice might affect other scientific disciplines for which NASA also had responsibility.

In the meantime, the Great Observatories were beginning to obtain support from SESAC, which recommended to Edelson that he make a significant commitment to the AXAF mission, so that technically critical long-lead items, such as its High Resolution Mirror Assembly, might be constructed without further delay.

The President's Science Advisor

Bill Graham became the President's Science Advisor in late 1986. During his brief tenure as NASA Acting Administrator, he had exhibited a keen interested in the Great Observatories. On December 23, 1986, usually a relaxed time in Washington just before Christmas, Harvey Tananbaum and I visited him in the Old Executive Building at the White House. We wanted to see how best to promote further interest in AXAF. But Graham said he would also enjoy a discussion of the current problems in astrophysics and cosmology. This was an opportunity to have George Field participate, and on February 27, 1987, George, Harvey, and I again met with Graham and this time also Richard G. Johnson on his staff. George gave a great presentation and we all had an enjoyable time. As I wrote George a few

---

[e] A *new start* conveys NASA's authorization for advancing a mission from the design phase to *Phase C/D* – construction and eventual launch into space.

days later, "That we got asked back was an excellent boost. That Dick Johnson cared enough to ask us to lunch and gave us good advice may be even more significant."[53,54,55]

International Cooperation

With the difficulties the recovery from the *Challenger* disaster posed, international cooperation to advance the astrophysics program was also coming up in discussions. At the August 1986 meeting of the Astrophysics Council, I had been asked to write Dr. Roger-Maurice Bonnet, Scientific Director at the European Space Agency (ESA), to see whether he might be willing to attend an upcoming Council meeting. Bonnet agreed and participated in the February 1987 meeting of the Council, along with Burt Edelson, who was soon to retire. Bonnet emphasized the strong impact that cancelations and uncertainties at NASA were having on ESA programs. It was a somber but constructive exchange.[56,57,58,59,60]

At a December 12, 1986 meeting, Council members had raised the question of finding a launch for the German-led ROSAT mission. Unable to launch AXAF soon, the U.S. X-ray community was eager to participate in this German-led venture. By mid-1987, Charlie informed the Council that negotiations were underway to obtain either the earliest Delta or earliest Atlas/Centaur launch for ROSAT. ROSAT was eventually placed into orbit on a Delta-II Heavy rocket on June 1, 1990, and successfully operated until February 12, 1999. At this same mid-1987 meeting, Charlie also revisited the possibility of combining SIRTF with ESA's proposed Infrared Space Observatory (ISO).[61] But this did not lead anywhere. By then ESA was already far advanced in its plans for ISO, and the U.S. infrared astronomy community preferred to work on its own mission design for SIRTF.

A Changing Atmosphere at NASA

By April 1987 the atmosphere at NASA was also changing. In May 1986, James C. Fletcher, a former NASA Administrator, had been brought back to permanently replace Jim Beggs. And on April 6, 1987, Lennard Fisk, a theorist and founding member of the Institute for the Study of Earth, Oceans and Space at the University of New Hampshire, succeeded Burt Edelson as Associate Administrator of OSSA. His style could not have differed more from that of Edelson. Where Edelson had attempted to reach decisions through consensus that was difficult to find within NASA or the scientific community served by NASA, Fisk was analytical and decisive. He willingly considered advice, but also was guided by a structured long-term vision for the tasks OSSA should be pursuing.

The good relations Pellerin had long enjoyed with both Fletcher and Fisk also helped to advance the prospects of the Great Observatories. When the Astrophysics Council first became a permanent fixture in the Astrophysics Division, and its membership needed to be broadened, one of the first letters of invitation

Charlie had sent out had gone to Jim Fletcher, whose advice as former NASA Administrator would have been invaluable.[62] Fletcher never did accept, but the invitation was symptomatic of the relationship between the two that had sprung up during Fletcher's earlier term as NASA Administrator. Fisk also was a long-term friend. In the 1970s, when Len Fisk and Charlie had both been conducting energetic particle studies at the Goddard Space Flight Center, Fisk had often been a helpful mentor to the younger Pellerin then working toward a PhD while a full-time researcher at NASA.[63]

Len Fisk set up a coherent long-term science and applications program that included the Great Observatories, but also missions the other OSSA directorates were promoting. And the President's Office of Management and Budget (OMB), as well as the Congress responded with admirable support. The changed atmosphere at NASA led to a more favorable view of the Great Observatories and resulted, over the summer of 1987, in NASA's advocating AXAF as its prime science mission for a new start in FY89, to be included in the President's budget submission in early 1988.[64]

Liaison with the Solar System Exploration Division

Meanwhile, Charlie was seeking closer ties with, and support from, the Solar System Exploration Division within OSSA, headed by Geoffrey Briggs. He asked me to work with Jim Head of Brown University, my counterpart in the planetary community working with Briggs. We were to come up with a recommendation on common interests for both SIRTF and the planned Stratospheric Observatory for Infrared Astronomy (SOFIA), designed to carry aloft a 2.5-meter telescope aboard a Boeing 747 aircraft.

On July 24, 1987, Jim and I co-chaired a small meeting at Brown University in Providence, RI, that included Bob Brown and Warren Moos, both members of the Astrophysics Council; and David Morrison, Bradford Smith, and Jurgen Rahe, representing planetary sciences and Geoff Briggs's Management Council. By the end of the day, we formulated a letter to Charlie and Geoff, in which we cited a number of areas of potential mutual interest to astrophysics and planetary science and recommended the establishment of a small group that could regularly meet to seek closer ties.[65]

Stepping Down from the Council

In early summer 1987, I was offered the Directorship of the Smithsonian Institution's National Air and Space Museum in Washington, DC. With considerable regret, I told Charlie I felt I needed to leave the Council to take on this new job which would require my full attention.[66]

Half a year after my starting work at the museum, during the lunch hour on February 1, 1988, the phone rang. It was the President's Science Advisor, Bill

Graham. "I thought you would like to know," he began, "that the President's FY89 budget went public today, and AXAF is in it slated for a new start." That was fantastic! And Bill obviously was as delighted as I was. I later learned that the new start for AXAF had been in considerable doubt at the last minute.[67] Perhaps Bill Graham's personal enthusiasm for the mission had carried the day.

At any rate, with AXAF now officially started, and with interest for SIRTF established in both the Astrophysics and Solar System Exploration Divisions at NASA, it looked as though the Great Observatories concept was being accepted throughout the relevant bureaucracies. There would be years of hard work ahead, but at least the past three years had established acceptance of a new way of planning space missions.[f]

Effective Leadership in Science

A number of conclusions may be drawn from this particular phase of the Great Observatories' history:

- First, Charlie Pellerin's single-mindedness, supported by a community which accepted that a measure of discipline and a willingness to compromise was necessary to advance the common good of the field, made for a model worth pursuing.
- Second, as Lennard Fisk demonstrated, government officials need to be able to formulate science priorities on their own, based on rational scientific advice. But, as we shall see in Chapter 15, Burt Edelson's quest for unanimity among the many factions shown in Figure 11.2 was far more important than most us realized at the time.
- Third, the many existing science advisory bodies should work together to offer dispassionate, coordinated, coherent, and consistent long-term advice. In advancing the Great Observatories, this requirement was not fulfilled until sometime in 1987.
- Fourth, science advisors and administrators should recognize that past performance is not an infallible forecast of the future. Areas that in previous decades were at the forefront of a discipline often need to give way to others promising to yield substantial advances, as astrophysics promised to do in the mid-1980s. It takes insight, an open mind, and a great deal of scrutiny to recognize new, promising areas that should be pursued and replace others

---

[f] This did not mean that turmoil and divisions would suddenly disappear. The extent to which these can persist throughout the detailed design and construction of such complex missions has been depicted with insight by Robert W. Smith in his book on the Hubble Space Telescope.[68]

that were in recent vogue. This may be the most difficult task an administrator or a science advisory committee faces.

- Fifth, Presidential directives intended to meet other national priorities should not lightly promise that they will also lead to scientific advances. The Space Station may have served the purposes of forging improved international cooperation on joint projects in space, which certainly constituted an important consideration. It may also have served to advance space engineering efforts. But care must be taken not to promise scientific worth, when the primary motivation for a project lies elsewhere. Advocating that a project will advance a range of sciences, when it will not, merely reduces investments in science by redirecting funds elsewhere.

By the time the Space Station was completed and its final components were borne aloft in 2011, it is difficult to believe that even President Reagan would have been happy. The Station's final cost has been estimated as approximately $150B, of which the U.S. contributed about $100B, far above the original cost estimate of $8B, not including ancillary launch and Space Station program costs.[69,70,71]

That the assurances of the Space Station's promise for science had been tenuous was shown by withdrawal of the Station's support for astronomical missions as soon as the Station's cost began to escalate. The fleet of Shuttles that had provided U.S. astronauts access to the Station was also retired once the Station was completed. Fortunately, by then a range of astronomical programs had shown that missions operated more stably in the absence of astronauts, and were less costly due to lower safety requirements. Few scientists were sorry to see the Space Shuttles go. For most purposes, powerful unmanned rockets were serving the community equally well. Previously discontinued to make way for Shuttle launches, the diverse family of rockets was returned to service only after the *Challenger* disaster had shown how high the cost of Shuttle launches could be, particularly in the loss of human lives.

* * *

Lastly, it is worth asking how well the Astrophysics Council performed.

We certainly misjudged situations, made mistakes, and suffered under misconceptions. We felt constrained by President Reagan's political directives to work with the Space Station. Looking back a quarter century later, I find it amazing how seriously we tried to tie the fates of the Great Observatories to the Space Station and the co-orbiting platforms that were to accompany it so that astronauts could refurbish our planned observatories and keep them permanently active and productive in space.

None of that ever transpired! What saved us from ourselves, in the end, were two events and one miscalculation:

First, when budgetary realities sank in, the co-orbiting platforms were quickly abandoned by the Space Station project. Second, with the *Challenger* disaster and the many-months-long hiatus of launches that followed, it became clear that, where possible, it was better to launch science missions on unmanned rockets, rather than on the Shuttle. Although HST and AXAF – later renamed "Chandra" – were still launched on the Shuttle, and HST was repeatedly serviced by Shuttle astronauts, none of the Great Observatories required Space Station operations. In the end, all this worked out well for the scientific community, but it goes to show how political directives can lead to engineering solutions that in hindsight are bizarrely complex – when far simpler technology will actually do.

Finally, our miscalculation was that the Great Observatories should be permanent, or at least long-lived. It never occurred to us that the cost of mission operations, and support for the large scientific community required to handle the high data rates produced by these extremely powerful missions, would be unaffordable if the life expectancy of the observatories was too long: The NASA Senior Reviews, which later were conducted every other year, have shown that we need to phase out some observatories to pay for others that either complement them or provide new, superior capabilities.

Termination of an observatory, once its primary mission has been completed, however, may not be the most cost-effective available option. In Chapter 12 which, among other matters, touches on the need for repeating observations that may have been cast into doubt, and Chapter 15, which confronts issues of cost, I examine some alternatives to irreversible termination of missions – operational modes that may serve the community better.

The question of when a given observatory has peaked in its scientific yield and should be phased out is complex and needs careful scrutiny. Some observatories are worth continuing for a long time, whereas others appear ready sooner for replacement by missions capable of providing information totally inaccessible to date. These decisions are always wrenching. But, back in 1985, I think we failed – even as we loudly advocated the need for them – to see how immensely powerful the Great Observatories would turn out to be and how quickly they would fulfill their initial promise of yielding novel insights on the workings of the Cosmos.

Few of us also took into account that there will always be a limit on how much society can afford to expend on science. Other societal concerns tend to trump launching observatories into space. One way or another, society keeps this in mind as it funds astronomical efforts.

Afterthought

At the start of this chapter I mentioned that I had been unaware, in 1984–5, that Frank Martin and Charlie Pellerin had found my article on 'The Number of Class A Phenomena Characterizing the Universe' and the book 'Cosmic Discovery' useful in arriving at a strategy for achieving the astronomical community's goals with a family of four Great Observatories.[72,73] I became aware of this only in 2012, when I read their respective interviews with Renee Rottner.

I believe I had met Frank only once, before 1985, a brief encounter sometime in 1979 or 1980, during a coffee break at some meeting or other in Washington. I recall his strolling up, introducing himself, and remarking that the article I had written must have been intended for just one or two people, most probably just him, because so much of what the article proposed would require telescopes out in space – the domain for which he, the director of NASA's Astrophysics Division, was responsible. At the time, Frank's statement surprised me. Surely, I thought, there would be more than just one or two people interested in the article, which I had enjoyed writing because it also provided a new way of assessing how many more cosmic phenomena might be awaiting discovery.

An article is usually thought to have been influential if it has been cited several hundred times. A few phenomenally influential articles in astrophysics may garner more than a thousand citations. In 2012, when I came across Frank's 2009 interview with Renee Rottner, in which he recalled his assessment that the article had probably been written for just one or two people, I thought I'd check how many people might have found the article sufficiently useful to cite in the 37 years since its publication in 1975. Because it appeared to have influenced the planning for the Great Observatories I thought the article would probably have been quite widely cited. But Frank had been right! There were just five citations, three among which had one and the same lead author. How the article had come to Frank's attention, and why he and Charlie Pellerin had decided it could be useful to them in planning the Great Observatories, and thus the way that the major astronomical efforts in space would be conducted for the 25 years to follow, still puzzles me. How had the article somehow happened to find its most receptive readers, when only three other astronomers world-wide appeared to have found it worth citing? Who was the now-forgotten person at NASA who had "walked down the hall and handed" the paper to Frank, thinking he might find it interesting?

Notes

Unless Otherwise Designated, All Documents Cited Below can be Found in the Cornell University Kroch Library Archives, Division of Rare and Manuscript Collections: (14/7/2402 Martin Harwit Papers, Boxes 9, 10 and 11)

1. *The Space Telescope: A Study of NASA, Science, Technology, and Politics*, Robert W. Smith, with contributions by Paul A. Hanle, Robert H. Kargon & Joseph N. Tatarewicz. Cambridge University Press, 1993.
2. http://books.google.com/books/about/Outlook for space.html?id=yxUgAAAAIAAJ
3. *Making the invisible visible: A history of the Spitzer Infrared Telescope Facility (1971–2003)*, Renee M. Rottner & Christine M. Beckman, Monographs in aerospace history, NASA-SP 4547, 2012, contains an interview of Franklin D. Martin conducted by the historian Renee Rottner in 2009, from which I have quoted here. I am indebted to Frank Martin for permission to quote him.

4. The Number of Class A Phenomena Characterizing the Universe, Martin Harwit, *Quarterly Journal of the Royal Astronomical Society*, 16, 378–409, 1975.
5. *Cosmic Discovery, The Search, Scope and Heritage of Astronomy*, Martin Harwit. New York: Basic Books, 1981.
6. Address Before A Joint Session of the Congress Reporting on the Sate of the Union, President Ronald W. Reagan, http://reagan2020.us/speeches/state_of_the_union_1984.asp
7. For a detailed description of the turmoil surrounding the community's belated backing of the HST see Robert W. Smith's book Ibid., *The Space Telescope*, pp.138–46.
8. Letter from George Field to The Honorable Jake Garn, March 8, 1984.
9. *How NASA Builds Teams*, C. J. Pellerin. New York: John Wiley & Sons, 2009, p. 196.
10. Ibid. *Making the invisible visible*, Rottner & Beckman, 2012. I thank Charlie Pellerin for permission to quote him.
11. Ibid., *Cosmic Discovery*, Harwit.
12. Letter from Pellerin to Harwit, December 21, 1984.
13. Minutes of the Astrophysics Program Coordination Group, January 3, 1985, p. 1.
14. Ibid. Minutes, p. 6.
15. Minutes of the Astrophysics Council Meeting, February 22, 1985, p. 4.
16. Ibid., Minutes pp. 2–3.
17. Letter from Richard L. Daniels to EZ/Director, Astrophysics Division, May 9, 1985.
18. Ibid., *How NASA Builds Teams*, Pellerin, p. 198.
19. *The Great Observatories for Space Astrophysics*, prepared under the auspices of the NASA Astrophysics Division, Dr. Charles H. Pellerin Jr., Director, by Martin Harwit, Cornell University and Valerie Neal, Essex Corporation in consultation with the Astrophysics Management Operations Working Group; graphic design and illustration by Brien O'Brien.
20. Letter from Riccardo Giacconi to The Honorable Ronald Reagan, June 12, 1985.
21. Minutes of the Meeting of the AMOWG, July 22, 1985, p. 3.
22. Speakers, Congressional Lecture Series, Space Science for the Non-Scientist. Announcements of the individual talks are filed under October 30, November 6, 14, and 21, December 2, and December 5, 1985.
23. Minutes of the Meeting of the Astrophysics Council, March 25 and 26, 1985, p. 3.
24. *Space Astrophysics in the 1990's: The Decade of Achievement*, filed under February 12, 1985 because of the designation A. F. P. 2/12/85 at the upper right. (The initials probably refer to Adrienne Pedersen of the BDM Corporation, which supported the Astrophysics Division at the time.)
25. Ibid., Minutes of the Meeting of the Astrophysics Council, March 25 and 26, 1985, p. 3.
26. Letter from Martin Harwit to Burton Edelson, April 5, 1985.
27. Minutes of the Meeting of the AMOWG (Astrophysics Council), September 20, 1985, pp. 3–4.
28. Letter from George Field to Martin Harwit, October 15, 1985.
29. Minutes of the AMOWG meeting, December 6, 1985, p. 1.
30. Postscript (P.S.) on a typed note dated December 24, 1985 and headed "For George Field"
31. Hand written notes some dated "3/1/85," some overwritten as "3/1/1986" (but all of them meaning January 3, 1986); see p. 5.
32. Letter from Martin Harwit to George Field, January 7, 1986; see the attached "Enclosure."
33. Ibid., Hand written notes some dated "3/1/85," some overwritten as "3/1/1986"; see p. 5.
34. Letter from Harwit to George Field, January 7, 1986.
35. Ibid., "Enclosure" attached to letter from Harwit to Field, dated January 7, 1986.
36. Letter from Martin Harwit to William R. Graham, January 20, 1986.
37. Letter from Martin Harwit to Burton I. Edelson, January 20, 1986.
38. Letter from Charles J. Pellerin, Jr. to A/Dr. Graham, Jan 22, 1986.

39. Minutes of the Astrophysics Management Operations Working Group April 17, 1986, p. 2.

40. Minutes of AMOWG meeting held on June 3, 1986, pp. 2–3.

41. Ibid., p. 4.

42. Minutes of the August 15, 1986 AMOWG meeting, p. 2.

43. Ibid., Minutes of the August 15, 1986 AMOWG meeting, p. 3.

44. http://www.nytimes.com/1987/04/04/us/president-scales-back-plans-for-space-station-over-costs.html

45. Ibid., Minutes of the August 15, 1986 Meeting of the AMOWG, p. 6.

46. Letter from Harvey Tananbaum to Martin Harwit, September 22, 1986.

47. *New Windows on the Universe: The NASA Great Observatories*, slide set and text prepared by C. Jones, C. Stern, and W. Forman, Harvard-Smithsonian Center for Astrophysics © 1987 SAO.

48. Astrophysics Council Membership 1985–1986 (revised 3/6/85).

49. Letter from Edelson to Harwit, dated July 26, 1985.

50. Memorandum from Martin Harwit to Claude Canizares, Warren Moos, Sabatino Sofia, Stephen Strom, and Michael Turner, dated November 27, 1985.

51. Letter from Claude Canizares to Martin Harwit, December 12, 1985.

52. Ibid., Minutes of the Astrophysics Management Operations Working Group April 17, 1986, p. 2.

53. Letter from George Field to Dr. William Graham, 31 December, 1986.

54. Minutes of the AMOWG Meeting of February 24, 1987, p. 2.

55. Handwritten letter from Martin to George, dated March 1, 1987, final paragraph.

56. Ibid., Minutes of the August 15, 1986 Meeting of the AMOWG, p. 6.

57. Letter from Martin Harwit to Roger Bonnet, August 22, 1986.

58. Letter from Martin Harwit to Roger Bonnet, November 26, 1986.

59. Letter from R. M. Bonnet to Martin Harwit, December 11, 1986.

60. Ibid., Minutes of the AMOWG Meeting of February 24, 1987, p. 4.

61. Minutes of the June 23, 1987 AMOWG meeting, p. 3.

62. Letter from Charles J. Pellerin, Jr. to James C. Fletcher, March 14, 1985.

63. Conversation of Charles J. Pellerin with Martin Harwit, at Pellerin's home, Boulder, CO, May 26, 2008.

64. Letter from James C. Fletcher to the Hon. Edward P. Boland, Chair Subcommittee on HUD-Independent Agencies, Committee on Appropriations, House of Representatives, October 14, 1987 (NASA Headquarters Archives, AXAF-1 File 5604); a copy of this letter is also in Martin Harwit's files.

65. Letter from James W. Head III and Martin Harwit to Charles Pellerin and Geoffrey Briggs, July 24, 1987.

66. Ibid., Minutes of the AMOWG Meeting of June 23, 1987, p. 4.

67. *Revealing the Universe, The Making of the Chandra X-ray Observatory*, Wallace Tucker and Karen Tucker. Cambridge, MA: Harvard University Press, 2001, pp. 102–4.

68. Ibid., *The Space Telescope*, Smith.

69. WhatItCot, http://historical.whatitcosts.com/facts-space-station.htm

70. Space Station: U.S. Life-Cycle Funding Requirements, Testimony Before the Committee on Science, House of Representatives, June 24, 1998.

71. Shuttle programme, Roger Pielke & Radford Byerly, *Nature*, 472, 38, April 7, 2011.

72. Ibid., The Number of Class A Phenomena, Harwit.

73. Ibid., *Cosmic Discovery*, Harwit.

12

Cascades and Shocks that Shape Astrophysics

In earlier chapters, I noted that the acceptance of new scientific views was usually determined by a loosely defined leadership, which Ludwik Fleck had called the community's *thought collective*. A clearer picture of how this acceptance comes about has gradually emerged through an improved understanding of the ebb and flow of influence within the scientific community. When the weight of mounting influence triggers a cascade, widespread acceptance becomes inevitable. Once set in motion, however, cascades are almost impossible to control; misinformation can spread as readily as more reliably documented data. High priority thus needs to be assigned to containing the spread of error.

Transforming Belief

The inflationary theory of cosmology envisioned by Alan Guth, and extended most persistently by Andrei Linde, proposed that at the birth of time the Universe comprised a high-density vacuum permeated by chaotic fluctuations on all scales. Almost at once, the vacuum explosively expanded at a rate that exponentially increased. During this inflationary phase, which may have lasted no more than 10^{-33} seconds, distances across the Universe separating individual fluctuations dramatically increased by factors of order $\sim 3 \times 10^{27}$ or more. This left the fluctuations isolated, able to seed the formation of clumps of radiation and matter as the inflationary phase subsided.[1,2]

This astonishing theory was widely adopted almost as soon as it was proposed. It accounted for the homogeneous and isotropic appearance of the Universe; explained why space is seemingly flat, meaning that light propagates along straight lines rather than curved paths as it traverses the Universe; and also suggested why the Cosmos appeared to be devoid of *magnetic monopoles*.

By the end of 1981, 15 refereed journal articles had already cited Guth's January 1981 paper; another 75 refereed articles referred to it in 1982, rapidly rising to 150 published in 1985!

256

When we look back at this quick adoption of the inflationary theory, one of its most surprising criteria for adoption was its accounting for the absence of magnetic monopoles. Yet only a handful of astrophysicists could even have known that a monopole problem existed, or that monopoles were an integral part of fundamental particle theory at the extremely high temperatures that Guth proposed to have existed at the very earliest primordial epochs.

Magnetic monopoles have never been verifiably observed in Nature, have never been detected at the high energies generated in the most powerful particle accelerators, nor has any trace of them appeared among the far-higher-energy cosmic rays lighting up Earth's upper atmosphere from time to time.[a] So, even members of the high-energy particle-physics community might have been surprised that absence of magnetic monopoles had played a prominent role in leading Guth to invent the theory.

Recollections on the Acceptance of Inflation

Fortunately, considerable insight on the astrophysics community's acceptance of the inflationary theory can be gained from a published collection of interviews, amounting to 400 pages of transcribed conversations between 25 leading astrophysicists of the late 1980s and the MIT astrophysicist and essayist, Alan Lightman.[4,b]

In approaching the topic of inflation, most of Lightman's interviews began with a question regarding the flatness of the Universe, a problem raised by Robert Dicke of Princeton, which Guth's inflationary theory was designed to explain. That the Universe appeared to be quite *flat* – that is, that light, on average, travels along straight lines, rather than along curves, in traversing the Universe – was a problem which had been around a long time; but most of the astrophysicists interviewed indicated that they had not considered it particularly pressing. On the contrary, Sandra Faber of the University of California at Santa Cruz, Margaret Geller of Harvard, and Jeremiah Ostriker of Princeton, all pointed out, in different ways, that the Universe appeared not to be as flat as the inflationary theory demanded. The best estimates in the 1980s were that the cosmic mass – energy density was an order of magnitude too low to agree with the predicted flatness. As Sandra Faber

---

[a] On the night of February 14, 1982, a year and a half after Guth had submitted his paper to the *Physical Review*, an experiment set up by the physicist Blas Cabrera at Stanford University registered an event that appeared to indicate detection of a magnetic monopole. No similar event ever recurred. Cabrera cautiously reported "A single candidate event consistent with one Dirac unit of magnetic charge, has been detected during five runs totaling 151 days. These data set an upper limit of 6.1×10^{-10} cm^{-2} sec^{-1} sr^{-1} for magnetically charged particles moving through the earth's surface."[3]

[b] Lightman's interviewees also included Guth and Linde, who had proposed the original theory; but the questions they were asked naturally differed somewhat from those posed to the 25 others.

understandably commented, this bothered her "intensely." The discrepancy was no minor matter.[5,c]

Others also were of two minds. Recalling his first impressions on inflation, Dennis Sciama of the International School of Advanced Studies in Trieste recalled how initially,"[It] all fell into place and I saw how potentially important it was." But then,

> Well, in the end I think it's turned out a bit disappointing. It was a marvelous idea. It had various difficulties. It's now in what I call a Baroque state. There are so many variations, and there is no formalism, there is no reasonable grand unified theory and cosmological formalism that gives a scheme that really does all that is required of it. Half a dozen people in the field have produced their own variations. Perhaps this is the nature of scientific research. I'm not saying therefore the idea is wrong, but it's a mess at the moment....[6]

Roger Penrose of Oxford felt that "The union of particle physics and cosmology is a magnificent thing, and what one can learn about the early universe from both ways is fascinating. That's fine as long as the particle physicists appreciate the problems of general relativity... fundamental problems we have argued over endlessly... in trying to quantize [general relativity]". He worried that the particle physicists were ignoring these problems.[7]

Martin Rees of Cambridge University felt that "The idea of inflation clearly offered an important new insight, suggesting a possible explanation. It meant that one would not necessarily have to go right back to quantum gravity to get a solution [to the problem of why the Universe seemed close to flat]. Prior to that time, I had supposed – in an entirely ill-informed way – that it was kind of a question that only quantum gravity could answer." On the other hand, he added, that the theory, "didn't make a tremendous impact on me when I first heard about it.... Over the years since then, I've come to think that there's almost certainly something in the general idea of a de Sitter phase, but not necessarily anything closer to Guth's first formalism than to Starobinsky's first formalism." Here Rees was referring to two earlier papers Aleksei Starobinsky in the Soviet Union had published, some features of which had anticipated portions of Guth's work.[8,9,10]

Don Page of the University of Alberta recalled that he was skeptical of the inflationary universe when it was first proposed. He commented that, to have a direction of time defined, "special low-entropy conditions were required even to

---

[c] Not until the mid-1990s did a cosmic mass – energy density roughly ten times higher than previously estimated become apparent. Attributed to a new *dark energy*, this mass – energy density, alongside an improved value of the Hubble constant, guaranteed that a general relativistic model of the Universe had to be much closer to being perfectly flat.

have inflation ... I thought that homogeneity and isotropy might come as a consequence of whatever it was that solved this bigger problem of the second law of thermodynamics."[11]

Robert Wagoner of Stanford University also was skeptical. "I'm still not a believer in inflation. ... I don't think its necessary. ... [U]ntil we have a theory of quantum gravity, we don't know that there's a horizon problem. I'm perfectly willing to live with inflation. I'm just not an apostle of inflation."[12]

Joseph Silk of Oxford University thought that "[Y]ou need to specify particular conditions in order for inflation to work. If the universe is too inhomogeneous initially, then inflation may not work. That's one of the worries, and there are other problems, too. Now it may be that we don't yet have a satisfactory theory of inflation. I think one is still waiting. If you really need to push inflation back to the Planck epoch, as some people believe, then you need to know the theory of quantum gravity at the same time." However, he also acknowledged that, "The inflationary advocates make persuasive arguments. They're still having some trouble getting the right amplitude of the fluctuations, but they get the spectrum, ... the Harrison–Zel'dovich-Peebles spectrum, ... the distribution, the sizes to come out remarkably well, close to what we think we see."[13]

P. J. E. (Jim) Peebles of Princeton recalled that he initially had been skeptical about how the inflationary phase could have made a transition to a Friedmann-Lemaître phase so that all parts of the Universe moved coherently. But once he reconciled himself on this point,

> I could see how beautifully the inflation model solved what seemed ... to be the essential puzzle – how did the universe get to be so homogeneous. I ... am still not convinced that it *has* to be the way the universe started, but I certainly had to agree that it was a wonderfully elegant idea and so should be pushed harder.

Asked why inflation had so rapidly caught on, Peebles suggested,

> In the absence of any other idea, a good idea will capture the field. ... [I]t doesn't mean that the idea is right. It means that we didn't have any options. ... [W]e should be careful. ... [T]ere is a reasonable chance we've been led down the wrong path. It certainly has happened before.[14]

Several of those questioned by Lightman thought inflation "extremely clever and a big surprise," as Edwin Turner of Princeton put it.

> I would *not* have expected what did happen, which is that it became the foundation of a whole renaissance in the study of the early universe. ... [It is] a model that allows one to do lots of cute calculations. It's a theorist's gymnasium, so to speak, where one can go, and there are lots of nice problems to do and to be worked out. It wasn't obvious to me when the theory was originally described ... But I think that does tend to happen

> when some new idea comes up that has lots of problems or aspects of it that can be worked out. Naturally people do. There's nothing wrong with that. But I think it gives the *impression* that the theory is likely to be right, or that the field is more important than maybe it is.[15]

In his interview with Lightman, Hawking had reservations only about the original form of inflationary theory, which had predicted much greater variations in the microwave background radiation than observed. He commented:[16] "A better model, called the chaotic inflationary model, was put forward by Linde in 1983.... [It] does not depend on a dubious phase transition, and it can moreover give a reasonable size for the fluctuations in the temperature of the microwave background that agrees with observation."[17] Hawking and Ian G. Moss had made a number of theoretical contributions to inflationary theory as early as the beginning of 1982, so that Hawking certainly was fully familiar with the theory's development over the years.

Among the 25 senior astrophysicists Lightman interviewed, David Schramm of the University of Chicago probably was the most enthusiastic advocate for inflation. Schramm had been only 35 years old in 1980, but by then already was a leading astrophysicist at Chicago.

In the late 1970s, Schramm, Gary Steigman of Ohio State University, and various collaborators had written a series of papers in which they established upper limits on the number of neutrinos and superweak particle species the Universe could contain and still come up with the observed abundances of primordially produced helium – ^{4}He – and other light isotopes. This work involved the use of sophisticated particle theory.[18,19]

Later, Schramm and the young University of Chicago postdoc, James N. Fry, had written a paper on the monopole problem in 1980, and so Schramm was also familiar with the work of Alan Guth and Henry Tye.[20] Schramm recalled,[21]

> I was quite up on what [Guth and Tye] were doing. Then I heard about Alan's inflation idea, and suddenly everything just fit together. It immediately hit me that he had found it. There was the solution, and it all fit together. I was an immediate convert, I guess. There are still a lot of nagging problems with the fluctuations, but I remember right away, I was enthusiastic about it.

The interviews from which I have extracted the cited opinions were all conducted between September 1987 and August 1989, six to eight years after Guth's original paper had appeared. By then, several hundred papers on inflationary theory had referred to Guth's paper and to those of others working on inflation. Yet, judging from the opinions of the senior astrophysicists Alan Lightman had interviewed – a representative number among whom I hope I have quoted fairly – most were ambivalent about the theory's future, even if they thought it held promise.

Aside from Schramm and Stephen Hawking, both of whom had been active in developing aspects relevant to inflation theory, most of the astrophysicists interviewed appeared to be watching from the sidelines. To them, it may have seemed that there were reasonable alternatives to the inflationary scenario – for example, a variety of oscillating models of the Universe, in which the Cosmos was thought to have cyclically expanded and contracted throughout history. In each successive cycle the Cosmos rose phoenix-like from its own ashes as its temperature first reached great heights during collapse, erased every trace of its past, and then expanded again – repeatedly following the same sequence of expansion followed by collapse.

One of Lightman's interviewees, Robert Dicke, had long been a proponent of an oscillating universe. Although conceding that he had no detailed theory, he thought that a large number of these cycles stretching back into the distant past might have helped to establish homogeneity across great distances. A sufficiently large Universe of this kind could also have assured low enough curvature to make its space appear flat. An oscillating model of the Universe could have served the astronomical community of the early 1980s just about as well as inflationary theory. Each appeared to have associated problems but also advantages. Oscillating models originally proposed by Alexander Friedmann in 1922 had long been cited in standard textbooks dating back many decades.[22]

The Intervention of High-Energy Theorists

So, if only a few senior astrophysicists, like David Schramm and Stephen Hawking, were sufficiently expert in high-energy-particle physics to fully understand and work on the theory, why was the inflationary model so rapidly and universally accepted, whereas alternative models were effectively sidelined?

I believe the rapid acceptance came about because a wave of young high-energy-particle theorists became fascinated by the idea. They were motivated by two features of the theory. A strong impetus came from the insight of Sidney Coleman at Harvard and his model for a vacuum – empty space – that could exist in two different states. A *false vacuum*, at a higher energy could jump into the lower energy state, the *true vacuum*, with the release of energy.[23] The implication was that the *true vacuum* then would correspond to a stable medium that no longer could yield further energy.

High-energy-theory physicists also had another reason for welcoming inflation. Their *standard model* had proved itself highly successful, as far as it went; but it did not yet incorporate all the elementary forces of particle physics, the strong and weak nuclear forces and the electromagnetic force in a unified approach. In 1974, Howard Georgi and Sheldon Glashow at Harvard had proposed a symmetry scheme going by the technical name SU(5) which, at extremely high energies – far in excess of any one could ever hope to reach with available accelerators – would combine all three forces into a single force predicted by what were called grand unified theories (GUTs).[24] Energies as high as those contemplated by the GUTs could have existed nowhere except at very early times when the Universe was highly compact and extremely hot. Among competing GUTs, the Georgi–Glashow theory had a particular simplicity and thus an inherent elegance and appeal for the high-energy physics community. Like all other GUTs, however, the theory predicted a high cosmic abundance of magnetic monopoles that seemed not to exist.

Guth's inflationary phase, based on Coleman's and Georgi and Glashow's theories, expanded the Cosmos so greatly that any magnetic monopoles initially present would essentially vanish. None, or no more than a small handful, could have remained in even as vast a volume as the entire Universe our telescopes now survey. This explained why no magnetic monopoles had ever been found, even though they were an inevitable consequence of grand unified theories.

The explanation was welcome news to physicists who otherwise would have had to revise their theories drastically. Welcome also were Guth's views on the physical conditions in the early universe. They promised high-energy physicists a virtual laboratory in which to study physics at energies far higher than their accelerators could ever produce.

That inflation also explained the observed homogeneity of the Universe, which until then had been largely ignored, both by astronomers and physicists, was an additional feature that favored Guth's approach.

Inflationary theory thus became a cosmological theory invented by high-energy theorists largely for their own purposes. Most of those writing on inflation were young physicists publishing in journals devoted to high-energy physics rather than astrophysics. Later, when the arXiv preprint site became popular in the 1990s, papers on inflation were routinely deposited in the section devoted to high-energy physics theory, *arXiv hep-th*, rather than the section dealing with astrophysics, *arXiv astro-ph*.

Initially, inflation appears to have been accepted by traditional astronomers more or less passively. They questioned some of its aspects but, by and large, were quite pleased to have the help of high-energy physicists who undoubtedly would bring with them new tools to investigate the origins of the Universe and potentially other problems, including why the Cosmos consists largely of matter to the exclusion of antimatter.

Within a few years, the inflationary theory had become astrophysical lore. The only remaining question was how this rapid embrace had been triggered.

Shocks and Cascades Traversing Networks

The cascade theory of Duncan Watts mentioned at the end of Chapter 9 provides a plausible explanation.[25] When he first proposed his theory, Watts wanted to obtain insight on more mundane questions, such as,

> Why do some books, movies, and albums emerge out of obscurity, and
> with small marketing budget, to become popular hits, when many *a pri-*
> *ori* indistinguishable efforts fail?…Why does the stock market exhibit
> occasional large fluctuations that cannot be traced to the arrival of
> any correspondingly significant piece of information?…. How do large,
> grassroots social movements start in the absence of centralized con-
> trol or public communication?…These phenomena are all examples
> of…*cascades*

Watts portrays such instances of rapidly rising consensus as cascades triggered
by respected peers. Astrophysicists faced with a decision too complex to be readily
fathomed tend to take their cues from influential colleagues. If a threshold fraction
of these appear to favor a particular theory, a decision by even one additional
colleague in favor of the theory may persuade an individual to act similarly, thus
triggering a cascade involving all others who previously were just as undecided.

Processes triggering a cascade differ from those responsible for an infection.
An individual may become infected with enthusiasm for a particular theory while
conversing with any immediate colleagues. But he or she will join a cascade of
believers only when some threshold fraction of influential colleagues adopts the
theory.

Watts depicts decision-making colleagues as nodes in a random network. The
mean number of edges joining the nodes – that is, the mean number of channels
linking colleagues – and the fraction of colleagues on the verge of a decision, then
determines the probability of initiating a cascade that shocks the system into
adopting a new stance.

For astrophysics, the key concept of cascade theory is that if a new way of viewing the Universe is accepted by a significant fraction of knowledgable colleagues,
it will quickly trigger near-unanimous embrace. Acceptance of the theory grows
not because every member of the community understands it, but because a few
members whom others trust demonstrably approve of the theory. They express
their commitment to it by lecturing on the topic, or publishing papers that accept
the new approach as a reasonable, if provisory, way to move forward.

* * *

We now need to ask whether any of the earmarks of a cascade were present as
inflation caught on?

As he was developing his theory Alan Guth was a postdoctoral fellow at the
Stanford Linear Accelerator (SLAC). Interviewed by Lightman, some years later, he
offered[26]:

> When I first came up with the idea, I was at SLAC. That year, Sidney Cole-
> man [normally at Harvard, also] was at SLAC and Lenny Susskind was at
> Stanford, spending a good deal of time at SLAC. When I gave the semi-
> nar, when I first talked about inflation, Coleman and Susskind were in

the audience. Both of them got very excited about it and felt right away that it was a good idea. At least initially, as far as the spread of information, I think they were both instrumental. Both of them went around and talked about it a lot. I was just a lowly postdoc. If I had gone around talking about (the inflationary model), no one would have listened for quite some time.

Sidney Coleman would understandably have been happy with Guth's theory, which gave his proposed two-phase model of vacuum serious backing; and Alan Guth's admirable frankness seems to confirm that the main requirements for triggering one of Duncan Watt's cascades were present through the active support of Coleman and Susskind, both influential theorists whose enthusiasm provided Guth's theory the required threshold credence.

Coleman and Susskind's influence on the young theorists at SLAC was matched by Schramm's influence at the Fermi National Accelerator Laboratory, Fermilab. There, Schramm convinced Fermilab's director, Leon Lederman, that setting up an astrophysics group to work with Fermilab's high-energy physicists and with the University of Chicago astrophysicists could spark an entirely new line of research. To this end, Lederman and Schramm obtained NASA support and, in 1982, "Schramm moved to a house on the Fermilab site (an hour's drive from the Chicago campus) to start up the astrophysics activities at Fermilab." As Michael S. Turner later recalled, "[Schramm's] presence and a series of high-profile astrophysics seminar speakers began to introduce the two cultures to one another."[27]

Schramm and Lederman then hired Edward "Rocky" Kolb, who had been at Los Alamos as an Oppenheimer fellow, to join Michael Turner, whom Schramm had attracted to the University of Chicago in 1980. Together, Kolb and Turner were to lead the newly created NASA / Fermilab Astrophysics Center (NFAC) with a dedicated group of young researchers working on particle physics and cosmology. In May 1984, Kolb and Turner and this newly created group organized a spectacular conference 'Inner Space / Outer Space – The Interface between Cosmology and Particle Physics' at Fermilab attended by more than 200 scientists from all over the world – astronomers, theoretical astrophysicists, high-energy experimentalists and theorists, low-temperature physicists, relativists, and cosmic ray physicists. The array of about 90 talks marking the conference was subsequently published in a densely packed 600-page volume.[28] With this initiative, the investigation of cosmology from a viewpoint of high-energy-particle physics was in full swing.

The acceptance of inflation by particle physicists Coleman and Susskind; by Stephen Hawking, who wrote some of the early papers examining inflation; and by the young and energetic David Schramm, who was willing to invest his time and career in promoting high-energy-particle cosmology, appears to have been critical. Schramm's influence in setting up the Fermilab research center devoted to the topic, and help in organizing the 1984 conference led by Kolb and Turner, which gave its 200 participants the opportunity to exchange ideas on the subject, thus precipitated a cascade to jump start and establish this new direction of investigation.

* * *

Similar cascades may be revealed in the quick acceptance of three other physical and astrophysical theories.

The first of these was the acceptance of Einstein's 1905 theory of special relativity. During the winter semester of 1905–6 Max Planck, by then already recognized as a leading theorist of his time, presented the theory in the physics colloquium in Berlin. Through his own research, Planck established the transformation laws relating momentum to energy. As early as 1906, experimentalists also were becoming interested in the mass – energy relation, through investigations of the relationship between the velocity and energy of β-rays – energetic electrons.[29]

By 1908, Max Planck had demonstrated that relativity elucidated many previously vexing problems of thermodynamics, and – most important – that his eponymous Planck constant, h, remains invariant under relativistic transformation.[30] That same year, 1908, Hermann Minkowski announced his novel mathematical insight that a four-dimensional world view combining the previously separate concepts of *space* and *time*, into a unified *space–time*, was a fruitful new mathematical approach to theoretical physics. Minkowski's semi-popular address in late 1908, to the 80th assembly of German scientists at Cologne, introduced this theory to a broad scientific audience.[31] Both of these influential men, one an eminent theoretical physicist, the other a leading mathematician, had clearly been willing to invest their time in using Einstein's theory to view the most varied areas of physics from new perspectives.

A similar cascade promoted the acceptance of general relativity. The announcement, in 1920, by the two leading astronomers in Britain, Arthur Stanley Eddington and the Astronomer Royal Frank Watson Dyson, that their solar eclipse expedition of 1919 had verified Einstein's predicted bending of light past the Sun, similarly launched confidence in the theory.[32] It initiated an outpouring of public enthusiasm, followed by further attempts by astronomical observers to challenge or verify Einstein's predictions on the bending of light and the gravitational redshift.[33] Attempts by theorists, notably by Alexander Friedmann in Leningrad, and later by Georges Lemaître in Belgium, to pursue Einstein's and de Sitter's early attempts at establishing a general relativistic cosmology also were significant signs of acceptance.[34,35,36]

Finally, the rapid acceptance of Hans Bethe's theory of energy generation in stars, as we saw, was set in motion by Henry Norris Russell's whole-hearted endorsement of the work. Russell quickly publicized Bethe's theory in three of his widely read columns in the *Scientific American*. These introduced Bethe's work, both to a lay audience and to American astronomers. As Bethe was pleased to acknowledge, a more powerful advocate and propagandist for his theory could not have been found. Russell's approval initiated a cascading acceptance of Bethe's theory, the details of which only a few astronomers would have been capable of mastering, but to which Russell's unabashed acceptance gave strong support.[37,38]

Astrophysical Cascades and Instabilities

Unfortunately, as we shall see, cascades are not a universal boon. They also generate a variety of instabilities in astrophysics. Defining these, and dealing with the difficulties they pose, are topics to which I now turn.

The astrophysicist's landscape we first encountered in Chapter 9 has no memory. It records only what we know, here and now. Today's landscape erases yesterday's, which is then lost from memory. The erasure is useful because an astronomical exchange of ideas has to be based on mutually accepted vocabularies on precisely what we mean by words denoting facts or concepts, and this requires agreement on what those facts and concepts currently are.

As it erases recently accepted landscapes, the community also loses a sense for the stability or instability of astrophysics. If no memory remains of how the Universe appeared to us yesterday, last year, or a century ago, how can the stability or instability of the field be reliably judged?

Most astronomers strongly resist any suggestion that astrophysics is simply buffeted from one unstable state to another, possibly by one mistake following another. We prefer to think of the field as steadily advancing from strength to strength, in a stable progression, even if this progression is seen as a series of gradual advances punctuated by occasional leaps forward.

But does twentieth century history substantiate this view?

An Unruly Landscape

Natural disasters, financial crashes, national policy directives, or wars, all predictably alter the course of science. However, intrinsic discontinuities brought about by novel scientific findings can have the same effect, creating shocks that cascade through astrophysics, redirecting research. If these were better understood, a more deliberate way to pursue astronomy might emerge:[d]

In 1917, when Einstein created his general relativistic cosmology he, like everyone else, imagined a static immutable universe. In modern parlance, the rate of cosmic expansion, the *Hubble constant, H*, was zero.

---

[d] The word *shock* used here is adopted from the terminology developed for the study of financial markets, where the failure of a single bank may trigger a shock that cascades through the entire system and leads to the collapse of sizable portions of the system. The mathematical approaches used to study the propagation of financial shocks can be applied equally well to studying the propagation of shocks through other social systems, including the network of astrophysical disciplines.

By 1931, even Einstein was persuaded that the Universe was expanding. The Hubble constant measured at the time was $H \sim 530$ km/sec-Mpc.[39]

Two decades later, in 1952, Walter Baade, by then working with the new 200-inch telescope at Mount Palomar, revised this downward to ~ 250 km/sec-Mpc. He had uncovered a mixup of two types of *Cepheid variables*, whose luminosities had been used to calculate the galaxies' distances. Seldom in a hurry to publish, Baade provided a more detailed exposition on this topic four years later in 1956.[40]

That same year, Milton L. Humason, Nicholas U. Mayall, and Allan R. Sandage, further lowered the expansion to $H = 180$ km/sec-Mpc, partly by correcting a second earlier mistake, the misidentification of ionized nebulae as luminous hot stars.[41]

Only two years later, in 1958, an even more thorough investigation of the same misidentifications, led Allan Sandage to revise this value downward by another factor of ~ 2.5, yielding $H \sim 75$ km/sec-Mpc, albeit with a possible uncertainty of a factor of 2.[42]

By the mid-1980s, Sandage had lowered the Hubble constant to a value closer to $H = 50$ km/sec-Mpc; but today's best estimates, obtained by a variety of means, cluster around 70 km/sec-Mpc.[43]

Taken at face value, the difference between $H = 530$ and $H = 70$ km/sec-Mpc led to a drop in the number density of galaxies in the Universe by a factor proportional to H^3 or about 430, on the basis of standard estimates at the time. The mass density of galactic matter in the Universe would drop in proportion to H^2 or by a factor of ~ 57.

Factors as large as this have repercussions on the most basic aspects of the Universe, among them the density of atomic matter. An error of 57 in estimating the atomic density alters calculations on the primordially produced ratio of helium to hydrogen.[44] The ratio of the abundances of these two elements observed in stars today would then affect not only our estimates of the additional helium produced through hydrogen burning in stars, but also the expected ages of stars. Those ages, in turn, determine the history of cosmic heavy element production, as judged, for example, by the relative abundances of the elements in Solar System meteorites, whose chemical abundances have remained frozen for nearly five billion years since the birth of the Sun. Those abundances then also constitute much of the evidence we have for the relative abundances of elements deep in the interior of giant planets.

These chains of interrelations make astrophysics a coherent science, even if the field superficially appears to comprise an array of more or less autonomous disciplines, such as cosmology, stellar energy production, star formation, or planetary structure.

The Hubble constant, and other measures like it, are not just measures of interest to a single discipline like cosmology. When such parameters drastically change, as the Hubble constant did between 1930 and the year 2000, their repercussions ricochet through all of astrophysics and, at least temporarily destabilize it, as specialists in different disciplines scramble to seek coherence between previously established parameters and the changes newer observations now demand.

* * *

The history of the Hubble constant is not an isolated instance of major conceptual changes marking twentieth century astrophysics. Nor were those changes confined largely to the century's early decades.

Despite much searching, the question of whether planets exist around other stars eluded an answer until the 1990s. A first hint was provided by Aleksander Wolszczan and Dale A. Frail. Working at the Arecibo radio observatory in Puerto Rico in 1992, they noticed that the pulsar PSR1257+12, a rapidly rotating neutron star emitting a sharp pulse with great regularity every 6.2 milliseconds, exhibited a slight periodic anomaly. They traced this to two small planets, not much more massive than Earth, orbiting and gravitationally tugging the central pulsar hither and yon, respectively with periods of 98.2 and 66.6 days.[45] Two years later, in 1994, Wolszczan noted the influence of yet a third planet orbiting the pulsar.[46]

Pulsars, of course, are rather special, and so these findings did not cause a great stir. But the following year, 1995, measurements by Michel Mayor and Didier Queloz of the Geneva Observatory in Switzerland showed the radial velocity of a solar-type star to be measurably affected by the orbital motion of an inferred Jupiter-sized companion.[47] This set into motion a blizzard of further planetary discoveries by a rapidly growing community of astronomers who were dropping whatever they had been doing to study planets orbiting other stars. The work of Mayor and Queloz had triggered a cascade drastically altering the structure and interests of the astronomical community.

Initially, the adopted method of searching for planets indicated the existence of systems very different from our own. Jupiter-sized, or even larger planets were being found in orbits far closer to their parent stars than Jupiter's distance from the Sun. This was rather misleading. The initial method used to discover planets around other stars simply worked best for discovering this particular class of planetary systems.

By 2012, the use of a variety of other techniques exhibited a much richer mix of planetary systems. Roughly a thousand planets, large and small, had been discovered around other stars. Some planetary systems were quite similar to ours. Others were strikingly different. Over a period of 20 years, we had gone from wondering whether our planetary system was totally unique, the only such system in our Milky Way or possibly in the entire Universe, to realizing that the number of planets in the Galaxy lies in the hundreds of billions, comparable to and perhaps exceeding the number of stars!

These findings have implications going well beyond just the study of planets. They tell us that the formation of stars generally entails the accompanying formation of a stable circumstellar gaseous disk within which planets form. Just how massive these disks must be is now being further pursued, and will provide us with new information about the internal structure of the giant molecular clouds within which most stars are currently thought to form.

If the existence of life elsewhere hinges on the availability of suitable planetary habitats, the rich abundance of newly discovered planetary systems makes the likelihood that life can exist around other stars far higher as well.

An entire network of disciplines is changing its research pursuits in response to a discovery made in just one of its member disciplines.

* * *

Whereas the collapse of stars that had used up their nuclear fuel was a concept heatedly debated in the late 1930s, the discovery of pulsars, 30 years later, made clear that pulsars are rapidly rotating neutron stars that could have originated only in such a collapse. But what about the possibility that stellar black holes might exist? This was a more difficult question to answer because so little was known about the extent to which nuclear matter could resist severe compression and thwart a star's further collapse to form a black hole.

By the mid-1960s, a variety of Galactic X-ray-emitting sources were known. On a rocket flight in 1964, Herbert Friedman's NRL team had detected a powerful new X-ray source in the constellation Cygnus. It is now designated Cygnus X-1.[48] Intense study of this source at many wavelengths soon revealed it to be part of a binary system composed of a blue supergiant and an invisible companion. In 1971, Louise Webster and Paul Murdin at the Royal Greenwich Observatory in England inferred the mass of the companion from the supergiant's *orbital velocity* and its 5.6-day periodic Doppler shift. They concluded, "The mass of the companion probably being larger than about $2M_\odot$, it is inevitable that we should also speculate that it might be a black hole."[49]

This appears to have been the first plausible observational hint that massive stars could overcome the mutual repulsion of a neutron star's nucleons, to collapse even further to form black holes!

By late 1972, Clifford Rhoades and Remo Ruffini at Princeton had determined, on the most general theoretical grounds, that the repulsion among nucleons in a neutron star would never suffice to resist collapse into a black hole if the star's mass exceeded $3.2M_\odot$.[e] But this did not rule out that collapse might readily occur also at significantly lower masses.[50]

Perhaps even more striking than Cygnus X-1, was an odd-looking source discovered in the late 1970s. Its spectrum exhibited a bizarre set of changes. Known by its listing in a catalogue assembled by C. B. Stephenson and N. Sanduleak, the source referred to as SS433, was identified first by David Clark in England and Paul Murdin in Australia as a peculiar object with properties similar to those of a known

---

[e] To set this upper limit they needed to postulate only that Einstein's general relativity is correct; that an increased hydrostatic pressure always increases density; and that resistance to collapse would need to be causal, meaning that the speed of sound, that is, the speed at which pressure differences propagate across a star, could not exceed the speed of light.

variable X-ray and *radio star*, Circinus X-1, possibly associated with a supernova remnant.[51] This finding piqued the curiosity of Bruce Margon of the University of California at Los Angeles and that of his colleagues. Their spectral observations identified the source as a new type of phenomenon – later termed a *microquasar* – ejecting beams of excited atomic hydrogen at speeds of order 70,000 to 80,000 km sec^{-1}.[f] It resembled a quasar except that its mass might be just a few solar masses, whereas the masses of quasars could amount to those of a billion suns.[53,54,55]

Recent analysis indicates this microquasar to likely be a stellar-mass black hole surrounded by a supernova remnant. The black hole formed in the collapse of a more massive predecessor about to give rise to a supernova explosion has a binary companion star from which it tidally strips matter. An X-ray-emitting accretion disk surrounds the compact star. Matter stripped from the companion falls onto the disk, which precesses around the compact star with a period of ~164 days. As it does this, the oppositely directed beams emanating from the compact star precess as well, redirecting the beam pattern with the same periodicity of ~164 days.[56]

The insights gained from studies of microquasars may advance our understanding of precisely how massive stars undergo collapse as they give rise to at least some types of supernovae, and how companions to those massive stars appear to survive those explosions. Microquasars may also provide opportunities to further our understanding of quasars and their birth early in the history of the Universe.

The concept of stellar collapse, so strongly rejected by Eddington in the 1930s, has become an active area at the forefront of research!

* * *

From planetary systems, to collapsed stellar-mass black holes, to cosmology, whether early in the twentieth century or late, the astrophysical landscape reflecting the basic set of tenets of the astrophysical community was continually being buffeted by observations that required drastic conceptual changes.

Unrequited Expectations and Sporadic Success

The arrival of new tools, or other reasons to believe that a new direction of research might be promising, can also induce astronomers to veer off along new paths, many of which might never entirely fulfill their promise, although some may.

For a while, in the late 1950s, plasma physicists and engineers were confident that magnetohydrodynamically controlled fusion would soon provide unlimited sources of energy for human consumption. The versatile Princeton astrophysicist Lyman Spitzer, Jr., had become the director of Project Matterhorn, a first attempt

---

[f] The term *microquasar* was coined by Felix Mirabel and colleagues, who identified a similar source 1E1740.7–2942, in 1992.[52]

to build a fusion machine to generate energy by converting hydrogen into helium. Declassified by the U.S. Atomic Energy Commission in 1958, Spitzer's *Stellarator* concept raised expectations that magnetohydrodynamics would soon explain many astrophysical processes as well.

In 1942, Hannes Alfvén in Sweden had predicted the existence of what he initially called *electromagnetic-hydrodynamic waves* – now often referred to as *hydromagnetic* or *magnetohydrodynamic* waves – which he thought might be discovered in the surface layers of the Sun.[57] Alfvén's ideas took some time to be accepted, but by the late 1950s they had gained considerable credence. A book Spitzer wrote in 1956, another by the British astrophysicist Thomas G. Cowling published the next year, and a book by Chandrasekhar published in 1960 all signaled a heightened interest in the roles that magnetohydrodynamics and plasma physics might soon play in astrophysics.[58,59,60] Ultimately, however, the complexity of magnetohydrodynamic theory and the lack of quick progress on fusion machines led many astrophysicists to look for simpler ways to advance their science, even though some stayed on to ply this work further.

* * *

Another wave of optimism then swelled that supernovae might intimately control star formation, gradual enrichment of heavy chemical elements in the Universe, cosmic ray production, and galaxy evolution – all leading to a similar burst of enthusiasm propelling astrophysics in new directions. Soon this too was superseded, this time by a shift of attention to supermassive black holes, some highly active in quasars and *Seyfert galaxies*. These might be even more powerful generators of cosmic rays and influences on star formation and the evolution of chemical abundances.

* * *

Many who were graduate students or young postdocs in the late 1960s and early 1970s still recall fondly *the golden age of black hole physics*, which rapidly yielded insight on rotating and charged black holes, their interaction with electromagnetic and gravitational radiation or incident neutrinos, and the generation of energy that might be extracted from giant black holes in the nuclei of galaxies.

A legacy of the golden age, which appeared to end just as quickly as it had started, was the deduction of the basic laws of black hole thermodynamics. Then, inevitably, a less heady pace set in as the community began to tackle more difficult problems. It took several decades for the requisite techniques to arrive but, early in the new millennium, it became possible to model the mutual interactions of black holes and their potential mergers in giant outbursts of energy – events to search for in γ-ray bursts or in the nuclei of interacting galaxies.

* * *

Starting in the 1990s, a new vogue arose. It involves massive computations simulating the formation of stars, quasars, and galaxies in the first few hundred million years in the life of the Cosmos, around the time of redshift $z \sim 20$. It tries to determine whether the earliest structures to emerge in the Universe would have been stars, or quasars, or galaxies. And how can one be sure? Such computational attacks will continue at least until more powerful observational tools come along and show how reliably the computational models simulate actual events.

Speeding Neutrinos – An Astrophysical Debate

Some past misunderstandings, such as those that arose from errors in deriving the Hubble constant, take decades to sort out, and are generally considered to have arisen from weighty challenges. Other misjudgments may be ferreted out in a few months. It may be useful to still recount one of these here, because rapidly discovered errors are sometimes derisively dismissed as follies, whereas in reality they provide a measure of how subtle many aspects of astrophysics can be. The incident I shall describe also shows how an organized sharing of information among specialists in many different areas of physics and astrophysics may ultimately help in confronting an important impasse.

On September 23, 2011, a team of high-energy physicists working deep underground in the Gran Sasso mountain in Italy made an announcement in which they sought the advice of the science community.[61] The team, known by its acronym OPERA, had been measuring the speed of neutrinos in 2009, 2010, and 2011, steadily improving their procedures throughout. To their surprise, pulsed bunches of neutrinos generated 730 km away in Switzerland at the CERN accelerator in Geneva, appeared to be arriving 60 nanoseconds – 60×10^{-9} seconds – earlier than a simultaneously generated pulse of light could have.[g] This violated a long-accepted view that nothing can travel faster than the speed of light – a tenet of relativity to which all of modern physics is solidly anchored. After checking everything they could think of that might have gone wrong, the team notified the rest of the scientific community of their finding, but noted:

> Despite the large significance of the measurements reported here and the stability of the analysis, the potentially great impact of the result motivates the continuation of our studies in order to investigate possible still unknown systematic effects that could explain the observed anomaly.

This note effectively invited colleagues to examine their methods and determine whether anything untoward might have been overlooked.

The striking result and the evident care the team of more than 170 scientists, engineers, and other specialists had taken in analyzing the arrival times of some 16,000 muon neutrinos at Gran Sasso spontaneously propelled the physics and astrophysics communities into action. Over the following six months, more

---

[g] CERN, Centre européenne pour la recherche nucléaire, is the European Organization for Nuclear Research.

than 200 articles were posted on the physics and astrophysics communities' *arXiv* websites, where others could read them and respond as early as the next day. The website became a virtual roundtable around which experts from different areas were arrayed, bouncing their thoughts across to each other to determine how to cope with the problem. Some advised on the intricacies of Global Positioning System (GPS) timing measurements over distances of 730 km; others commented on potential electronic circuit analyses; others yet wondered whether the neutrinos might have found a shortcut by traveling through an as yet unknown dimension of space, or attempted to see whether the laws of relativity could somehow be made to accommodate the observed results, should they turn out to be secure.

Right from the start, something seemed at odds with astronomical measurements. 6×10^{-8} sec may not appear to be much; but at the speed of light, the distance of 730 km is covered in 2.4 milliseconds. This meant that the neutrinos were racing along at a fractional rate of $\sim 2.5 \times 10^{-5}$ in excess of the speed of light.

Most astrophysicists knew that this was extremely unlikely. Two decades earlier, in February 1987, astronomers had observed a supernova explosion in the Large Magellanic Cloud, a distance of nearly 170,000 light years from Earth. Theorists had predicted that a supernova explosion should be preceded by a catastrophic stellar collapse terminating in a pulse of neutrinos and antineutrinos triggering the explosion. Such a pulse would then precede, by a few hours, the rise in the star's luminosity as its hot outer layers explosively expanded, causing the supernova to brighten rapidly.

Indeed, the brightening of the supernova was observed just three hours after a neutrino and antineutrino pulse had arrived at Earth after a journey of those 170,000 years. This meant that their speed could not have exceeded that of light by more than three hours after all those years, or a fractional amount of about 2×10^{-9}, roughly 10,000 times less than the OPERA experiment claimed. The OPERA team was well aware of these observations and specifically cited them. Nevertheless, they felt that their experimental data might be revealing some new, unanticipated physical phenomenon.

The energies of the muon neutrinos generated at CERN were ~ 17 GeV, about 500 times higher than the highest energy neutrinos observed from the supernova, ~ 35 MeV. Moreover, the OPERA experiment was carried out with muon neutrinos, whereas the astrophysical detections were sensitive solely to electron neutrinos and antineutrinos. These differences made an experimental velocity measurement worthwhile. On the other hand, the velocities of the supernova neutrinos and antineutrinos had been essentially independent of their energies. Over the observed 5 to 35 MeV range, all the neutrinos had arrived within a span of 12.5 seconds, meaning that their speeds over this energy range were identical to within at most a fractional amount of $\sim 2.5 \times 10^{-12}$. Even if the difference between neutrino speeds and the speed of light had increased as the square of the energy – in violation of all relativistic expectations – 17 GeV neutrinos could not have exceeded the speed of light by a fraction higher than $\sim 10^{-6}$, still an order of magnitude lower than the OPERA team claimed.

Be this as it may – eventually, as a result of the intense exchange between hundreds of physicists, engineers, and astrophysicists, two subtle sources of potential experimental error were identified, both of which at least cast doubt on the measured neutrino speeds. Perhaps the only way this could have been established so quickly after so many previous months of checking and rechecking for any potential errors in the course of 2009–11 was for the best and brightest theorists, electronics experts, accelerator physicists and engineers, and astrophysicists to weigh in and explain their respective findings to each other and home in on a rational analysis. By July 12, 2012, less than ten months after their original announcement on the *arXiv* site, the OPERA team had not only pinpointed the source of difficulties and revised their findings, but also had come out with a new analysis of their data. Their measured difference in the arrival times of the muon neutrinos and light at Gran Sasso now was no greater than 6.5 nanoseconds, well within their experimental uncertainties.[62]

Although a significantly superluminal velocity of neutrinos was now cast in doubt, a matter about which no uncertainty at all could exist was that the community of physicists and astrophysicists had just witnessed a massive social cascade setting in motion a beneficial communal debate.

* * *

Such buffeting by new fashions; the availability of increasingly powerful observational or computational tools promising the solution of previously intractable problems; major new discoveries, such as those of dark matter or dark energy, which suggest the opening of entirely new vistas; outright mistakes that nevertheless are subtle and need to be carefully analyzed before they are rejected; or, more prosaically, governmental research policies and availability of funding, all play a critical part in shaping astrophysics.

Work in astronomy thus advances as though perpetually driven by destabilizing thrusts and cascades to which the field has to respond.

An Intrinsic Potential for Error

Why is it that astrophysics is so prone to error or the lure of promising new findings, that the field forever staggers from surprise to surprise?

Astrophysics comprises many areas of investigation that have historically emerged as distinct *disciplines*. The physical processes at work in each of these may occur on vastly different scales. Physical dimensions in astrophysics range from microscopic grains of interstellar dust, to kilometer-sized *asteroids* and millions of times more massive giant planets, stars and stellar groupings, entire galaxies and their clusters, and finally to the entire Cosmos.

Temperature regimes of interest range from 2.7 K for the microwave background radiation; to thousands of degrees Kelvin on the surfaces of stars; tens of million degrees at the centers of main sequence stars; billions of degrees at the center of stars collapsing before exploding as supernovae; and higher temperatures still in the early Universe.

Across these wide ranges, different physical laws come into play with changing temperature, density, and size.

Because of this diversity, research in each discipline is conducted by its own dedicated experts, with different areas of expertise and specialized toolkits for solving problems. Planetary astronomy, star-formation studies, stellar structure, galaxies and their evolution, and cosmology all have their own specialists. Even within these disciplines, astronomers often dedicate their careers to narrower sub-disciplines. Planetary astronomers may devote themselves to Solar System planets or, since they were first discovered in significant numbers in the mid-1990s, to the study of planets orbiting other stars. Within the Solar System community, some might dedicate their careers to investigating comets and how they may have evolved since the Solar System first formed.

Nonetheless, the various disciplines are not entirely isolated. To determine, say, the rise in the abundances of the most massive chemical elements, over the eons, we may find it useful to pursue data based on a variety of approaches: (i) calculations of nucleosynthesis early in the evolution of the Universe; (ii) calculations of nuclear processes at work in massive stars collapsing today; (iii) spectroscopic analyses of the chemical composition of supernova ejecta; (iv) spectra of stars formed when the Universe was much younger than today; (v) chemical and isotopic analyses of meteorites formed as the Solar System came into being; (vi) the chemical composition of cosmic rays impacting Earth, and (vii) a variety of other calculations and observations. These different investigations are carried out by members of distinct disciplines, specialists who otherwise work on barely related research topics.

Although the various astrophysical disciplines are thus loosely connected through such areas of shared interest, the diversity of the disciplines and the differences in adopted research tools lead to concepts and vocabularies that may take on disparate meanings in the different disciplines and lead to misunderstandings. Information exchanged among disciplines may further be misinterpreted through lack of familiarity with the tools and methods by which data in unfamiliar disciplines have been assembled.

All these factors can inadvertently precipitate cascades triggered by misunderstandings of technical complexities arising in new observations, differences in trusting previously established results, and a potential misapprehension of the relevance of significantly different approaches.

Fragmented Communities, Shocks, and Cascades

A feature that Duncan Watts emphasized in his paper on cascades precipitated by shocks to social networks is that these networks can be both *robust* and *fragile* – a finding that might initially appear contradictory. He writes[63]:

> Although they are generated by quite different mechanisms, cascades in social and economic systems are similar to cascading failures in physical infrastructure networks and complex organizations in that initial failures increase the likelihood of subsequent failures that … are extremely difficult to predict even when the properties of the individual components are well understood. … [A] system may appear stable for long periods of time and withstand many external shocks [be robust], then suddenly and apparently inexplicably exhibit a large cascade [become fragile].

As the network of astrophysicists splits into discrete disciplines, it tends to destabilize. The instabilities express themselves in episodic crises in which a new finding initiates a cascade that propagates from discipline to discipline, cycling through the entire field with shocks and aftershocks, just as changes in the Hubble constant changed not just cosmological conclusions but also those of a wide range of seemingly unrelated disciplines.

This should not be surprising. Large, loosely coupled networks are by now known to be inherently unstable. The root causes leading to sudden imbalances were first noted and analyzed in ecological systems. But they are equally prevalent in social and industrial networks. Widely studied examples of instability in the industrial world are the periodic brownouts or blackouts in the major electric grids spanning the United States. The global banking crisis of 2008 reflected analogous instabilities.

Watts points out that, following each industrial cascade, the temptation arises to blame failure of some minor identifiable component for precipitating the entire calamity. But, as he convincingly argues, this misses the point that cascades are endemic to highly complex networks with myriad components. If one specific component does not fail, some other inevitably will. The problem lies with the network, not the individual components.

* * *

Shocks propagating through astrophysical networks may have two causes. The first is natural; a novel finding in one discipline affects considerations in neighboring fields to which the discovery is conveyed. Those fields must then readjust their own findings based on revised calculations, insights, and conclusions and convey those further through the network to yet other disciplines relying on them. This process is natural because that is the way science should advance. Astrophysics is not meant to remain static.

However, there is a corresponding instability that follows the same pattern but has less fortunate consequences. It arises because the separation between disciplines and their vocabularies, the languages they speak, diverges. The various disciplines then no longer understand each other well. Data exchanged in the absence of full understanding of their provenance and limitations can then lead to misunderstandings. Mutual reliance and communication based on partial information thus threaten to break down the astrophysical network. Yet, the different disciplines need to keep in touch; otherwise astrophysics splinters and we lose a coherent cosmology.

The sole way of avoiding such misunderstandings and the resulting cascades of errors is through far clearer channels of communication between disciplines than astrophysics has established to date. The resolution of the problem raised by the speeding neutrinos came about through just such a clarification. It identified a subtle misunderstanding in the originally reported neutrino measurements that invalidated the apparent findings.

We might be tempted to dismiss the speeding neutrino affair as due to some blunder that should never have occurred. But this would be missing the essential feature of cascades. After the fact, the initiating cause of any given failure can often be traced to some minor defective component – in this instance to the way GPS timing was linked to neutrino pulse arrival. But the actual fault was a lack of clear communication across the network of experts – physicists, engineers and astrophysicists – who might have spotted the potential for error rather earlier. For the astrophysical network, one lesson to be learned is that improved communication between disparate disciplines is paramount. Unless we can assure ourselves of reliable communication across the entire network, we will never genuinely understand the workings of the Universe.

Yet, any system that expands in size, whether it be a country's electrical grid, or an international scientific enterprise, places an increasing premium on reliable parts. Parts that rarely fail may be completely adequate for a small system; but as the system grows, the probability of failure, somewhere, keeps increasing, until a point is reached where occasional failures become routine. To avoid failure, the quality of the parts must then be significantly increased, raising cost and threatening to make the system unaffordable.

More important, however – even if none of the individual parts fail – are the ways in which the parts are mutually linked. The design of the communication links that join them can equally lead to failure, as we shall now see.

The Behavior of Large Dynamic Systems

How complex networks of interacting communities or machines behave has been intensely studied in recent years, particularly in the wake of costly banking failures. The results of these studies show how the interactions of varied organizations, among them the mutual interactions of the various astrophysical disciplines, may be understood.

Network theory reveals that entities exchanging noisy or ambiguous information tend to exhibit catastrophic instabilities caused by random mutual interactions rather than by component failure. Models developed to understand the stability of natural ecosystems or to assess the stability of financial markets and banking consortia have shown that systems consisting of individuals, communities, or organizations, referred to in network theory as *modules* or *nodes*, can instigate mutual instability when permitted to interact, even when each isolated node may be perfectly robust.

Describing how such instabilities arise is relatively straightforward and may help in devising ways to avoid them. But network studies are still rather young, and we have no assurance as yet that catastrophic failures will always be foreseen and avoided. Human ingenuity, enthusiasms, and hubris often succeed in circumventing buffers emplaced to avoid disaster. The unintended consequences of a tightly meshed network may then lead to major meltdowns.

Studies of large interacting systems date back to 1970, when Mark R. Gardner and W. Ross Ashby at the Biological Computer Laboratory at the University of Illinois in Urbana made a seminal computational discovery.[64] They wondered whether a system of interconnected nodes, such as those depicted in Figure 12.1, would be stable? Would the interconnection of nodes – for example, systems of interconnected computers – make the resulting network unstable even if each node, when isolated from all others, was perfectly stable?

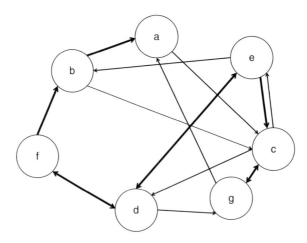

Figure 12.1. A network with seven nodes. Here each node is meant to represent a different discipline of astrophysics. The nodes are connected by directed links with varied coupling strengths, called *weights*. Stronger links among disciplines are indicated by darker arrows. Links along which exchange occurs equally in either direction are marked by two-headed arrows.

Gardner and Ashby modeled such a system by assuming the individual nodes to start out in an inherently stable state. This initial state would change in some fashion once random stimuli could be exchanged among the nodes. Their simple model involved linear behavior, that is, a response of any node was directly proportional to the signal the node received from some other node. They examined what would happen if even just a few of the nodes were interconnected. For a fraction C of interconnected nodes the interaction strengths were randomly assigned values evenly distributed over a range that normally would not directly destabilize some other computer. Self-stimulation by each node was restricted to stabilizing feedback.

For different sets of nodes consisting of networks with 4, 7, or 10 nodes, a computer program then generated many successions of such interactions, constrained only by specified values of the *connectance C* – the percentage of the nodes that were connected. For each of the simulated interactions, the final state of the system was considered stable if, and only if, every node then still operated within specified bounds from its original state. Figure 12.2 illustrates how a system of sound amplifiers might operate under such conditions.

Figure 12.3 shows the probability of Gardner and Ashby's system of n nodes ending up in a stable state for systems with $n = 4$, 7, or 10 nodes and connectance C.

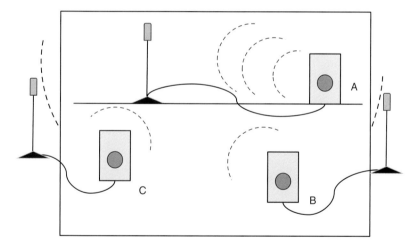

Figure 12.2. Unstable networks. Microphones amplifying sound excessively can pick up tones emitted by their speakers, amplifying these successively until a loud whistling or wailing sound fills an entire auditorium. To prevent such instabilities microphone amplifiers are built to reject feedback from their own speakers. This makes a system A, at top, stable if completely isolated from all others. But if remote microphones B and C can pick up part of speaker A's output, amplify it so that sound from their speakers B and C reach microphone A, A's feedback rejection may no longer work, making the combined network, A, B, C, unstable. Computer networks considered by Gardner and Ashby (see text) tended similarly to destabilize their coupled computers.

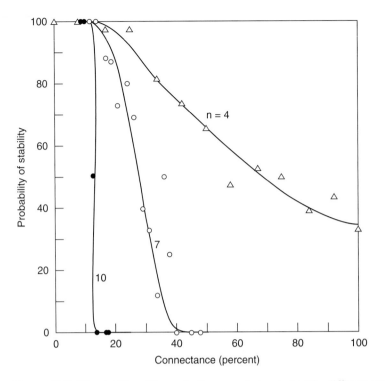

Figure 12.3. The stability of dynamically connected systems. The different curves show the probability that systems with $n = 4$, 7, or 10 nodes will remain stable when connected with other nodes. The *connectance C* is the percentage of nodes connected to each other. (*Reproduced drawing reprinted by permission of Macmillan Publishers Ltd from: 'Connectance of Large Dynamic (Cybernetic) Systems: Critical Values of Stability,' Mark R. Gardner & W. Ross Ashby, Nature, 228, 784, 1970 ©1970 Nature Publishing Group*).

For a system of 10 nodes the probability for stability dropped to zero at a connectance even as low as $C = 13\%$. For a system of 7 nodes, a connectance just above 40% would predictably induce instability.[65] The hypothetical system of Figure 12.1 would likely be unstable.

Interaction Among Species

In 1972, two years after the appearance of Gardner and Ashby's work, the ecologist Robert M. May used analytic arguments to extend these findings to larger networks.[66] Although his investigation was mathematical, May's ultimate interests lay in biology. He asked how stable the ecology of n populations of interacting species would be if restricted to a small neighborhood of their equilibrium point, where a displacement from the postulated equilibrium would be directly proportional to the exerted stimulus.

May postulated that, left to itself, each specie if perturbed away from equilibrium would return to equilibrium within some characteristic time. For simplicity he assumed these damping times to be identical for all the species. Thereafter, he followed a procedure essentially identical to that of Gardner and Ashby in assigning random values to all the interactions, specifying only that the mean value of the interaction strength $\langle w \rangle$ be zero and its mean square $\langle w^2 \rangle$ have some specified value α.

Again, the criterion for stability was that values of the nodes had to remain unchanged within specified bounds. With these provisos, May was able to reach back into the mathematical literature to cite the result that, for mean square interaction strengths $\alpha < (nC)^{-1/2}$ the system was stable with near certainty, whereas for $\alpha > (nC)^{-1/2}$ it was nearly certain to be unstable.

The fewer and weaker the interactions, and the smaller the number of interconnected species, the more stable the system would remain. With frequent, strong interactions, any sufficiently large system became unstable. A system with strong interactions but low connectance would have the same stability as one with weak interactions but high connectance as long as the product $C\alpha^2$ remained the same.

May then showed that his analytically obtained results, aimed at depicting the stability of large systems, differed from those of Gardner and Ashby's by no more than 30% even for systems as small as their system of $n = 4$ discrete interconnected species.

The Network of Astronomical Disciplines

To determine how stable our current ways of describing and understanding the Universe may be, we can model astronomical studies as divided into N different disciplines designated $j = a, b, c, \ldots$, each of which is a node in a network within which nodes may be connected to each other along linking edges with varied strengths as in Figure 12.1. The strength may be measured in terms of a weight w ranging anywhere between $w = 0$, meaning that there is no connection, to $|w = w_{max}|$, implying a strong interdependence between disciplines. The sign of the interaction, $+$ or $-$, indicates influence, respectively exerted by a node or acting on it. The mean squared characteristic weight of the interaction is equivalent to May's mean square interaction strength $\langle w^2 \rangle = \alpha$. Arrows indicate the direction of the influence.

The populations of astronomers active in any given discipline correspond to the different *species* in May's ecological perspective. We need to think of n populations of different species of astronomers – for example, planetary, galactic, or stellar-interior specialists – each interacting with some number of other species n with a connectance C. The ecological setting is all of astrophysics.

Taken as separate entities the different astronomical disciplines often are stable. If astrophysics were modular, with no connection between disciplines, an error in one discipline would be contained without propagating to others. Usually it might soon be discovered and corrected. However, the primary aim of astrophysics is to unify insights gained in the various disciplines in order to obtain an overview of how the Cosmos has evolved and continues to develop. This linkage of disciplines into a grander network of understanding enables both beneficial and harmful shocks to propagate, domino fashion, from one discipline to the next, initiating instabilities throughout.

Following May, the system is then intrinsically unstable whenever α exceeds $(nC)^{-1/2}$. The catch in astrophysics is that some level of instability is desirable to permit transmission of a new finding in one discipline to one or more others. We want astrophysics to change significantly when new observations or calculations herald a significant advance, but if the transmitted information is tainted, a system of detecting this is needed to reject the information before excessive harm is done. This is a balancing act that productive disciplines continually face.

Around the turn of the millennium, a particularly powerful shock propagating through a broad set of astrophysical disciplines arose from the discovery that the light received from distant supernovae of type SN Ia was systematically fainter than their observed redshifts would have led us to believe. This suggested that the exploding supernovae were more remote than their redshifts indicated. A plausible explanation was that the expansion of the Universe had recently been accelerating, and that this was likely to have happened only if there was an unknown form of energy, now known as *dark energy*, with a cosmic mass – energy density significantly exceeding that of baryonic and dark matter combined. For decades, the rate of cosmic expansion had been thought to be decelerating. Now, in the late 1990s, a countervailing acceleration was coming into vogue. This new interpretation also implied that previous estimates of the cosmic baryon density had been too high, and raised new questions about the nature of vacuum. Virtually all astrophysical disciplines were affected, one way or another, as were the very foundations of physics, arising from questions about the true nature of empty space – the vacuum. We are still trying to understand this shock's potential implications. Will it provide deeper insight on a new form of energy, or merely indicate that we are looking at Nature from the wrong perspective, as physicists had, a century earlier, when they were seeking to understand the aether they were sure was pervading space?

Propagation of Shocks

Astrophysical shocks may be modeled, at least in simplified ways, using techniques advanced by evolutionary biologists and, more recently, through studies of interacting financial systems.

Shocks in financial systems have been illustrated in many studies by now, usually in deliberately oversimplified schemes. These simplifications are appropriate in any young set of investigations, and consequently also suffice for a first attempt to illustrate the vulnerability of astrophysics to shocks. Much of what I present here borrows directly from financial models.[h]

---

[h] In banking, catastrophic failures have been described by Andrew G. Haldane.[67] Models of failures have been summarized by Haldane and May.[68] More detailed models describing

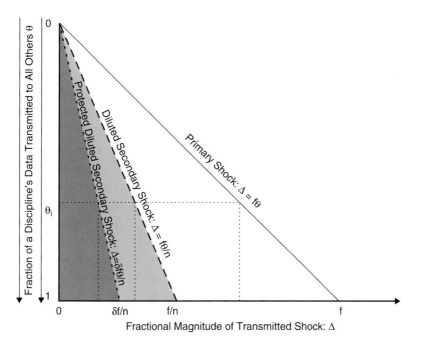

Figure 12.4. Fractional loss of information Δ due to a fraction f of transmitted information tainted by error. The transmission is taken to flow from an initiating discipline to n other linked disciplines. The failure f characterizes the fraction θ of the initiating discipline's data that is actually transmitted. Each of the n linked disciplines is assumed to acquire only a diluted fraction $1/n$ of the transmitted data, making its fractional contamination n times lower than that actually transmitted. If a discipline has adequate contamination protection, the magnitude of the shock may decrease to a further reduced fraction δ, as indicated by the lowest diagonal line. The tainted information assimilated by a typical discipline is indicated by the abscissa of the intersection point of θ_i, the initial value of θ, and the labeled diagonal lines. (See text.)

A particularly instructive illustration, based on the financial modeling by Robert May and Nimalan Arinaminpathy, is shown in Figure 12.4.[71] We may consider the damage inflicted by a harmful shock on a network of $N = n + 1$ disciplines, one among which is struck by a shock that reduces the value of its transmitted information by a fractional amount f.

For simplicity in illustrating the propagation of astrophysical shocks, we may assume that each discipline is identical to all the others, both in the value of its total assets of information and in its ability to check for errors and contamination. We now assume that the magnitude of the shock suffered by the initially trans-mitting discipline is restricted to a fraction $\Delta = f\theta$ of the discipline's total asset of

the erosion of banking systems by shocks, and the spread of contagion across financial systems have been published by Fabio Caccioli, Matteo Marsili, and Pierpaolo Vivo,[69] and by Prasanna Gai and Sujit Kapadia.[70]

information, where θ is the fraction of the discipline's total information actually transmitted.

We further assume that the transmitted shock is equally distributed among the other n disciplines, so that each of the others receives erroneous material amounting to only a fraction $1/n$ of the entire fraction of tainted transmitted data. If even this reduced shock exceeds the receiving discipline's error detecting and removing ability, and a fractional amount of the erroneous material δ still takes foothold, the assimilated shock will amount to a fractional amount $\delta f \theta / n$, which may then be transmitted further to tertiary disciplines, some of which may not have been directly linked to the initiating discipline.

In Figure 12.4 the triangular area $(1, 0, f)$ below the solid black line indicates the extent to which the fractional magnitude of the initial shock of erroneous information $S = f\theta$ taints all of the information from the initiating discipline. When $\theta = 1$, meaning that the entirety of the initiating discipline's information is transmitted, the fractional magnitude of the original shock is $S = f$.

In general, without the ability to reject tainted transmission, the shock assimilated by each of the n linked disciplines then lies below the dashed line $\Delta = f\theta / n$, intersecting the abscissa in Figure 12.4 at $\Delta = f / n$. With the ability to reject all but a fraction δ of the tainted material, an even steeper decline intersecting the abscissa at $\Delta = \delta f / n$ may be anticipated. For smaller fractions of tainted material, $\theta_i < 1$, the intersection points move to the left in the figure, to fractional levels $\theta_i \Delta$.

These estimated losses may, however, underestimate actual damage. In astrophysics, much of the transmitted material may reach and affect most if not all disciplines to which the initiating discipline is linked. All of these linked disciplines may then be equally interested in and become contaminated by the entire transmitted material – meaning that $n = 1$. Further, it may often be difficult to detect tainted or erroneous material despite considerable precautions, and so the fraction δ of retained tainted material may be close to 1. Shocks may thus be transmitted widely, with significant repercussions, particularly if the receiving disciplines propagate the tainted information further.

The parallels drawn here between the linkages binding financial institutions and those connecting different disciplines of astronomy indicate the importance of more clearly understanding the extent to which astrophysics may be liable to instabilities similar to those afflicting financial markets.

* * *

A lesson that astronomy may learn from banking systems is that mandated buffers, especially in the form of monetary or other available liquidity can help counter the effects of shocks. If a key foundation on which much of a discipline rests is unexpectedly cast into doubt and begins to threaten the collapse of linked disciplines, liquidity – funding or facilities – to quickly reexamine the causes of doubt and correct any deficiencies may limit the spread of damage to other astronomical disciplines.

This is the lesson to take away from the case of the speeding neutrinos. A finding that had immense potential repercussions on many different disciplines was quickly checked through in the course of six months, even though no budgets or facilities had been planned for the intense efforts that eventually discerned the sources of error. The necessary liquidity was available largely through volunteer efforts to come to terms and avoid further propagation of the tainted results.

Liquidity is a particularly scarce asset in astrophysics, which strongly depends on observations often accessible solely by means of space telescopes whose operational lifetimes seldom exceed a decade. As we saw in Chapter 11, the high cost of operations in space generally forces the termination of a mission once it has met its primary goals. Keeping it operational longer saps funding available for launch of other missions already in queue.

> • At current funding levels maintaining continuous and simultaneous access to observations from space in all astronomical disciplines is unaffordable. However, ways might be sought and found to enable some of the most powerful and most expensive among these observatories to enter lengthy states of hibernation from which they could be reawakened for limited periods, from time to time, to undertake essential observations inaccessible by other means.

A hibernation phase is not a feature normally included in the design of a space mission. Intelligently designed into some missions, it could provide liquidity and flexibility to a field that otherwise may be too rigid to face unanticipated contingencies: In 2005, the Submillimeter Wave Astronomical Satellite (SWAS) was called out of retirement to help analyze the surface material of comet 9P/Tempel 1 by observing the plume of water vapor ejected as NASA's high-speed Deep Impact probe was crashed into the comet. Although not specifically designed for such contingencies, SWAS was successfully revived for this task before again returning to retirement.[72]

Infection and Its Speed of Transmission

The ease with which harmful shocks infect a network of disciplines depends on the ease of escaping detection and rejection. A discipline that lacks means for detecting transmitted errors or contaminating misinformation is readily infected and susceptible to propagating shocks. A discipline with a functioning error- or contagion-detecting mechanism damps out shocks.

Shocks can spread like infectious diseases. Two factors are involved. One is the availability of transmission links; the other is the existence of sites susceptible to infection. Like diseases, shocks can propagate both over frequently available short distances, and less often encountered links to some remote discipline that shares few common interests. Short hops quickly spread information locally. Transmission to remote links quickly affects the entire network. Once a shock reaches a remote node, it readily infects tightly linked neighboring disciplines. This is the way computer viruses as well as infectious diseases are globally transmitted. The same patterns of transmission infect linked astrophysical disciplines.[73]

A complexity, here again, is that many shocks actually benefit and are essential to the conduct of astrophysics. Every new discovery needs to propagate swiftly through the system. Incoming information cannot be rejected merely because it threatens to transform existing knowledge. A discipline needs to be willing to receive potentially beneficial new data. But this leaves it vulnerable to undesirable errors and misunderstandings, so that some system of monitoring the flow of influence from one discipline to another remains essential.

In Chapter 9 we noted the existence of a giant component linking about 90% of the members of the astrophysical community through jointly published papers. No more than 14 steps separated any one member of this set of loosely linked co-authors from any other. The existence of the giant component makes certain that shocks will propagate readily through the ecology of astrophysical disciplines because both long-range and short-range paths appear abundant.

The question we need to ask is, how best to detect and control potentially harmful shocks?

Two approaches may be considered. One is based on a model for controlling infectious diseases, the other on practice in the experimental sciences.

Consider the strategy on which the prevention and control of epidemics is based. It has two aspects: The first is to prevent carriers of diseases from traveling. The second is to immunize individuals in potential contact with carriers of infectious diseases.

In astrophysics, the first of these methods, a partial bulwark, already exists in the refereeing system employed by all major astrophysical journals. It will not prevent all misconceptions, or ferret out every observational or computational error, but it generally is helpful.

Regrettably the refereeing system is all too easily circumvented. Many astrophysicists each day download unrefereed preprints posted overnight on the publicly available *arXiv astro-ph* site. This provides access to new findings, months before they appear in a journal. The danger in this is the adoption of apparently new findings on faith. Until those findings have been sifted for errors by a knowledgeable referee, they may contain avoidable misinformation. Careful refereeing is time consuming and a thankless task; most authors don't appreciate a referee's criticism and grumble about the added work a referee may recommend. But a thorough referee renders the community invaluable service by identifying potential errors, insisting on clarity, and thereby reducing the chances of spreading misinformation.

In astrophysics, the second means of preventing epidemics of misinformation – the equivalent to immunization – is by ensuring that transmitted information will be appropriately checked and understood. This requires the payment of increased attention to connotations of language and nuanced concepts in different disciplines. A clear sense of vocabulary can prevent the spread of misinformation and misunderstandings and the resulting confusion as information is exchanged among disciplines. Establishing clear vocabularies permits easier identification of misinformation – tantamount to immunization against its spread.

I will develop this notion further in Chapter 15, but only after I have described how astronomers from different disciplines discuss and persuade themselves about new findings, and how this variegated astronomical community is organized and functions.

Notes

1. Inflationary universe: A possible solution to the horizon and flatness problems, Alan H. Guth, *Physical Review D, 23*, 347–56, 1981.
2. A New Inflationary Universe Scenario: A Possible Solution of the Horizon, Flatness, Homogeneity, Isotropy and Primordial Monopole Problem, A. D. Linde, *Physics Letters B, 108*, 389–92, 1982.
3. First Results from a Superconductive Detector for a Moving Magnetic Monopole, Blas Cabrera, *Physical Review Letters, 48*, 1378–81, 1982.
4. *Origins – The Lives and Worlds of Modern Cosmologists*, Alan Lightman & Roberta Brawer. Cambridge MA: Harvard University Press, 1990.
5. Ibid., *Origins*, Lightman & Brawer, pp. 273, 336, 371.
6. *Ibid. Origins*, Lightman & Brawer p. 148.
7. Ibid., *Origins*, Lightman & Brawer, p. 429.
8. Spectrum of relict gravitational radiation and the early state of the universe, A. A. Starobinsky, *Soviet Physics, JETP Letters, 30*, 682, 1979.
9. A New type of Isotropic Cosmological Models without Singularity A. A. Starobinsky, *Physics Letters B, 91*, 99–102, 1980.
10. Ibid., *Origins*, Lightman & Brawer, pp. 154–69.
11. Ibid., *Origins*, Lightman & Brawer, p. 411.
12. Ibid., *Origins*, Lightman & Brawer, pp. 179–80.
13. Ibid., *Origins*, Lightman & Brawer, pp. 193–94.
14. Ibid., *Origins*, Lightman & Brawer, pp. 224–25.
15. Ibid., *Origins*, Lightman & Brawer, p. 315.
16. Supercooled Phase Transitions in the Very Early Universe, S. W. Hawking & I.G. Moss, *Physics Letters B, 110*, 35–38, 1982.
17. Ibid., *Origins*, Lightman & Brawer, pp. 398.
18. Constraints on Cosmology and Neutrino Physics from Big Bang Nucleosynthesis, Jongmann Yang, David N. Schramm, Gary Steigman & Robert T. Rood, *Astrophysical Journal, 227*, 697–704, 1979.
19. Cosmological Constraints on Superweak Particles, G. Steigman, K. A. Olive & D. N. Schramm, *Physical Review Letters, 43*, 239–43, 1979.
20. Unification, Monopoles, and Cosmology, J. N. Fry and D. N. Schramm, *Physical Review Letters, 44*, 1361–64, 1980.
21. Ibid., *Origins*, Lightman & Brawer, pp. 444–45.
22. Über die Krümmung des Raumes, A. Friedman, *Zeitschrift für Physik, 10*, 377–86, 1922.
23. Fate of the false vacuum: Semiclassical theory, Sidney Coleman, *Physical Review D, 15*, 2929–36, 1977.

24. Unity of All Elementary-Particle Forces, Howard Georgi & S. L. Glashow, *Physical Review Letters*, 32, 438–41, 1974.

25. A simple model of global cascades on random networks, Duncan J. Watts, *Proceedings of the National Academy of Sciences*, 99, 5766–71, 2002.

26. Ibid., *Origins*, Lightman and Brawer, p. 476.

27. David Norman Schramm 1945–1997, A Biographical Memoir, Michael S. Turner, Biographical Memoir, National Academy of Sciences of the USA, 2009.

28. *Inner Space / Outer Space, The Interface between Cosmology and Particle Physics*, edited by Edward W. Kolb, Michael S. Turner, David Lindley, Keith Olive, & David Seckel, University of Chicago Press, 1986.

29. *Subtle is the Lord – The Science and the Life of Albert Einstein*, Abraham Pais, Oxford University Press, 1982, pp. 150–53.

30. Zur Dynamik bewegter Systeme, M. Planck, *Annalen der Physik*, 26, 1–35, 1908.

31. Raum und Zeit, H. Minkowski, *Physikalische Zeitschrift*, 10, 104–11, 1909.

32. A Determination of the Deflection of Light by the Sun's Gravitational Field, from Observations at the Total Eclipse of May 29, 1919, F. W. Dyson, A. S. Eddington, & C. Davidson, *Philosophical Transactions of the Royal Society of London*, 220, 291–333, 1920.

33. The Relativity Displacement of the Spectral Lines in the Companion of Sirius, Walter S. Adams, *Proceedings of the National Academy of Sciences*, 11, 382–87, 1925.

34. Ibid., Über die Krümmung des Raumes, Friedman.

35. Über die Möglichkeit einer Welt mit konstanter negativer Krümmung des Raumes, A. Friedmann, *Zeitschrift für Physik*, 21, 326–32, 1924.

36. Un univers homogène de masse constante et rayon croissant, rendant compte de la vitesse radiale des nébuleuses extra-galactiques, G. Lemaître, *Annales de la Société scientifique de Bruxelles, Série A*, 47, 49–59, 1927.

37. My Life in Astrophysics, Hans A. Bethe, *Annual Reviews of Astronomy and Astrophysics*, 41, 1–14, 2003.

38. *Henry Norris Russell – Dean of American Astronomers*, David H. DeVorkin, Princeton University Press, 2000, pp. 254–55.

39. A Relation between Distance and Radial Velocity among Extra-Galactic Nebulae, Edwin Hubble, *Proceedings of the National Academy of Sciences*, 15, 168–73, 1929.

40. The Period-Luminosity Relation of the Cepheids, W. Baade, *Publications of the Astronomical Society of the Pacific*, 68, 5–16, 1956.

41. Redshifts and Magnitudes of Extragalactic Nebulae, M. L. Humason, N. U. Mayall, & A. R. Sandage, *Astronomical Journal*, 61, 97–162, 1956.

42. Current Problems in the Extragalactic Distance Scale, Allan Sandage, *Astrophysical Journal*, 127, 513–26, 1958.

43. The Hubble constant as derived from 21 cm linewidths, Allan Sandage and G. A. Tammann, *Nature*, 307, 326–29, 1984.

44. Updated Big Bang Nucleosynthesis Compared with Wilkinson Microwave Anisotropy Probe Observations and the Abundance of Light Elements, Alain Coc, et al., *Astrophysical Journal*, 600, 544–52, 2004.

45. A planetary system around the millisecond pulsar PSR1257+12, A. Wolszczan & D. A. Frail, *Nature*, 355, 145–47, 1992.

46. Confirmation of earth-mass planets orbiting the millisecond pulsar PSR B1257+12, A. Wolszczan, *Science*, 264, 538–42, 1994.

47. A Jupiter-mass companion to a solar-type star, Michel Mayor & Didier Queloz, *Nature*, 378, 355–59, 1995.

48. Cosmic X-ray Sources, S. Bowyer, E. T. Byram, T. A. Chubb & H. Friedman, *Science, 147*, 394–98, 1965.

49. Cygnus X-1 – a Spectroscopic Binary with a Heavy Companion? B. Louise Webster & Paul Murdin, *Nature, 235*, 37–38, 1971.

50. Maximum Mass of a Neutron Star, Clifford E. Rhoades, Jr. & Remo Ruffini, *Physical Review Letters, 32*, 324–27, 1974.

51. An unusual emission-line star/X-ray source/radio star, possibly associated with an SNR, David H. Clark & Paul Murdin, *Nature, 276*, 44–45, 1978.

52. A double-sided radio jet from the compact Galactic Centre annihilator 1E1740.7-2942, I. F. Mirabel, et al., *Nature, 358*, 215–17, 1992.

53. New H-alpha Emission Stars in the Milky Way, C. B. Stephenson and N. Sanduleak, *Astrophysical Journal Supplement Series, 33*, 459–469, 1977.

54. The Bizarre Spectrum of SS 433, Bruce Margon, et al., *Astrophysical Journal, 230*, L41–L45, 1979.

55. The Quest for SS433, David H. Clark, Viking, 1985.

56. Inflow and outflow from the accretion disc of the microquasar SS 433: UKIRT spectroscopy, Sebastian Perez & Katherine Blundell, *Monthly Notices of the Royal Astronomical Society, 397*, 849–55, 2009.

57. Existence of Electromagnetic-Hydrodynamic Waves, H. Alfvén, *Nature, 105*, 405–06, 1942.

58. *Physics of Fully Ionized Gases*, Lyman Spitzer, Jr. New York: Interscience, 1956.

59. *Magnetohydrodynamics*, T. G. Cowling. New York: Interscience, 1957.

60. *Plasma Physics*, S. Chandrasekhar, University of Chicago, 1960.

61. http://arxiv.org/pdf/1109.4897v1.pdf

62. http://arxiv.org/pdf/1109.4897v4.pdf

63. *Six Degrees – The Science of a Connected Age*, Duncan J. Watts. New York: W. W. Norton, 2003.

64. Connectance of Large Dynamic (Cybernetic) Systems: Critical Values of Stability, Mark R. Gardner & W. Ross Ashby, *Nature, 228*, 784, 1970.

65. Ibid., Connectance, Gardner & Ashby.

66. Will a Large Complex System be Stable?, Robert M. May, *Nature, 238*, 413–14, 1972.

67. Rethinking the financial network Andrew G. Haldane, http://www.bankofengland.co.uk/publications/speeches/2009/speech386.pdf

68. Systemic risk in banking ecosystems, Andrew G.Haldane & Robert M. May, *Nature, 469*, 351–55, 2011.

69. Eroding Market stability by proliferation of financial instruments, Fabio Caccioli, Matteo Marsili, and Pierpaolo Vivo (2009), *European Physics Journal B, 71*, 467–79, 2009.

70. Contagion in financial networks, Prasanna Gai & Sujit Kapadia, *Proceedings of the Royal Society of London A, 466*, 2401–23, 2010.

71. Systemic risk: the dynamics of model banking systems R. M. May and N. Arinaminpathy, *Journal of the Royal Society Interface, 7*, 823–38, 2010.

72. Submillimeter Wave Astronomy Satellite observations of Comet 9P/Tempel 1 and Deep Impact, Frank Bensch, et al., *Icarus, 184*, 602–10, 2006.

73. Ibid., *Six Degrees*, Watts, chapter 6.

13

Astrophysical Discourse and Persuasion

Given the unruly onset of cascades described in the last chapter, how can we ever be sure whether our investigations are yielding a credible picture of the Universe? What warning signs should we look for to determine whether our pursuit may merely lead to a *construct*, potentially accounting for all available observations but nonetheless a mere caricature of the Cosmos – rather than a true portrait?

Language and Its Role in Persuasion

Astrophysics progresses through persuasion. Astronomers need to convince each other of the reality of a new finding. A discovery that fails to persuade colleagues goes nowhere. Once the astronomical community is convinced, it has to also persuade funding agencies of the discovery's importance.

Neither step is easy. Researchers investigating planetary systems, the structure and evolution of stars, the chemistry of interstellar dust and gases, the formation of galaxies, and the birth and evolution of the Cosmos all express themselves using slightly, but nevertheless significantly differing vocabularies. A clear exchange of ideas among the members of different disciplines, however, requires those vocabularies to be well defined and mutually understood. In practice this is rare. Words assume new meanings as novel findings in a discipline enrich a concept. The connotations of identical expressions used in different disciplines then gradually diverge, so that misunderstandings easily arise and dialogue suffers.

Questions about the language of science are not new, but they are seldom discussed by astrophysicists in the daily pursuit of their work.

In his book 'The Web of Belief' written in 1970, the Harvard philosopher and logician Willard Van Orman Quine referred to Kurt Gödel's fundamental mathematical proof and its implications for natural language and exchange of information.[1] Four decades earlier, Gödel had proved the impossibility of setting up a mathematical system that is entirely self-defining and complete. To Quine it was clear that if mathematics, the most formal of all languages, was incomplete in this sense, then languages of every other kind must suffer the same shortcoming. This by itself was a warning that mutual persuasion might not always be easy in science.

A decade after Quine, Douglas R. Hofstadter, one of the most imaginative proponents of artificial intelligence, examined the same question in a comparison of artificial and human intelligence. This is particularly relevant because a robot can respond correctly only when its instructions are unambiguous. Similarly, colleagues in different astronomical disciplines can converse and respond to each other correctly only if they can unambiguously relate novel findings in their respective disciplines. This explains how stringent the requirements on clarity must be, particularly as language continually evolves in response to ever changing scientific findings.

Hofstadter made clear that for artificial intelligence to function, the robot, the machinery that needs to process information and respond, has to be programmed to follow specific rules. These have to be made increasingly complex for artificial intelligence to come up to par with human intelligence. At some point, the two may approach each other sufficiently for the robot's use of language to be indistinguishable from that of humans.

In an astrophysical context, this also defines a minimum level of sophistication a language must meet for an astrophysicist in any chosen discipline to understand fully what a colleague from another discipline is stating.

Hofstadter lists some of the essential abilities that robotic intelligence should include:

> to respond to situations very flexibly;
> to take advantage of fortuitous circumstances;
> to make sense out of ambiguous or contradictory messages;
> to recognize the relative importance of different elements of a situation;
> to find similarities between situations despite differences which may separate them;
> to draw distinctions between situations despite similarities which may link them;
> to synthesize new concepts by taking old concepts and putting them together in new ways;
> to come up with ideas that are novel.[2]

If we consider how all of these capabilities apply to language for discussing astrophysical topics among members of different research communities, those that are most demanding are the development of concepts, and thus new words in the language, that find "similarities between situations despite [their] differences," and simultaneously also "distinctions between situations despite [their] similarities," particularly if we do not know whether we already have all the necessary information in hand or whether a considerable body of factual information may still be missing and beyond reach.

Astronomers and astrophysicists cannot do without distinct sets of concepts to discuss and debate new findings; but to the extent that these concepts are incomplete and often evolve, they can also lead to confusion. Clarity of language becomes indispensable if astrophysics is to evolve as a coherent field in which the findings of different disciplines can be compared to determine just how compatible and how significant they are.

* * *

Clarity of language is critical also to astronomical planning. In Chapter 11 we saw how members from different astronomical disciplines had to be brought together to establish consensus on how a major astrophysical undertaking, the establishment of the Great Observatories, should be structured and implemented, what its primary aims ought to be, and how its different components should relate to each other. Eventually it became clear that these observatories would need to meet the aspirations not only of Galactic and extragalactic investigators, but also of planetary scientists. Members of all the various disciplines in both these communities had to agree to work together, understand each other's requirements, and reach consensus.

Members of the astronomical community participating in the decadal reviews recommending research plans and expenditures for an upcoming 10-year period also need to understand each other clearly. They represent a wide range of research disciplines each with its own priorities, but need to express themselves in ways that will persuade participants from all the represented disciplines. Bridging language across different disciplines then becomes paramount.

Communal Precepts

Language alone, however, is not the sole impediment to coordinated progress. More important to accept is the need, at critical junctures, to adopt new and different ways of thinking and working. Ludwik Fleck documented this.[3] A drastic change in the *Denkstil* – the thought convention by means of which we express an idea or formulate a logical argument, the very vocabulary we use to convey complex concepts – constitutes a barrier that the community often finds impossible to transcend. Its acceptance ultimately hinges on the backing of the administrators of the scientific language and dictionary who belong to Fleck's *thought collective*, guardians of the bounds of permitted scientific thought.

This leadership may not be well defined: For some purposes it might be the elected members of academies, for others it might be referees deciding whether a scientific paper is worthy of publication. Advisory committees to organizations sponsoring research or directors of research institutes can play similar roles. A few individuals may serve in all four of these capacities as well as others that may influence the conduct of science. Together, these influential community members constitute Fleck's thought collective.

The thought collective is not some sinister Orwellian invention. Its members tend to be outstanding scientists – competent and often thoughtful with genuine concerns about the welfare of their discipline. But research at the forefront of science is ever changing, and even scientists who established distinguished credentials no more than a decade or two earlier, may be out of touch. The thought collective may then become a damper on progress, leading to strains between the collective and young scientists who have not yet become victims of the thought convention, and who see the Universe from new, often more fruitful perspectives.

As we saw earlier, Eddington and Einstein may have had good reasons in the 1930s for objecting to Chandrasekhar's and Oppenheimer's work on catastrophic gravitational collapse, but the Universe has its own way of reasoning about such matters, and this always trumps anything we – the astronomical community – can offer.

The history of science records innumerable disputes arising over new concepts. They show how difficult it is for established scientists to accept a new way of thinking, a novel perspective on the world, a new thought convention that, just like its predecessors might again be based on somewhat arbitrary motivations, including extraneous religious beliefs, political constraints, or budgetary imperatives, as well as purely scientific reasoning and preferences. Examples illustrating this abound:

Once the nature of galaxies had come to be properly understood in the first half of the twentieth century, every astronomer took for granted that the masses of galaxies were dominated by their stellar content with additional contributions from interstellar gases. But in 1973, the Princeton astrophysicists Jeremiah P. Ostriker and P. J. E. (Jim) Peebles postulated that galaxies might have unseen massive spherical haloes that could account for the apparent stability of the thin disks of stars observed in spiral galaxies. To keep the disks of galaxies from breaking up the masses of these haloes had to be comparable to, and perhaps higher than, the galaxies' stellar masses.[4]

A few years later, Vera Rubin and Kent Ford, of the Carnegie Institution of Washington, and Morton Roberts of the National Radio Astronomy Observatory (NRAO) made the apparent existence of such massive haloes particularly clear by deducing them from the rate at which interstellar gases orbit their parent galaxies at increasing radial distances from their centers.[5]

Vera Rubin, Figure 13.1, had been working on galaxy rotation since the mid-1960s, when she, E. Margaret Burbidge, Geoffrey Burbidge, and Columbia University's Kevin Prendergast had derived the rotation and mass of the inner parts of the galaxy NGC 4826.[6] As observing techniques improved, the rotation curves for the fainter, outer portions of galaxies came within reach. If most of a galaxy's mass was concentrated around its central nucleus, the rotational speed of matter at large radial distances from the center should decline. As Figure 13.2

Figure 13.1. Vera Rubin, at the Carnegie Institution of Washington in 1974, charting the radial velocities of matter in galaxies. (*Courtesy of the Carnegie Institution of Science*).

shows, by 1980, Vera Rubin and her associates had found instead that, beyond the central regions of a galaxy, the rotational velocities of stars and interstellar gas were almost independent of radial distance, or perhaps slightly rising.[7] If our laws of gravity were correct, this meant that matter must be gravitationally bound to its parent galaxy not just by the observed stellar matter, but by some much larger mass of matter emitting no perceptible radiation – *dark matter*.[8]

Sixteen years earlier, in 1964, Fritz Zwicky and Milton Humason at the Mount Palomar Observatory had similarly noted that galaxies were traversing their clusters at speeds of hundreds of kilometers per second. At these high velocities they should long ago have escaped the cluster, dissolving it altogether. Yet the structure and spectra of the galaxies suggested that the clusters must be very old. Some *missing mass*, as they called it, appeared to be providing the gravitational forces keeping the cluster intact.[9]

Zwicky had already anticipated these results three decades earlier in a 1933 paper published in the Swiss journal *Helvetica Physica Acta*. There, Zwicky had applied the *virial theorem* to the spread of recession velocities for the seven

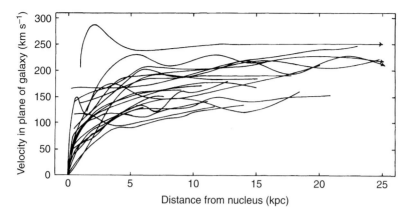

Figure 13.2. Superposition of 21 rotation curves of spiral galaxies obtained by Vera C. Rubin, W. Kent Ford Jr., and Norbert Thonnard in 1980. As the figure shows, the velocities of the ionized hydrogen regions about the centers of their galaxies either leveled off or kept rising out to radial distances at which they became too faint to observe. The smallest galaxy whose rotation curve is shown had a radius of only 4 kpc. The radius of the largest was 122 kpc. (*Reproduced from 'Rotational Properties of 21 Sc Galaxies with a Large Range of Luminosities and Radii, From NGC 4605 (R = 4 kpc) to UGC 2885 (R = 122 kpc), Vera C. Rubin, W. Kent Ford, Jr. and Norbert Thonnard, Astrophysical Journal, 238, 480, 1980. Reproduced by permission of the American Astronomical Society).*

trustworthy radial velocities for Coma Cluster galaxies available at the time – 6600–8500 km/sec – to derive a cluster mass far in excess of the luminous mass he had estimated to exist in the cluster. He concluded, "Should this prove to be true, the surprising result would emerge that dark matter exists in far greater density than luminous matter."[a;10] However, in 1933 the Hubble constant was thought to be 500 km/sec-Mpc, whereas by 1964, it appeared to be closer to ~70 km/sec-Mpc. This would have lowered his earlier ratio of dark to luminous matter by a factor of about 7. The 1964 result of Zwicky and Humason, therefore, was considerably more accurate.

These mutually reinforcing sets of observations led to the tentative acceptance of the novel concept of *dark matter* binding stars to their galaxies and preventing galaxies from escaping the cluster to which they belong. The unambiguous observational evidence could no longer be ignored. That an unobserved type of matter, referred to as *dark matter*, might exist, soon became accepted.

We do not yet know whether the observations on galaxies and clusters will eventually come to be understood in terms of some exotic form of matter or, possibly, a new theory of gravitation that holds sway over large distances – or something else altogether. The essential point is that the increasing weight of observational anomalies at some juncture could no longer be ignored and opened

---

[a] Falls sich dies bewarheiten sollte, würde sich also das überraschende Resultat ergeben, dass dunkle Materie in sehr viel grösserer Dichte vorhanden ist als leuchtende Materie.

a preparedness to consider new entities or laws governing the motions of stars and galaxies on scales measured in hundreds of thousands and millions of light years.

No formal logic defines just how much observational or experimental evidence must be collected to change the minds of fellow astrophysicists. Nor does logic take account of a potential psychological reluctance to accept new points of view. If anything, the hostility that often greets apparently frivolous novel ideas – frequently dismissed as *crank ideas* – forbids a logical growth of understanding. This is why the ways in which new understanding is gained are so difficult to depict: We lack a predictive theory of how scientific insight grows or how knowledge of the Universe will continue to evolve. No universal theory of knowledge competently touches on this central topic.[11]

* * *

Ludwik Fleck points out that even where we are able to look back on the history of a successful scientific advance, it is difficult and perhaps not even possible to correctly record its evolution. Often the history would have to be presented in terms of multiple, crisscrossing, mutually interacting evolutionary lines of thought, which all would have to be followed individually as well as in relation to their mutual development.

Fleck characterizes this as akin to trying to faithfully record a conversation between several people all simultaneously talking with and past each other while, despite the apparent confusion, a single idea suddenly crystallizes from the conversation.[12] We would have to continually interrupt the portrayal of these thought trajectories in order to introduce a newly developing line of thought; to halt discussion in order to explain the various connections; and to omit a great deal in order to maintain sight of some main emerging trend.

A historical sequence then emerges as a more or less artificial thread that replaces the lively interchange, memory of which this history simultaneously erases because the path along which the new agreement was reached no longer seems important. A new, more stylized justification for the new viewpoint is invented instead, which is more fit, more acceptable for presentation to colleagues or for inclusion in scientific publications or text books.

It is this extinction of memory of how a new perspective emerges that largely explains why astronomers so firmly – if mistakenly – believe that our depiction of the Universe, the astrophysical landscape discussed in Chapter 9, stands on its own, isolated from extrinsic influences such as money or politics.

* * *

A similar erasure of memory can be found in the astrophysical community's acceptance of thought conventions.

Chapter 8 recounted how, in the late 1940s, Hermann Bondi, Thomas Gold, and Fred Hoyle proposed that the Universe had existed for all time and was continually creating new matter to replace existing stars and galaxies as the Universe expands and galaxies escape beyond the cosmic horizon.

The Steady State theory thus removed the apparent contradiction between a rapidly expanding universe that appeared to be younger than the Solar System and globular-cluster stars. The theory was generally opposed and sometimes derided by physicists of the 1950s, who maintained that you cannot just form something out of nothing! They objected that it violated the law of conservation of mass and energy, the bedrock on which all of physics rested.

An early example is an exchange in the journal *Nature* between Max Born, one of the leading pioneers of quantum mechanics, and Fred Hoyle. Born protested, "if there is any law which has withstood all changes and revolutions in physics, it is the law of conservation of [mass–energy]."[13]

Hoyle responded that, despite its continual creation, the amount of matter and energy enclosed within the cosmic horizon was precisely constant for all time.[14] But this global energy conservation did not satisfy physicists who were more concerned with conservation of energy on smaller scales.[15] How would one account for energy conservation in the creation of individual hydrogen atoms as the Universe expanded?

There was a twist of irony in all this.

The creation of matter that the Steady State theory required was provided by a quantity formally quite similar to the *cosmological constant* of general relativity, which Einstein had employed in 1917 to model the Cosmos.

Although the Steady State theory is by now recognized to conflict with observations, the physics and astrophysics communities have come full circle to embrace a *dark energy* pervading the Cosmos and accounting for its accelerating expansion.

A rational explanation for accepting the potential self-regeneration of a universe can be found by returning to Chapter 3 and reexamining Lemaître's equation 3.4. The equation holds true, whether or not a cosmological constant is at play and, in particular also, if all or part of the mass–energy density of the Universe, ρc^2, includes or even is dominated by a cosmological constant. If the mass–energy density is to remain constant as the Universe expands, that is, if $\dot{\rho} = 0$, equation (3.4) simply tells us that the relation between pressure, P, and mass–energy density must be:

$$P = -\rho c^2. \tag{13.1}$$

If the mass–energy density is due to a positive cosmological constant, then the pressure must be negative – and vice versa. In everyday life, we are not used to thinking in terms of negative pressures or negative mass–energy densities, which is why Max Born and physicists of the 1950s thought the Steady State theory violated energy conservation. But, setting everyday intuition aside, we have no formal, no objective, reason for ruling out a universe in which negative pressures exist.

Why do physicists now see no conflict between this continually regenerated energy and the laws of conservation of mass and energy?

In part it may be because they were not thinking of the Steady State theory in the early 1980s, when they invented inflation to account for the apparent flatness of space, the homogeneity of the Universe out to the greatest distances, and the absence of magnetic monopoles. In his interview with Alan Lightman conducted in early 1989, the cosmologist Dennis Sciama, who had worked on aspects of the Steady State theory in the 1950s, recalled a chance conversation with Alan Guth soon after 1981 when he had first proposed his theory of inflation. "Guth came to the Royal Society in London for some meeting. He spoke, and at lunch I remember saying to him, 'Do you realize that your inflationary epoch is just the Steady State theory?' And he said, 'What is the Steady State theory?' He hadn't even heard of it. So that is just one of many reminders about culture gaps, or time gaps and culture gaps."[16]

The new vacuum, which remained after inflation had subsided and may account for the dark energy existing today, exhibits an energy density that pales when contrasted to that of the inflationary vacuum. Nevertheless, it has the same ability to regenerate itself and its dark energy, and may continue to do so for all time.

Today, few physicists would deride either the existence of different types of vacua or their ability to regenerate their energy densities out of nothing. Where the idea of an indefinitely renewable vacuum energy comes from no longer seems to matter. Far from seeing any contradictions in not just one, but two sets of vacua that can regenerate themselves and their energy contents, physicists now embrace the vacuum as a vibrant medium. Its full set of properties still remains unknown and is a topic vigorously investigated. A concept that once seemed unthinkable is now widely accepted.

Erasure from memory of such historical debates shares similarities with the erasure of discovery tools mentioned earlier. Once the astronomical community accepts a novel discovery, mention of the observing apparatus that made the discovery possible no longer seems relevant. Once a new theory is adopted, the rationale originally developed for accepting it seems unimportant. Once the community establishes a coherent astrophysical account, this in itself justifies and supports the newly reconfigured landscape.

Thought Conventions, Paradigm Shifts, and Network Cascades

The meandering cosmological history just recounted illustrates two of Ludwik Fleck's early observations on a scientific community's thought conventions worth recalling here, albeit in a more modern astrophysical setting:

1. *A contradiction to an accepted set of beliefs becomes unthinkable.* The Steady State theory ran headlong into this mindset: To the community of physicists creation of matter out of nothing violated conservation of energy, a proposal

that in mid-twentieth century appeared outrageous. Not until 30 years later could such an apparent violation be accepted.

2. *However, once a concept has entered the scientific vocabulary, it becomes part of accepted knowledge.* Whether right or wrong makes no difference; once formulated, the concept cannot be erased. Often an erroneously held view that may rightly have been discarded is resurrected in a new setting and found to be of use. Scientists will unearth and rehabilitate it if it is handy. The creation of matter and energy out of nothing is just one example of this historical trend.[17]

In his book, 'The Structure of Scientific Revolutions,' Kuhn recognized that normally any branch of science is pursued under two conditions: first, the existence of a recognized set of legitimate problems and tools for solving them; and, second, a sense that the field is sufficiently open-ended to permit all sorts of problems to be tackled.[18] The combination of these two enables the discipline to attract a new generation of adherents who otherwise might engage in other scientific activities.[19]

In Chapter 12 we saw Edwin Turner depicting the attraction of inflationary theory in somewhat similar terms, "a model that allows one to do lots of cute calculations. It's a theorist's gymnasium, so to speak, where one can go, and there are lots of nice problems to do and to be worked out."

Kuhn's paradigms guide what he called *normal science*. He wrote, "by choosing [the term paradigm] I mean to suggest that some accepted examples of actual scientific practice – examples which include law, theory, application, and instrumentation together – provide models from which spring particularly coherent traditions of scientific research."

Today, astronomical disciplines such as cosmology, X-ray astronomy, relativistic astrophysics, the search for and investigation of extrasolar planetary systems, all might be considered areas in which the scientists involved share a commonly accepted set of approaches, tools, and criteria for judging success.

However, as Kuhn pointed out, the fate of paradigms is that they ultimately generate anomalies that might at first be ignored but then are found in increasing number, until workers in the field become aware of a crisis they never anticipated when they initiated their investigations. Kuhn writes[20]:

> Confronted with anomaly or with crisis, scientists take a different attitude toward existing paradigms, and the nature of their research changes accordingly. The proliferation of competing articulations, the willingness to try anything, the expression of explicit discontent, the recourse to philosophy and to debate over fundamentals, all these are symptoms of transition from normal to extraordinary research.

Kuhn calls transitions of this type a *paradigm shift* accompanying a *scientific revolution*. He takes these to be[21]

[T]hose non-cumulative episodes in which an older paradigm is replaced in whole or in part by an incompatible new one. . . . [s]cientific revolutions are inaugurated by a growing sense . . . that an existing paradigm has ceased to function adequately in the exploration of an aspect of nature to which that paradigm itself had previously led the way. . . . the sense of malfunction is prerequisite to revolution.

Today, Kuhn's revolutions might more appropriately be thought of as a particular subclass of the network cascades depicted by Duncan Watts.[22] Astrophysicists, like members of any other social network, are continually looking over their shoulders to see what their peers are doing. When they perceive a critical aggregate within the *giant component* of their community tending to seek new directions, a change of mind by even just one colleague may convince them to also join in the search for a new communal mind set, a new thought convention, a new direction for their research, a new paradigm. This results in a cascade, which stabilizes only once a sufficient number of colleagues agree on a new way forward.

Associating astrophysical revolutions of all kinds with Watts's cascades makes good sense. At least revolutions and cascades share one significant trait: A significant membership of a scientific community changes its research directions or topics.

In the course of the twentieth century such astrophysical cascades were more frequently precipitated by totally unanticipated findings, such as the discovery of neutron stars, black holes, or galaxies that radiated most of their energy as radio waves or X-rays, rather than in the visual domain. Each of these unanticipated discoveries led whole communities of astronomers to drop what they were doing and start work on one of the newly discovered phenomena. The new phenomenon appeared to offer more exciting prospects!

Note that none of these observational discoveries had been preceded by Kuhn's troubling anomalies or crises, or a community's doubts about the paradigm it was pursuing. The observational discoveries came out of the blue. They were exciting! The astronomers involved wanted to shift to something new.

This is not to say that either Fleck or Kuhn were wrong in their perceptions about the prevalence of changing thought conventions or paradigms. However, to suggest that all the revolutions cited by Kuhn – notably the discovery of X-rays or radioactivity, which he included among his revolutions – had been preceded by confrontations between factions holding different beliefs would be difficult to defend. Both discoveries were unexpected. Neither involved an agonizing readjustment, even if considerable research was required to understand it fully. But both discoveries decisively changed the course of science.

Cascades of many different kinds are marked by mass migrations of scientists, from one range of activities to another. But these can and do have a variety of different causes, which should be recognized and distinguished. Clarity of vocabularies is useful not just in science, but also in depicting how the community of scientists works and behaves. The word "revolution" is too easily applied to cascades that may have quite different origins. Sometimes it might also be applied to significant advances that show no signs of having triggered a cascade. Developing a clearer vocabulary for these different types of behavior could provide greater insight on how science advances.

Social Constructs

Some of the many twists and turns of astrophysics arise from constructs vigorously defended because they seem so closely aligned with everyday intuition, or are an apparently clearcut consequence of observations that have survived so many decades of scrutiny that they *must* be true.

Must?

It helps to be specific. What exactly is an *astrophysical construct*? Here are three brief examples:

In 1917, Einstein tried to show how general relativity, whose structure he and David Hilbert had just divined with great effort, could provide a suitable model of the Universe. As everyone else agreed at the time, the model had to be static – unchanging for all time. But general relativity could not prevent a static universe from collapsing under the mutual gravitational attraction of its stars. Isaac Newton had already worried about this problem two centuries earlier. So Einstein just added a cosmological constant to general relativity. This fixed it. It kept the universe from collapsing.

Fourteen years later, when Einstein was persuaded that the Universe is expanding, he ruefully realized he had manufactured the cosmological constant only to keep the Universe from collapsing. General relativity was far better off without it. The model of the Universe he had created, he realized, had been a construct, an edifice based solely on his preconception that the Cosmos must be static.

* * *

In those same decades, Henry Norris Russell and Arthur Stanley Eddington were beginning to model stars. As everyone else agreed at the time, chemical abundances in stars had to be identical to those on Earth. The strong spectral features of heavy elements in stellar spectra indicated as much. On this expectation, Eddington created a beautiful theory of the structure of stars composed of heavy elements, deriving how radiation is transferred from the stars' interior out to the surface, and deducing a relation between the masses of stars and their luminosities – an apparent theoretical triumph.

When the Sun and stars stubbornly refused to cooperate, and Cecilia Payne's measurements clearly exhibited a predominant abundance of hydrogen, Russell demurred. In a letter to *Nature*, he and his Princeton colleague Karl Compton had already made clear that a straightforward analysis "would demand an absurdly great abundance of hydrogen" if taken at face value.

Why? Apparently, to maintain the notion that the chemical composition of the stars had to be identical to that of Earth. The model Eddington and Russell had created was definitely a construct.

* * *

In 1948, Bondi, Gold, and Hoyle created the Steady State theory to explain how stars could be older than the apparent age of the Universe judged by its expansion. To do this, they needed to account for the continuous creation of matter, and its rapid conversion into stars – all at the precise rate required to maintain the appearance of the Universe, and all of the stars and galaxies it comprised, constant for all time. When quasars were discovered 15 years later, and found to have existed in higher abundance at high redshifts – at earlier times – the theory encountered its first observational challenge. The subsequent discovery of the microwave background radiation, for which Steady State theory had no apparent explanation, was a further blow. The model Bondi, Gold, and Hoyle had created was a construct created to conform to an error-riddled Hubble constant.

* * *

Given these examples, one needs to worry whether the inflationary universe, created to explain why space is homogeneous, isotropic, and flat, and why magnetic monopoles can't be found, might be a construct as well. Like the Steady State theory, it also requires the constant regeneration of a highly energetic vacuum, out of nothing, during its inflationary phase.

Is dark energy a construct? Its explanation similarly is based on the notion that vacuum energy can regenerate itself out of nothing, for all time, by virtue of a hypothesized negative pressure. Has its existence been postulated only to avoid challenges to Einstein's general relativistic theory of gravity?

While neither inflation, nor dark matter, nor dark energy need be constructs invented to salvage theories we feel too important to abandon, the possibility that they may well be constructs should not be hastily dismissed.

The true Universe may be quite different!

Social Pressures, Productivity, and Rewards

One more question: Why are some astrophysical advances that are obviously important in hindsight not more thoroughly taken to heart when they first appear?

In Chapter 8, we encountered Ralph A. Alpher and Robert C. Herman, two young theorists who in 1948–9 predicted that the Universe should be permeated by electromagnetic radiation at a temperature roughly 1 to 5 degrees above absolute zero.[23] In 1965 this radiation was accidentally discovered by Arno Penzias and Robert W. Wilson at the Bell Telephone Laboratories in New Jersey.[24] Its temperature is now known to be 2.73 K, well within the predicted range. But by 1965, the work of Alpher and Herman had long been forgotten. Robert Dicke and his co-workers at Princeton University had in the meantime reinvented essentially the same theory, and Alpher and Herman were given little of the credit that prediction of a major new phenomenon is usually accorded.[25]

The simplest explanation for the neglect of such predictions may be that they seldom can be pursued quickly. The tools often are lacking and may take decades to develop, often only in response to unrelated commercial or national demands. Prudent scientists might then just ignore such predictions and go about the business of producing publishable results.

Steady scientific publication ensures survival as a scientist and shapes the work a scientist produces. Work within a well-established system of thought exposes a scientist to few risks. Publishing results that confirm widely held beliefs antagonizes nobody and yet can be useful to the community.

More difficult for an author is a paper that challenges the beliefs of fellow scientists, casting their work in doubt. Journal referees then ask for higher standards of proof than required of publications that conform and confirm. A journal's referee may even recommend the rejection of an audacious new paper on the grounds that it is probably wrong and at least needs considerably more backing.

Other practical factors are also at play. Most astronomers and astrophysicists are focused on projects they are committed to complete. Their work has been sponsored by an agency that will provide continuing support only if promised research is concluded satisfactorily. This requires avoiding distractions in order to obtain publishable results. Anomalies that arise can only be briefly explored before one must return to the main problem at hand. Where some nagging puzzle persists, one figures some explanation may eventually turn up; most likely, it will be nothing significant. In the meantime there are more important things to do....

This difficulty is emphasized in a posthumously published autobiographical sketch Einstein wrote shortly before his death.[26] He recalled his own good fortune in having obtained an appointment to the Swiss Patent Office where, once he had fulfilled his duties at work, he could bury himself in favorite problems without having to fear that his efforts might not succeed.[b]

> Through this, I was freed from financial worries in the years of greatest creativity, 1902–9. Quite apart from this, the work on the ultimate formulation of technical patents was a veritable blessing for me. It demanded multifaceted thinking, and stimulated physical reasoning. In the end, a practical occupation is actually a blessing for people like me. For, the

[b] Dadurch wurde ich 1902–9 in den Jahren besten productiven Schaffens von Existenzsorgen befreit. Davon ganz abgesehen, war die Arbeit an der endgültigen Formulierung technischer Patente ein wahrer Segen für mich. Sie zwang zu vielseitigem Denken, bot auch wichtige Anregungen für das physikalische Denken. Endlich ist ein praktischer Beruf für Menschen meiner Art überhaupt ein Segen. Denn die akademische Laufbahn versetzt einen jungen Menschen in eine Art Zwangslage, wissenschaftliche Schriften in impressiver Menge zu produzieren – eine Verführung zur Oberflächlichkeit, der nur starke Charaktere zu widerstehen vermögen.

academic career forces a young person into a bind to produce scientific papers in impressive numbers – an enticement to superficiality that only a strong character can resist.

Trust in the Leaders of Science

In Chapter 2, I described the system of justification and acceptance of new ideas in astrophysics. It suggested that scientists are persuaded through logic. But this makes for an incomplete depiction.

A less often cited, though very powerful type of persuasion is one that was frequently heard in the United States when the highly gifted theoretical physicist Richard Feynman was still alive. It went roughly like this: "I had lunch with Feynman yesterday, and he said...."

Feynman's influence was enormous. When he made a statement, physicists of all types took note. Statements by lesser if more numerous trusted scientists, as we saw earlier, can similarly persuade colleagues by launching a cascade leading to widespread acceptance.

Scientists do not always use objective means to persuade each other, but invoke unabashedly the authority of someone as brilliant as Feynman, even while expressing disdain for authority. Although this generally is only a temporary circumstance, and ultimately stronger proof is required, the influence of greatly respected scientists is often profound, and even lesser leaders of the scientific establishment can exert impressive influence through their work on advisory boards.

Given the importance of communal forces that shape acceptance of new concepts and theories, how can we ever be sure about whether our astrophysical and cosmological theories are sound, rather than mere *constructs* we may have inadvertently fabricated because they appear so plausible? Are there ways in which we can disabuse ourselves when we have constructed false theories?

A construct has two characteristics. First, it is authoritarian, deriving legitimacy mainly because the full authority of the scientific community stands behind it. Second, it agrees with most if not all established scientific findings; but it makes few specific predictions that could be verified experimentally or confirmed through observations. Where it does make such predictions it may soon become outdated when its predictions fall short.

Authoritarianism – the influence of Fleck's thought collective on the acceptance of a way of thinking – may, by itself, not be a sufficient reason to suspect that our understanding is a mere construct; it is only a warning that the community should beware.

Notes

1. *The Web of Belief*, W. V. Quine. New York: Random House, 1970, p. 29.
2. *Gödel, Escher, Bach – an Eternal Golden Braid*, Douglas R. Hofstadter, New York: Basic Books, 1979, p. 26.

3. *Entstehung und Entwicklung einer wissenschaftlichen Tatsache – Einführung in die Lehre vom Denkstil und Denkkollektiv*, Ludwik Fleck. Basel: Benno Schwabe & Co. 1935; Frankfurt am Main: Suhrkamp, 1980.) (English translation: *Genesis and Development of a Scientific Fact*, Ludwik Fleck, translated by Fred Bradley & Thaddeus, J. Trenn, edited by Thaddeus J. Trenn & Robert K. Merton. University of Chicago Press, 1979.

4. A Numerical Study of the Stability of Flattened Galaxies: or, Can Cold Galaxies Survive? J. P. Ostriker & P. J. E. Peebles, *Astrophysical Journal*, *186*, 467–80, 1973.

5. Extended Rotation Curves of High-Luminosity Spiral Galaxies V. NGC 1961, the Most Massive Spiral Known, Vera C. Rubin, W. Kent Ford, Jr., & Morton S. Roberts, *Astronomical Journal*, *230*, 35–39, 1979.

6. The Rotation and Mass of the Inner Parts of NGC 4826, V. C. Rubin, E. Margaret Burbidge, G. R. Burbidge & K. H. Prendergast, *Astrophysical Journal*, *141*, 885–91, 1965.

7. Rotational Properties of 21 Sc Galaxies with a Large Range of Luminosities and Radii, From NGC 4605 (R = 4 kpc) to UGC 2885 (R = 122 kpc), Vera C. Rubin, W. Kent Ford, Jr. and Norbert Thonnard, *Astrophysical Journal*, *238*, 471–87, 1980.

8. Rotation Curves in Spiral Galaxies, Yoshiaki Sofue & Vera Rubin, *Annual Reviews of Astronomy & Astrophysics*, *39*, 137–74, 2001.

9. Spectra and Other Characteristics of Interconnected Galaxies and of Galaxies in Groups and in Clusters III, Fritz Zwicky & Milton L. Humason, *Astrophysical Journal*, *139*, 269–83, & plates 11, 63, and 72, 1964.

10. Die Rotverschiebung von extragalaktischen Nebeln, F. Zwicky, *Helvetica Physica Acta*, *6*, 110–27, 1933; see p. 126.

11. Ibid., *Entstehung*, Fleck, p. 17.

12. Ibid., *Entstehung*, Fleck, p. 23.

13. Formation of the Stars and Development of the Universe, Pascual Jordan, with an introduction by Max Born, *Nature*, *164*, 637–40, 1949.

14. Development of the Universe, *Nature*, F. Hoyle, *165*, 68–69, 1950.

15. *Cosmology and Controversy – The Historical Development of Two Theories of the Universe*, Helge Kragh, Princeton University Press, 1996, 196–201, which has a more extensive discussion of the controversy.

16. *Origins – The Lives and Worlds of Modern Cosmologists*, Alan Lightman and Roberta Brawer. Cambridge MA: Harvard University Press, 1990, p. 147.

17. Ibid., *Entstehung*, Fleck, p. 31.

18. *The Structure of Scientific Revolutions*, Thomas S. Kuhn, The University of Chicago Press, 1962.

19. Ibid., *The Structure of Scientific Revolutions*, Kuhn, p. 10.

20. Ibid., *The Structure of Scientific Revolutions*, Kuhn, p. 90.

21. Ibid., *The Structure of Scientific Revolutions*, Kuhn, p. 91.

22. A simple model of global cascades on random networks, Duncan J. Watts, *Proceedings of the National Academy of Sciences of the USA*, *99*, 5766–71, 2002.

23. Remarks on the Evolution of the Expanding Universe, Ralph Alpher & Robert Herman, *Physical Review*, *75*, 1089–95, 1949.

24. A Measurement of Excess Antenna Temperature at 4080 Mc/s, A. A. Penzias, and R. W. Wilson, *Astrophysical Journal*, *142*, 419–21, 1965.

25. Cosmic Black-Body Radiation, R. H. Dicke, P. J. E. Peebles, P. G. Roll, & D. T. Wilkinson, *Astrophysical Journal*, *142*, 414–19, 1965.

26. Autobiographische Skizze, Albert Einstein, in *Helle Zeit – Dunkle Zeit, in memoriam Albert Einstein*, edited by Carl Seelig. Zürich: Europa Verlag, 1956, p. 12.

Part III The Cost of Discerning the True Universe

14

Organization and Functioning of the Astronomical Community

Whether we ultimately succeed or fail in determining the origin and early evolution of the Universe is likely to be determined by two competing factors. The first is the extent to which the high-temperature phases prevailing in the early Cosmos may have eradicated all memory of preceding epochs at the dawn of time; the second is the monetary cost of searching for shards of information that may have escaped erasure so we might recover and analyze the fragmentary surviving evidence.

The Larger Network in Which Astronomy Is Embedded

A part of the difficulty of accounting for the conduct and progress of astrophysics, even considering all the influences I have already cited, is that the field cannot be fully isolated from the far larger setting in which it is embedded.

Modern astronomy is expensive and competitive. It is expensive because powerful instrumentation is costly, whether it be telescopes or supercomputers. It also has to remain competitive because the cost of astronomical projects has to be justified at a national level where astronomy competes with other sciences for limited resources.

Seen from the perspective of an individual astronomer, these two factors lead to the need to continually justify the funding and potentially also the observing time required to initiate a project. First and foremost, this involves persuading the larger community of working astronomers to agree to the funding. Persuasion and its associated political activity within the field is part and parcel of the work of almost every established astrophysicist.

Proposals must be submitted to funding agencies. Obtaining funds not just for equipment, but also for graduate students and postdoctoral workers who will be needed to help carry out a project, is generally part of such an effort. Once funding is received, a separate set of persuasive proposals usually needs to be submitted to obtain observing time at an observatory or access to a sufficiently powerful supercomputer. Most of these facilities are heavily oversubscribed; eloquent persuasion is thus important even when funding has already been obtained. The network of persuasive steps is complex and outlined in simplified form in Figure 14.1.

Astronomers may also be asked to sit on advisory boards to a variety of funding agencies or to the National Academy of Sciences to assist in writing a community master plan. There they face the task of persuading fellow board members that a particular line of research should be assigned high priority.

As we saw in Chapter 11, the Astrophysics Council similarly was asking fellow astronomers to seek support for the proposed *Great Observatories* by visiting their representatives in the U.S. Congress, to explain the purpose and importance of this family of four observatories and ask for Congressional support.

High-level governmental decisions motivated by political concerns may occasionally also mandate new scientific initiatives. Some of these may be welcome. If fears of a potentially disastrous collision of a comet or asteroid with Earth leads the government to order a broad investigation into the nature and trajectories of these bodies, planetary systems researchers will appreciate the decision.

The process is complex but largely follows prescriptions envisioned in Vannevar Bush's 1945 report, 'Science – The Endless Frontier.'[1] Support for astronomy requires not only a scientific rationale, but also agreement with the priorities of a President and the Congress who, complicating matters further, do not always agree.

Funding for a project then vacillates with national elections, posing ever-present threats to long-term projects. Lean times can dictate delay or cancellation of missions. Military secrecy governing different classes of technical equipment may negate prospects for international collaboration on projects too costly for the United States to support by itself. The starts and stops on such projects are always unnerving to those who have spent years devoted to their success only to find those prospects threatened.

Frequently, top-down initiatives are motivated by nonscientific concerns – sometimes the need to maintain a national cadre of competent engineers, whose expertise might be required for some future defense efforts and who otherwise would have to be disbanded for lack of immediate work. To the affected working scientists these interventions may seem ill advised, but they may be essential for national security. If the government then assigns these engineers the construction of some powerful new facility that most astronomers might consider superfluous, those few willing to switch from whatever research they have been conducting may ultimately benefit from making the change.

In the United States, the Space Station, initially announced by President Ronald Reagan as important to astronomers and space scientists, is an example of a facility that required significant restructuring. Earlier, we saw how this Presidential initiative eventually restricted its support for astrophysical programs to avoid even higher cost overruns.

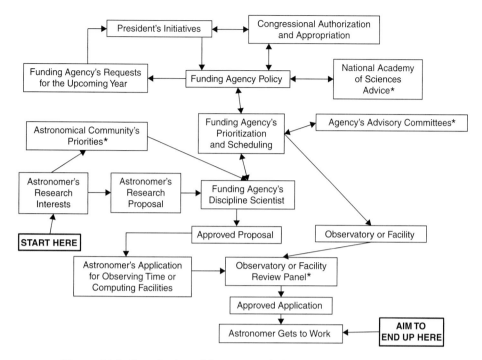

Figure 14.1. Organization of the astronomical community in the United States, as viewed by an individual observer wishing to carry out a novel observation. Depending on the cost of the proposed observation, and whether or not it falls into a research discipline that already is recognized as significant, substantial delays may be anticipated before even a well-reasoned proposal can find approval. Influential advisory boards convened every 10 years to evaluate major astronomical projects for the National Academy of Sciences and the National Research Council, which serves it; advisory committees to the National Science Foundation, NASA, or the Department of Energy; politically or economically motivated Presidential initiatives or Congressional legislation; all can sponsor potentially conflicting astronomical programs and greatly affect whether or not a project will receive approval and funding. Directors of observatories also have considerable discretion, subject to approval by observatory governing boards and time allocation committees that recommend how observing time should be distributed among astronomers submitting competing proposals. Asterisks * denote committees whose membership may appreciably overlap.

In a similar vein, two decades later on January 14, 2004, President George W. Bush proclaimed, "We will build new ships to carry man forward into the universe, to gain a new foothold on the Moon and prepare for new journeys to the worlds beyond our own." He offered to provide powerful new rockets with which astronomers could eventually place new observatories on the Moon.[2] This too wilted with time as the funding required could not be justified. But while the offer lasted, those who were persuaded to shift their research in the indicated direction were able to benefit. The government often persuades reluctant scientists with the funding it offers those willing to be flexible.

The Size and Structure of the Astrophysical Community
Utility and Funding

Scientific directions to pursue in a long-term program are paramount, but practical considerations come into play in determining the ways and the rates at which we should proceed.

We might, for instance, expect the rate at which astrophysical understanding accumulates to roughly correspond to the number of active investigators. But does it? And, if so, what are the limits that determine how many scientists might or should devote themselves to astronomy?

A primary and readily understandable limit on the size of the community of astronomers and astrophysicists, then, has to be funding that society can afford.

But other limits exist as well.

The Australian theoretical anthropologist Roland Fletcher has identified three such limits in his studies of the growth of human settlements – communities: villages, towns, and cities.[3] Remarkably, the same limits that apply to the growth and survival of Fletcher's settlements also apply to the prospects for success or failure of astronomical research communities and the development of the astrophysics community worldwide.

The Interference Limit

The first of Fletcher's limits arises when the density of a community becomes so high that people interfere with each other's activities. When that happens, the community ceases to grow. Its members move elsewhere. Fletcher calls this limit the interference limit and designates it by the letter I.

The mutual interference between astronomers is perhaps nowhere as apparent as in what the directors of major astronomical observatories euphemistically call their "oversubscription rates." A high oversubscription is considered a mark of an observatory's distinction and success. If four times as many astronomers are submitting applications to make use of the observatory than can actually be accommodated, the observatory must be offering a highly valued service. The flip side of this is that three out of every four astronomers who would like to make use of the observatory for their research are turned down.

Oversubscription rates of this magnitude are not uncommon, either for observatories or for funding agencies. And because the entire system of graduate student education depends on raising enough money or obtaining observing time, access to supercomputers, or other resources to support students' studies, most senior researchers involved in education cannot escape the necessities of competing for their share of support. Instead of devoting themselves to research, which is what they do best, they spend much of their time on entrepreneurial activities.

This is where the interference limit most surely applies. When astronomers spend most of their time writing grant applications and hardly any on research, we will have reached the limit where further growth is self-defeating. The field will be overcrowded.

The Threshold Limit

A young scientist may decide to enter a new rather than an established discipline because the newer realm, though far less certain in potential yield, may offer more freedom to flourish. On his arrival in Göttingen in the summer of 1928, for an opportunity to participate in the exciting new realm of quantum mechanical research, the 24-year-old George Gamow chose just this path to the future.

As we saw in Chapter 6, that summer, the Göttingen physicists were concentrating on the quantum theory of atoms and molecules. Gamow didn't want to compete in this melee. He chose to work quietly instead on a quantum theory of the nucleus – work that had not yet become fashionable. There he could work at his own pace, with no worries about interference or competition from others.

Venturing into a sparsely populated new discipline, however, is not without its dangers either. If the field attracts little interest, nobody will care much about what you do. Your publications will not be read and the funding required for your research will be hard to attract. No matter how intrinsically interesting the subject, the work is likely to find little appreciation, sometimes until long after you have left the scene when a new generation rediscovers the field, remains unaware of the earlier work, and repeats it without realizing that it had all been done before. This, as we saw in Chapter 8, was the fate of Alpher and Herman and their prediction of the microwave background radiation.

The scientific literature is full of advances that were ignored in their time only to be rediscovered later, usually with inadequate appreciation of the work of those who first carried it through. Pity the scientists ahead of their times and ignored. Their lives are unenviable.

Fletcher defines a threshold, designated T and illustrated in Figure 14.2, below which a human settlement cannot be expected to flourish. A field of science seeking to flourish requires a sizable community to provide interaction, discussion, debate, crosschecking, refutation, and ultimately acceptance. Unless the field succeeds in attracting a sufficient membership to show rapid progress, the larger scientific community will ignore it. The field will be considered unfashionable.

A vibrant community working on a fashionable problem readily attracts knowledgable colleagues able to provide constructive criticism. A lone astronomer or a small group working in a remote field of its own choosing may make an advance but will have few colleagues who fully understand its significance or who can voice an informed opinion. Workers outside the mainstream find themselves largely ignored, a group on whose work there is no consensus. Below the threshold limit they remain invisible. In the network of scientists, those detached from the giant

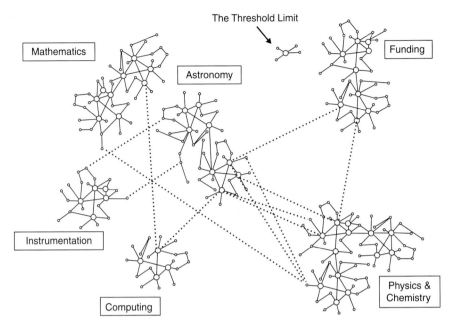

Figure 14.2. The threshold limit. Researchers who remain detached from the giant component connecting the vast majority of astrophysicists through co-authored publications may find their work largely ignored, even if it is original and interesting. A direct communication link to the larger research community may be needed if one's work is to gain attention – or obtain funding.

component in the upper right of Figures 9.8 and 14.2 fall below the threshold limit.

One productive scientist who often appeared to work just at, or somewhat below, the threshold level was the Estonian-born Ernst Öpik. Two factors led to his isolation. Most of his work was carried out at observatories outside the mainstream, initially at the University of Moscow during World War I, later at the Dorpat Observatory in Estonia; and, with Estonia occupied by the Soviet Union during World War II, at the Armagh Observatory in Ireland, where he worked until the end of his life in 1985. Öpik's interests ranged so widely that, in many of his investigations, he perpetually appeared to be an outsider. His main area of interest was the study of minor bodies in the Solar System – asteroids, comets, and meteorites. But he also emerged as a contributor to many other areas of astrophysics and cosmology.

As we saw earlier, in 1916, Öpik discovered the extremely high density of a star now known to be a white dwarf. He considered its density incredible but, in time, similarly dense stars came to be recognized and accepted as real.[4]

Six years later, in 1922, he derived a reliable distance to the Andromeda Nebula, establishing it as a stellar system equivalent to that of the Milky Way, but at great distance. From then on, other spiral nebulae could be safely taken to be independent galaxies as well.[5]

In 1938 Öpik published a 115-page-long article on 'Stellar Structure, Source of Energy, and Evolution.'[6] There, he noted that the "direct synthesis of the deuteron from protons (with the expulsion of a positron)" and also the merger of $^{12}C + {}^1H$ would be able to supply the radiation of the sun for 500 million years, at [a central temperature of] $T_c \sim 2 \times 10^7$. He did not anticipate the carbon cycle, which Bethe described the following year, but the stellar structure considerations he took into consideration made his work impressive.

In 1951, about the same time as Edwin Salpeter, Öpik also proposed the triple-α process for forming carbon ^{12}C in stars as their central temperatures rose.[7] He was not aware of the resonance level in beryllium, which Salpeter knew about through his affiliation with Fowler's Caltech group. But Öpik's effort nevertheless was impressive, particularly because the same paper also predicted the successive formation of oxygen, neon, magnesium, and argon through a sequence of α-particle assimilations as these more massive elements formed at increasing stellar temperatures.

In an autobiographical essay Öpik published late in life, he recalled receiving a letter from Gamow after his 1938 paper on 'Stellar Structure, Source of Energy, and Evolution' had appeared. Gamow was "underlining the importance of my work but reproaching me for publishing in such an 'obscure' place, wherefore in his opinion progress in the study of stellar structure must have been unnecessarily delayed."[8]

Öpik defended his custom of reporting on his work primarily in publications issued by the two observatories at which he had spent most of his professional life – initially the Dorpat Observatory in Tartu Estonia, and later the Armagh Observatory in Ireland. He praised the "centuries-old astronomical tradition of exchange of publications…[among]…all astronomical institutes in the world," and evidently felt that this was an honored tradition he was pleased to follow. But by the twentieth century, most astrophysicists were publishing in more accessible journals, so that much of Öpik's work was overlooked or discovered only too late to have an appreciable impact on evolving thought. Öpik tended to write lengthy, comprehensive articles, and noted that the cost of publication of his papers in the *Astrophysical Journal* "would have been absolutely prohibitive." For him and for the institutions at which we worked, the journal's page charges were unaffordable.

Working at or below the threshold is not a reflection on the quality of a scientist's contribution, nor on the importance of the work. A current endeavor of particular significance is the *Search for ExtraTerrestrial Intelligence (SETI)*. Its dedicated team works largely in isolation from the giant component of astronomers working on programs with a more assured yield. But if SETI should succeed in locating intelligent life elsewhere in the Universe, the finding would send tremors across the globe as astronomers rushed to pursue the discovery!

Astronomy needs observers and theorists willing to tackle difficult problems regardless of whether their quest soon flourishes!

The Communication Limit

Fletcher's third and final limit is the communication limit, designated C. Unless satisfactory lines of communication can be maintained across a field, it will not flourish. The field may then split into smaller disciplines that hardly communicate with each other. This is already true of the larger well-established field of physics. Note the nearly isolated clumps within physics, mathematics, and astronomy pictured in Figure 14.2 that are joined by only a single link.

At one time the American Physical Society published The *Physical Review*, a journal for professional physicists. This has now split into *Physical Review A*, which publishes articles on "atomic, molecular and optical physics"; *Physical Review B*, which specializes in "condensed-matter physics"; and so forth, through *Physical Review E* for "statistical, nonlinear and soft-matter physics." These journals' fields seldom intersect; but in order not to lose total oversight, the Physical Society does publish *Physical Review Letters*, which accepts short research accounts on all sorts of topics that may inspire colleagues active in other areas of physics. In 2011, the Physical Society also added *Physical Review X*, intended to serve as "an online-only, fully open access, primary research journal covering all of physics and its applications to related fields."

- Astronomy and astrophysics are now approaching a point where a similar lack of communication no longer permits most members of the field to maintain insight on work conducted in research disciplines other than their own. A new journal modeled along the lines of *Physical Review X*, but dedicated to astrophysics, could help to keep astronomy intact.

Sidestepping the Limits

As Fletcher points out, the interference and communications limits are not absolute. They can be overcome by taking suitable measures. In astronomy and astrophysics the tendency to avoid the interference limit is at least partially expressed in a steadily increasing joining of forces. Instead of competing against colleagues, you join them, often in an effort to accomplish more through joint rather than separate efforts.

Many complex projects, such as the construction and operation of astronomical observatories in space, require large collaborations of specialists and cannot be carried out by smaller groups. Increasingly, even purely theoretical papers now bear the names of many co-authors. In earlier decades, such articles might have been written by just one individual, or perhaps by a student and his or her professor. Today, a computer program required for theoretical simulations may be proprietary or may be so complex that only one member of a team is fully familiar with it. A higher number of co-authors then provides access to a larger suite of complex tools.

The larger number of co-authors on papers produces the increasingly dense links between authors as well as the emergence of the giant component of astronomers who publish jointly, establish mutual interdependencies, and perhaps end up thinking alike – a further way in which the thought collective evolves.

* * *

Although the interference limit may thus be sidestepped, albeit with still-uncertain net benefits, overcoming the communications limit may prove harder. Something needs to hold astrophysics intact; otherwise we will miss the grander features of how the Universe evolved.

The communications barrier is not the speed with which we can access information. Through searches on the Web, work on different subjects by any number of different authors is readily recovered. Much of the recent literature in astrophysics, and the entire contents of several of the premier astronomical journals, some dating back to their inception more than a century ago, are readily accessed. A daily download of new preprints, many of them articles submitted the previous day for publication in journals, is also available through the *arXiv astro-ph* website.

Access to fonts of information is not the real problem. Computerized archives already inundate us with anything we ask for. But we may have to find ways to absorb information more quickly, either by improving techniques for entering information into our consciousness, or through means to better extract key data from masses of information within which they are deeply embedded and easily missed.

Often the implications of earlier work are not fully recognized by astrophysicists until many years later. Advances made more than a few years in the past are quickly forgotten or presumed outdated, regrettably ensuring that identical work is unnecessarily reinvented. Improved means for locating and assimilating the large body of existing information would more effectively advance the field.

Fortunately, rapid extraction and visualization of information are key areas in which research is being aggressively pursued for many applications, ranging from computer games to displays of financial markets, where rapid comprehension of a continually changing landscape can make the difference between success and failure. The communication limit needs to be overcome by many communities; the community of astronomers is definitely among them.

Note how the limits to growth I have just described reflect the findings of the network theorists. The threshold limit corresponds to the portion of the network disconnected from the giant component shown in Figures 9.8 and 14.2. Community members that are part of the giant component remain ignorant of unlinked publications. The mutual interference limit inducing astronomers to work on ever larger collaborations is at least partially traced by the appearance of ever more numerous linkages across the network, giving rise to a giant component as people opt to collaborate rather than compete. Growth in the number of links between network nodes also may herald the approach of the interference limit.

The communications limit similarly makes itself partly apparent when a network begins to condense into densely connected clusters bridged by sparse links as disciplines split apart to concentrate on different problems and create separate dedicated journals. The networks reflecting publication patterns thus provide a way of tracing where astronomy and astrophysics are headed.

Who or What Runs the Universe?

The structure of the complex network of persuasive interactions, among scientists and governmental bodies, shown in Figures 11.2 and 14.1, and the interplay between the different communities of scientists sketched in Figures 9.8 and 14.2, may well lead one to ask, "Who's in charge of Astronomy?" – not just in one country, but internationally as well. It shows how complex the funding of astronomy can be and how closely the motivations of scientists, administrators, and political leaders need to be aligned before any major thrust can succeed.

In Chapter 12, we saw that an overly dense organizational network of any kind brings with it a threat of instability, partly as a result of mutual interference. But it also heralds a potentially unhealthy congeniality of thought. As Chapter 11 attests, in the early deliberations of how the Great Observatories might best be built, we astrophysicists convinced ourselves that acceptance of a close tie to the Space Station would be the best way forward – even though with hindsight we all realized how big a mistake this would have been.

It is these linkages arising within large collaborations which Andrew Pickering's work, in 1984, was pointing out. Within the astronomical community such links may lead to mutual acceptance of world models – initially at a purely scientific, but eventually also at political levels, as funding is sought and obtained to pursue a particular thrust further.

This is what led to Pickering's concern that science can lead to social constructs, rather than a gradual approach to some greater inherent truth about, say, the structure of the Universe.[9]

So we return to Pickering's thesis and ask, "Do our modern views about the Universe reflect inherent properties of the Cosmos, as is generally believed, or is our depiction of the Universe a social construct?"

In brief, we need to ask "Who or what actually runs – that is, determines the structure and evolution of – the Universe?" Is it some set of forces inherent in Nature? Is it the set of influential astronomers? Could it be the governments, which through the moneys they allocate determine the direction of research?

The answers to these questions are only partially clear: I think we can rule out religious organizations seeking harmony between scientific findings and traditional beliefs. Creationists, for example, may be motivated to influence scientific views, but probably do not have the technical means. On the other hand, as Paul Forman pointed out in his 1971 treatise, 'Weimar Culture, Causality, and Quantum Theory, 1918–1927: Adaptation by German Physicists and Mathematicians to a Hostile Intellectual Environment,' cultural forces do appear to encourage scientists to emphasize socially acceptable aspects of their findings.[10]

Governments may not be particularly motivated to influence astrophysical thought, but do have the technical means to do so through the support of scientific studies that take advantage of technologies devised for governmental or other societal priorities.

Finally, the most powerfully linked astronomers in the networks revealed by Mark Newman's studies do have not only the motivation, but also the opportunity and the means to fashion the Universe.[11] Their motivation may be a sincere wish to take astronomy into promising new directions. The opportunity and the means are provided by the boards on which these influential members of the community are invited to serve – boards asked to recommend which new observatories should be constructed, who will direct them, and how moneys should be distributed among the research projects they might carry out.

But, just as in a court of law, motivation, opportunity, and means are merely circumstantial evidence. The question is, "Do the most influential astronomers really run, or fashion, the Universe?"

My impression is that in areas where we have a significant accumulation of observational and experimental evidence, our astrophysical and cosmological theories do largely reflect a physical reality. Where evidence is scarce, however, the tightly knit community of leading astrophysicists may actually be constructing the Universe.

* * *

The extent to which an astrophysical theory is a construct can be determined only by checking its history as new observations or novel experimental data emerge. If most of the discoveries made after the theory is proposed require the theory to be augmented with new assumptions, themselves unsupported by – although quite possibly compatible with – existing scientific evidence, one should suspect that the theory is a construct, rather than a reflection of inherent properties of Nature.

We may hope that such a construct will ultimately be supplanted by a more powerful theory that similarly agrees with all extant data but requires less manipulation to fit further new data as these emerge over time. That way, constructs may eventually be unmasked as new evidence emerges and the construct is either shown to be wrong or has to be amended and supported by new assumptions that had earlier seemed unnecessary. Although it may then survive somewhat longer, one of two outcomes will eventually prevail. The construct will be supplanted by a theory that better reflects the true nature of the Cosmos, or it will remain in force solely because a search for further experimental or observational support has become unaffordable!

Regrettably, astrophysical tests are becoming increasingly expensive, often beyond society's means. The onus then is to find ways to keep research healthy, either by attempting to explore the Universe in smaller less costly steps, over longer periods, or by waiting for new technologies developed for medical, security, communications, environmental, recreational, or other purposes to catch up. Technological means may then emerge to provide us new tools that are financially affordable and enable us to move forward again.

How Will We Recognize a Social Construct?

The question then is, can we enumerate objective criteria that will prevent our deluding ourselves into accepting a social construct in place of the true Universe? I believe we can, but only as long as we can test the predictions the construct makes. If it makes no predictions, it is unlikely to be helpful and should be discarded. More serious is the alternative that the construct does make predictions but that testing them is unaffordable.

* * *

Social constructs are especially likely to arise in the search for the origin of the Universe. Several factors indicate this.

We know that the Universe was opaque to electromagnetic radiation for most of its first 400,000 years. It also was opaque to neutrinos for the first few minutes of its existence. The high opacity suggests that only indications of thermal equilibrium at the respective decoupling of neutrinos and radiation from the rest of the primordial mix will have survived from those early times. Thermal equilibrium is described by a temperature and a statistical set of thermal fluctuations; and these, by themselves, do not provide an abundance of information.

Gravitational radiation may also have decoupled from matter and radiation at earlier times. The hope, today, is that gravitational radiation may ultimately tell us more about the earliest moments in cosmic evolution. But even there, we may learn not much more than that these waves also reflect a time by which all the forms of radiation had significantly interacted and come into thermal equilibrium.

The epoch at which dark matter decoupled from radiation is unknown as well; but if this did not happen until fairly late in cosmic evolution, we also may not learn very much concerning the earliest moments of creation from the study of dark matter distributions.

The microwave background radiation does tell us about the distribution of mass, including dark matter, in the Universe, 400,000 years after the origin of the Cosmos. What we see there can, in part, also be attributed to conditions at the dawn of time, when temperatures may have been extremely high and fluctuations could have been chaotic. We now surmise that these fluctuations eventually seeded the formation of clusters of galaxies. We can study these suggestions in some detail with simulations, because the microwave map of the cosmic radiation background is so rich, and because so many millions of clusters of galaxies exist on which we can observationally test the validity of our theories. But none of this may suffice to reveal precisely what happened when the Universe was, say, one picosecond (10^{-12} sec) old – a question that could be important to particle physicists.

We may have to persist in gradually building more sophisticated instrumentation at increasing expense, yielding progressively less incisive information about the earliest cosmic times, because the information is likely to have been erased at the high densities and close interactions that led to bland thermal equilibrium states along the way.

A potential hope is that even the earliest times may have been punctuated by a series of major transitions, each accompanied by the decoupling from its physical environment of some *carrier of information* that still reaches us today. The cessation of primordial nuclear reactions through cooling of the Universe at an age of a few minutes, and the preservation of the ratio of hydrogen, helium, deuterium, and lithium isotope abundances from that epoch, still provides the most detailed insight we have into conditions in the early Universe. Quite possibly, tracing the temperature at decoupling of a succession of other entities – gravitational radiation, neutrinos, dark matter – all taken together would prove comparably informative. But it is not at all clear that the Cosmos will have been so generous to us as to convey insights on a complete string of such events by means of which we will be able to retrace cosmic history, step by step, back to its most remote past.

The energies of particles and radiation at the earliest times were far higher than any we can produce at even the most costly accelerators we can construct today. This will make difficult the simulation, in our laboratories, of physical conditions that prevailed at earliest epochs. Either more affordable accelerators will have to be built, or means may have to be found to derive high-energy information from the very few naturally occurring cosmic ray particles that have energies ranging up to 10^{19} to 10^{20} eV – far higher than any accelerator produces today. Given the very low number of these particles, and their relatively low energies compared to those thought to have prevailed in the era when the Universe was controlled by just one grand unified force, mediated by particles at about 10^{25} to 10^{28} eV, this too would probably leave many questions unanswered.

We also know very little about the laws of physics governing events separated by large distances across the Cosmos. General relativity may be the ultimate law, but we may have difficulty verifying this, particularly if the vacuum should have undergone some further phase transitions in the past 10^{10} years.

True understanding is marked by the ability to correctly predict the outcome of an ever increasing number and variety of experiments or observational studies. Right now, the many surprises we still encounter continue to show how little we know. They enrich our understanding but also demonstrate how incomplete our knowledge remains. Only when even the most far-sighted observations and laboratory experiments yield results fully anticipated by theoretical predictions are we likely to gain assurance that we are close to fully comprehending the complexities of the Cosmos, and that our theories may be more than just social constructs.

Alternatively, however, the fate of astronomical observations may approach that of complex experiments. In his book 'How Experiments End,' Peter Galison has aptly stated:

> Reading an article, one could conclude that an effect would follow from an experimental setup with the inexorability of logical implication. But lurking behind the confidence of the experimental paper lies a body of work that relies on a kind of subtle judgment…Only the experimentalist knows the real strengths and weaknesses of any particular orchestration of machines, materials, collaborators, interpretations and judgments…[E]xperiments are about the assembly of persuasive arguments…about the world around us, even in the absence of the logician's certainty.[12]

As a result, experiments reach an end when the instruments used have led to about as much information as their design will permit. Continuing the investigation further will yield few gains but will certainly raise cost.

The same fate governs astronomical observations. A variety of pressures may require termination of an expensive ground-based facility or a space mission as observations are judged to have culminated. Like a completed high-energy experiment, the observations then become frozen in history.

A Seamless Portrayal of the Universe

The prevailing thrust in astrophysics is to portray cosmic evolution from the birth of the Universe down to our times as a seamless process giving rise to galaxies, quasars, stars, and planets along the way, at epochs and through physical processes on which observational data and theory fully agree. Nowhere, perhaps, are the ambitions to produce such a tightly woven network more clearly spelled out, today, than in current cosmological investigations seeking quantitative agreement among myriad different observations and their inferences.

The most comprehensive of these approaches is an attempt to determine how the temperature, temperature fluctuations, and polarization of the microwave background radiation vary across the sky and correlate with (i) expected mass–energy fluctuations at the dawn of time; (ii) the clustering of galaxies and distribution of voids observed today; (iii) the X-ray emission from clusters of galaxies interacting with the microwave background; (iv) precise measurements of the Hubble constant; (v) the density of the dark energy originally discovered through observations of high-redshift supernovae of Type Ia; (vi) the equation of state of this dark energy; (vii) the primordial abundance of helium produced in the first few minutes of cosmic time; (viii) the number of different neutrino species and the individual masses of neutrinos belonging to each specie; and (ix) the foreground radiation emitted by the Galaxy that needs to be taken into account in determining the eight preceding correlations, as well as a further relation to local distributions of gas and dust, the formation of young stars, and other activities on Galactic scales. All these are topics tightly woven into a coherent story based largely on nearly a decade of uninterrupted observations with the Wilkinson Microwave Anisotropy Probe (WMAP), and a large number of independent observations carried out across the entire electromagnetic wavelength domain with ground-based telescopes and from space.[13,14]

The high degree to which these observations are interlinked makes our current cosmology particularly prone to instability. Not that the interlinking is undesirable. On the contrary, we would be troubled if a cosmological theory failed to reveal our Universe to be a coherent structure each part of which fits every other part through bonds the laws of physics ably explain.

The problem is only that a tightly formulated cosmology falls apart if just one of these bonds between its building blocks is found to be missing. Astrophysicists may then scramble to postulate the existence of some other bond that keeps the entire structure from collapsing; but this may just be a construct, a prop, that will continue to trouble conscientious theorists.

Today, despite the long list of satisfactory correlations that WMAP and complementing space and ground-based surveys have revealed, huge gaps in our knowledge persist. Building blocks with surrogate names, like *dark matter* or *dark energy*, prop up the cosmological structure to hide enormous blank spaces in our grasp of physics.

Nature may or may not agree to the fanciful edifice we have erected, which might just reveal itself as another of our human constructs.

Indeed, some cracks in the current cosmology may already be appearing, although whether they are meaningful is still far too early to tell. I mention them only because tests of the kind I cite usually provide the first signs of a theory's potential failings.

By now, nearly a decade of observations by the WMAP team indicate that the brightness distribution of the microwave background quite nicely matches the predicted scale-free scalar fluctuation spectrum, which Edward Harrison and Yakov Zel'dovich had postulated in the 1970s, and that the 1983 chaotic inflationary model of Linde also proposes. This match is confirmed at a level of $96.8 \pm 1.2\%$.[15,16,17] This amazingly strong agreement, nevertheless, slightly deviates from original predictions with a certainty of 99.5%. Other factors, the WMAP team suggests, must be at play that are not yet understood. In the meantime, however, a host of other types of fluctuations, in particular, fluctuations arising spontaneously in the primordial vacuum have also been suggested, and some of these may ultimately be found to provide better fits.[18]

Another finding of the WMAP mission is an indication that the effective number of neutrino species needed to provide the best fit to existing observations is $N_{eff} = 4.34^{+0.86}_{-0.88}$, albeit at a rather lower confidence level of only 68%. Because we currently recognize only three neutrino species, this too may be a worrisome finding.

Figure 14.3. Paul Adrien Maurice Dirac about the time he was prescribing a grand strategy for matching the physical world to corresponding mathematical forms. His prescription could be extended to encompass all of Nature and the Universe. (*Courtesy of the AIP Emilio Segre Visual Archives. By permission of the Master and Fellows of St. John's College, Cambridge*).

If such problems persist as we learn more, we will need to find ways to understand them. Eventually a new cosmology, a new view of cosmic structure may then emerge, and we may again be faced with the same question. Does the new cosmology accurately portray the actual workings of the Universe? Or is this too, just a convenient construct we have devised to smooth over loopholes in our understanding? Fortunately, studies of the microwave background radiation are continuing, aided by the powerful Planck space mission launched by the European space Agency with NASA participation in May 2009. Its first results are gradually coming into focus.

Dirac's Generic Long-Term Plan

We need to reconsider what the endgame of cosmology should be.

Once we have carried out all the tests we can think of, or at least all the tests that appear affordable, how certain will we be that we have finally unveiled the structure and history of the Cosmos, rather than that of some convenient construct, which happens to pass all the tests we can devise, but otherwise has no resemblance to the true Universe?

Perhaps the most appropriate approach available to us to decide such questions will be a strategy originally proposed by Paul Dirac in the 1930s.

In 1930, Dirac noted that the equations of relativistic quantum theory held equally well for electrons having positive energy and negative energy. He asked why electrons with negative energy did not appear to exist, and suggested that they might indeed exist, but that "all the states of negative energy are occupied except perhaps a few of small velocity. Any electrons with positive energy will now have very little chance of jumping into negative-energy states and will therefore behave like electrons are observed to behave in the laboratory."[19]

The holes constituting negative energy electrons, Dirac interpreted as positively charged particles with positive energy. When a hole and an electron combined, their charges would cancel and their energy and momenta would be carried off by radiation.

The only problem that Dirac faced and realized was that his theory did not quite fit if he considered the holes to represent protons. The mass of the proton did not match that of the electron. He thought that relativistic calculations might "lead eventually to an explanation of the different masses of protons and electrons."

As the historian of science Helge Kragh recalled half a century later, Dirac's claim of 1930 was quickly countered in separate publications by Hermann Weyl, Igor Tamm, J. Robert Oppenheimer, and Dirac himself. By the following year, 1931, they had shown that the positively charged particles would have to have the mass of electrons, if for no other reason than that collision of electrons and protons would otherwise annihilate too readily, inconsistent with the observed stability of matter.[20] Dirac now predicted that not only electrons but protons, as well, should have negative energy states. Protons with negative states and opposite charge he called "anti-protons".[21]

Some months after the appearance of Dirac's second paper, Carl David Anderson, then a 27-year-old postdoc in physics conducting cosmic ray experiments at the California Institute of Technology, found particle tracks in his cloud chamber photographs "indicating a positively-charged particle comparable in mass and magnitude of charge with an electron."[22] In a later article he called these particles *positrons*.[23]

This was the first antiparticle to be discovered and changed the course of physics![a,24]

In his 1931 paper, Dirac had also discussed a second set of symmetry arguments to predict that magnetically charged particles should exist in Nature. Analogous to electrons, which carry electric charge, magnetic monopoles would carry magnetic charges $\mu = \hbar c/2e$.[25] Although such particles have never been detected, they go by the name *magnetic monopoles*.[b]

---

[a] The antiproton was not discovered until 1959.

[b] The magnetic monopoles discussed in Chapter 10 are more massive variants of the monopoles Dirac had in mind.

The essential point in the research strategy Dirac proposed was his insight that,[26]

> [M]odern physical developments have required a mathematics that continually shifts its foundations and gets more abstract. Non-euclidean geometry and non-commutative algebra, which were at one time considered to be purely abstract fictions, ... have now been found to be very necessary for the description of ... the physical world ... The most powerful method of advance that can be suggested at present is to employ all the resources of pure mathematics in attempts to perfect and generalise the mathematical formalism that forms the existing basis of theoretical physics, and *after* each success in this direction to try to interpret the new mathematical features in terms of physical entities.

Roughly speaking, the research program Dirac was recommending was one in which successful formalisms of theoretical physics would be extended to investigate not only such systems as might exhibit, say, states of positive energies, or the charges of known particles, but also such peculiarities as particles in negative energy states, or particles called *tachyons* moving at speeds potentially exceeding those of light, or others that might carry predictable magnetic charges. Such a program would guide experimentalists and observers, and focus their attention on speculative but potentially ground-breaking lines of investigation.

Coincidentally, perhaps, Kragh's paper recalling Dirac's recommendations was written at just the time when Alan Guth was investigating the fate of magnetic monopoles in his inflationary model of the Universe. Looking back now, the research program Dirac had proposed half a century earlier may be seen to have shaped a cosmology that took the potential existence of magnetic monopoles seriously – and introduced inflation to avoid a glut of monopoles that would have violated observations.

* * *

Dirac's insight, that exploratory theories can emphasize novel mathematical representations based on new symmetries or topologies, or added abstract dimensions, has a strong appeal. Steps in this direction are likely to play an essential role in verifying whether a given theoretical framework is sufficiently specific to define the structure of the Universe, or whether a more general framework based on additional symmetries or more complex topologies is more germane: A well-understood theoretical structure may merely be a subset of a considerably more comprehensive edifice consistent with established experiments and observations, but able potentially also to account for newer findings leading to further discoveries.

Theoretical physicists have already been pursuing such a thrust for the past few decades in their search for string and brane theories that might ultimately provide a theory combining all the forces of Nature and thereby include some of the basic elements of cosmology.[27,28]

In Dirac's thinking, any potentially unexplored consequence of the equations of theoretical physics may well have an actual counterpart in Nature. Helge Kragh's 1981 study, which we encountered in Chapter 10, examined this broad research program and its spreading influence on the community of physicists, which in the waning decades of the twentieth century developed the attitude that anything not specifically forbidden by the laws of physics should be found to exist in Nature.[29] Thoughts along such lines can be traced back to Aristotle and are often expressed as a *principle of plenitude*.

Applied to astrophysics Dirac's proposals recommend transcending the physicists' search for a mathematical formalism correctly depicting high-energy particle physics, by extending it to a schema encompassing cosmology as well. Such a program promises a systematic search not only for new patterns revealing a more richly textured cosmos than previously realized, but also for uncovering false constructs that may have inadvertently crept into cosmological thought and prevent our perceiving the Universe in its true character.

* * *

Eventually, we may find ourselves persuaded that our search has come to an end. Two potential outcomes seem likely:

The first might be that we will ultimately find an acceptable scientific explanation, but that we will be unable to test it further – either because we will not know how to devise such a test, or because contemplated tests will be unaffordable. Chances are that the prevailing scientific explanation might then be a social construct. We just won't be able to recognize or even to deny it.

The second way might be that observations and experiments will ultimately yield a series of mutually consistent findings we can cross-check at will, under a wide range of physical conditions, to discover that the workings of the Cosmos obey a simple mathematical schema. This would not tell us why that particular schema, rather than some other, was ultimately favored by Nature, but nevertheless would be an impressive conclusion to a centuries-old search. Many might be persuaded then that we had advanced as far as will be possible, but chances are that some would persist in a search to learn more. And if they succeeded, a new era yielding even deeper insight might thus still emerge.

Notes

1. *Science – The Endless Frontier*, Vannevar Bush, reprinted by the National Science Foundation on its 40th Anniversary 1950–1990, National Science Foundation, 1990.
2. Transcript of Remarks on U.S. Space Policy, President George W. Bush, NASA release, Washington, DC, January 14, 2004.
3. *The Limits of Settlement Growth: A Theoretical Outline*, Roland Fletcher. Cambridge University Press, 1995.

4. The Densities of Visual Binary Stars, E. Öpik, *Astrophysical Journal*, 44, 292–302, 1916.

5. An Estimate of the Distance of the Andromeda Nebula, E. Oepik, *Astrophysical Journal*, 55, 406–10, 1922.

6. Stellar Structure, Source of Energy, and Evolution, Ernst Öpik, *Publications de L'Obervatoire Astronomique de L'Université de Tartu*, xxx, No. 3, 1–115, 1938.

7. Stellar Models with Variable Composition. II. Sequences of Models with Energy Generation Proportional to the Fifteenth Power of Temperature, E. J. Öpik, *Proceedings of the Royal Irish Academy*, 54, Section A, 49–77, 1951.

8. About Dogma in Science and other Recollections of an Astronomer, E. J. Öpik, *Annual Reviews of Astronomy and Astrophysics*, 15, 1–17, 1977.

9. *Constructing Quarks – A Sociological History of Particle Physics*, Andrew Pickering. University of Chicago Press, 1984.

10. Weimar Culture, Causality, and Quantum Theory, 1918–1927: Adaptation by German Physicists and Mathematicians to a Hostile Intellectual Environment, Paul Forman, *Historical Studies in the Physical Sciences*, 3, 1–115, 1971.

11. The structure of scientific collaboration networks, M. E. J. Newman, *Proceedings of the National Academy of Sciences of the USA*, 98, 404–09, 2001.

12. *How Experiments End*, Peter Galison, University of Chicago Press, 1987, pp. 244 and 277.

13. First-Year Wilkinson Microwave Anisotropy Probe (WMAP) Observations: Determination of Cosmological Parameters, D. N. Spergel, et al., *Astrophysical Journal Supplement Series*, 148, 175–94, 2003.

14. Seven-Year Wilkinson Microwave Anisotropy Probe (WMAP) Observations: Cosmological Interpretation, E. Komatsu, et al., *Astrophysical Journal Supplement Series*, 192, 18, 2011.

15. Fluctuations at the Threshold of Classical Cosmology, E. R. Harrison, *Physical Review D*, 1, 2726–30, 1970.

16. A Hypothesis, Unifying the Structure and the Entropy of the Universe, Ya. B. Zel'dovich. *Monthly Notices of the Royal Astronomical Society*, 1P-3P, 1972.

17. Chaotic Inflation, A. D. Linde, *Physics Letters B*, 129, 177–81, 1983.

18. Ibid., Seven-Year Wilkinson, Komatsu, et al., 2011.

19. A Theory of Electrons and Protons, P. A. M. Dirac, *Proceedings of the Royal Society of London A*, 126, 360–65, 1930.

20. The Concept of the Monopole. A Historical and Analytical Case-Study, Helge Kragh, *Historical Studies in the Physical Sciences*, 12, 141–72, 1981.

21. Quantised Singularities in the Electromagnetic Field, P. A. M. Dirac, *Proceedings of the Royal Society of London A*, 133, 60–72, 1931.

22. The Apparent Existence of Easily Deflectable Positives, Carl D. Anderson, *Science*, 76, 238–39, 1932.

23. The Positive Electron, Carl D. Anderson, *Physical Review*, 43, 491–94, 1933.

24. Antiproton-Nucleon Annihilation Process. II, Owen Chamberlain, et al., *Physical Review*, 113, 1615–34, 1959.

25. Ibid., Quantized Singularities, Dirac, p. 68.

26. Ibid., Quantized Singularities, Dirac, p. 60.

27. Large Extra Dimensions: A New Arena for Particle Physics, N. Arkani-Hamed, S. Dimopoulos & G. Dvali, *Physics Letters B*, 429, 263–72, 1998.

28. An Alternative to Compactification, L. Randall & R. Sundrum, *Physical Review Letters*, 83, 4690–93, 1999.

29. Ibid., The Concept of the Monopole, Kragh, 1981.

15

Language and Astrophysical Stability

Although most astronomers assign particular importance to the problems on which they are currently working, our understanding of the Universe will not advance satisfactorily unless the community can agree on a coherent research plan with a well-defined thrust. The plan cannot be too rigid; otherwise unanticipated initiatives leading to novel insight will be thwarted. Nor should changes in direction be opposed as we learn more and realize a deliberate course correction is needed.

These criteria seem mutually contradictory, so that care is required in respecting them. In Chapter 2 we saw how different scientists approach a given problem by disparate means, guided primarily by tools in whose use they have developed skill and confidence. Faced with a novel problem, they thus reach for distinct tools in their search for increased insight. But before the community can persuade itself that the use of a particular set of tools has indeed led to a significant advance, trusted experts may first need to explain to each other how the respective tools work and the findings to which they point. The present chapter shows how this mutual persuasion may most effectively be pursued.

How to Revive a Spacecraft Millions of Miles Away in Space

I introduce the problem of reviving a spacecraft because it stresses the overarching significance of language in shaping the way scientists and engineers manage to repair a complex system when it breaks down. The astrophysics community may benefit from adopting a similarly formal approach for weeding out errors thwarting the field's progress, formulating long-term communal plans for astrophysical research and archiving astronomical data so it may benefit future generations.

* * *

Although most scientists are unaware of it, a complex space mission tends to fail in major or minor ways, every week if not every day. The myriad interactions of the hundreds of thousands of discrete parts required to keep the mission going are so complex that bizarre failure modes cannot be fully anticipated.

This makes any complex space telescope inherently unstable and likely to fail. The impact of a dust grain traversing the Solar System at tens of kilometers per second may damage an onboard instrument. Highly energetic nuclear particles from a solar outburst may incapacitate the spacecraft's electronic systems.

With an ever present threat of setbacks, a major space project maintains experts on whom it can call to find ways of keeping the mission operating reliably. If one capability of the spacecraft malfunctions, work continues on essential programs that do not require this function. Meanwhile the experts determine what went wrong, how it may be fixed and, if it is irreparable, how an alternative mode of operation may restore the system to full health.

All this to fix a spacecraft a million miles from Earth!

* * *

Required first is a diagnosis of what went wrong. The detected failure might be due to any dozens of causes. Tests must be run to identify the root failure. Was the failure mechanical? Electrical? Could it be a software malfunction fixable through new sets of commands sent to an onboard computer?

To start, a panel of experts is assembled. The complexity of a spacecraft requires exceptional care in selecting each member and planning how the team is to work. Although each expert must be deeply familiar with how a particular family of spacecraft components functions, panel members must also be able to understand each other – share a language based on words that identify spacecraft parts, their function, the ways they operate, and their frailties.

Vocabulary is critical! Each spacecraft component, each spacecraft function has a name, often a complex designation contracted into an acronym. The dictionary of acronyms designates a minimal vocabulary that must be mastered to even begin to converse intelligently and solve a problem. The size of the vocabulary reflects the complexity of the spacecraft. To the expert each word of this private language conveys a maze of implications depicting the spacecraft's functioning.

The minimum requirement for meaningful exchange of information is that each expert be fluent in the vocabulary not only of his or her own area of expertise, but be equally fluent and able to understand thoughts expressed by at least two other panel members, each an expert on some other, mutually differing spacecraft function. This level of fluency is hard to find, but is critical to assembling a panel of experts able to solve an unprecedented problem.

* * *

Each space mission creates its own language. In the mid-1990s, the European Space Agency launched the Infrared Space Observatory (ISO), a billion-dollar astronomical mission. By the time ISO had been operating in space for two years and was reaching the end of life, a final list of roughly 1500 acronyms constituted its dedicated language.

Just as with words in everyday natural language, identical acronyms had arisen with diverse implications. Their meanings no longer could be understood except in context. Even the acronym ISO now stood not only for "Infrared Space Observatory," but also for the "International Standards Organisation." Similarly, PSS could be taken to mean "Pyramidal Support Structure," "Portable Software Simulator," "Power and Storage Subsystem," or "Power Supply Subsystem." The new vocabulary with its nouns (spacecraft parts, ground-based support equipment, related organizations), verbs (spacecraft activities and commands), and adjectives or adverbs (variants of nouns or verbs) was reaching the complexity and ambiguities of conventional English.

A dozen years after the end of the mission, its dedicated language has largely ceased to exist. Some of its vocabulary has been incorporated by succeeding missions, albeit with new shadings and ambiguities. Half a century from now, the language for understanding how ISO operated will be difficult to resurrect. Its experts will be gone; the language they created to communicate among themselves to build and operate the spacecraft and successfully complete its mission will no longer be spoken.

Cutting-edge technologies used in building the spacecraft and its instruments will have been superseded by other, entirely different techniques and might not even be found in working order in technical museums. If doubts should then arise about the implications of signals the telescope and its instruments may have registered, we will not know whether those signals conveyed significant astronomical information that was not fully appreciated in its time, or whether they merely constituted known quirks in the performance of the instruments.

Problem Solving

The 1500 words that constituted ISO's list of acronyms were only a start. The wealth of concepts handled by spacecraft experts is considerably richer than the vocabulary of acronyms. Each piece of equipment an expert handles may have many components, often distinguished by manufacturers' part numbers; each line of computer code has its assigned line number.

Acronyms are assigned only to equipment and concepts that need to be identified in conversation. Part numbers, although of critical importance as an investigation into problems proceeds, are generally not assigned acronyms. Lines of computer code may have meaning solely to those who wrote them, even if others can learn to follow their logic. The vocabulary and dictionary of acronyms constitute merely the foundation of language sufficing for the exchange of ideas. Like in English and other natural languages, highly specialized designations do not find their way into dictionaries or lists of acronyms.

Let us now look at how actual problems are solved by a panel of experts. To emphasize the principles involved, the scheme I describe is deliberately oversimplified.

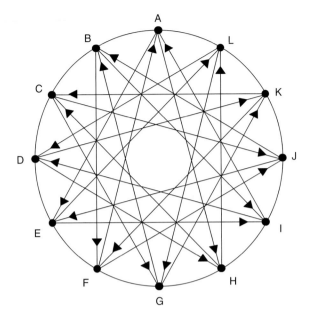

Figure 15.1. Effective roundtable discussion among twelve experts. With each participant an expert in one area and conversant with two others, each expert has four links to other experts – two incoming and two outgoing. An inquiry or its response can be transmitted to any other panelist in at most five steps, and is verifiable along a separate, but usually longer, transmission path. Note that each arrow, each communication link, points only in one direction.

We may imagine the team sitting around a large circular table, Figure 15.1. The project manager has asked the panel to determine what went awry yesterday, when the spacecraft failed to communicate data back to Earth between 12:33:07 and 15:46:18 Universal Time.

Expert A begins the investigation by asking whether this could be due to a certain procedure. But it isn't clear who among the other panelists can answer the question. Expert E understands the question raised by A and believes that Expert D might be able to answer it. D, however, is unsure of what Expert E is asking. E therefore explains the issue to Expert I, who then reformulates the question using vocabulary that D masters. The direction the conversation takes always progresses from an expert wishing to transmit a question to one who will understand its rephrasing. In Figure 15.1 this is indicated by the arrows that lead from A to E to I and then to D.

Having considered the question, D now wishes to answer A, but this will work only with help of clarification offered along the way by experts K and F, who share the necessary vocabularies to pass the message along in fully understandable fashion. For other members of the panel to also get the gist of D's response, A can further explain it to E and H who, between them can notify all the other panelists through different intermediaries. Each question and each response can be transmitted to all other members in a sequence of at most five steps, that is, through at most four intermediaries.

In a maximum of six sequential steps, each recipient can receive a copy of the initial message along at least two entirely distinct paths. All are able to compare these two versions to determine whether they concur. If they do not, a recipient may ask for further clarification until the messages arriving along the two paths agree. The message also returns to the initiator, A, in at most five sequential steps involving three partially overlapping paths.

If each node in this transmission – each intermediary – on average introduces an error of, say, 1% in passing the message along, the two received messages should have errors amounting to about 5 to 6% each, corresponding to the five to six steps required to convey the two messages. Moreover, expert A, the initiator of the original message, will be able to tell the extent to which the two returning messages correspond to, or even answer, the initially transmitted question. If they are found wanting, a more explicitly worded statement can be circulated to clarify potential misunderstandings.

This may appear to be an unnecessarily unwieldy stylization of how an actual panel discussion proceeds; but it encompasses what in essence transpires in a far more messy exchange among panel members before each member fully understands what the others are attempting to convey. It does this by providing a minimum number of cross checks to prevent the most likely potential misunderstandings.

Solving a complex technical question requires everyone on a panel to understand the problem fully in order to contribute to its solution. Anyone not able to follow the discussion in detail should probably not have been appointed to the panel.

A feature of this assembled roundtable that needs emphasis is the ability of each panelist to at least minimally verify that the transmitted information is error free. This is achieved by having each message relayed back to all members along two or more paths which, between them, traverse all the panelists. Verification ensures that every panelist adequately understands all others, even though the transmitted message was translated at each waypoint in its transit around the table.

Figure 15.2 shows a simpler model for a panel that can barely work, if need be, but with severely limited prospects for success. The simpler-looking seating arrangement illustrates pitfalls that need to be avoided in matching panel members with different areas of competence. In the configuration shown, each panelist is an expert in one area, as before, but conversant with solely one other area.

The only way the panelists' limited language skills can lead to a workable arrangement is for the flow of any conversation to progress consistently in an anticlockwise sense for the seating arrangement shown, in which each panelist can fully understand only the panelist on his left.

When each panelist understands the vocabulary of only one other expert, in addition to his or her own expert vocabulary, each question and each response has

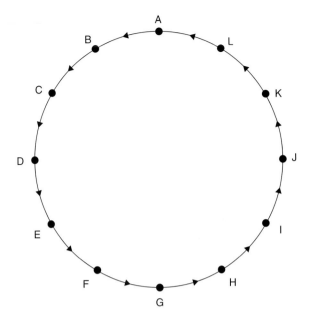

Figure 15.2. Thwarted roundtable discussion among experts. In this roundtable, each panelist, A to L, is expert in just one area and conversant with one other. Two links join each expert to others, an incoming link for queries from an expert whose language is familiar and an outgoing link for responses informed by the panelist's own area of expertise. This scheme of transmission can work albeit, as discussed in the text, with demonstrably less success than the arrangement illustrated in Figure 15.1.

to go through twelve successive steps before finally returning a suitable response to the originator. If we again assume the introduction, on average, of errors of order 1% per transmission step, the message returning to the originator will have roughly a 12% error. The originator of the message can compare the returning message to the query sent out, to see whether the transmission of information was satisfactory. None of the other panelists at the table have an independent way of checking whether the message they received contained errors, unless the messages make a second circuit around the table with the potential for intrusion of additional errors.

When each panelist understands the vocabulary of only one other expert, the prospects of verification that the entire panel fully understands a problem at hand are greatly diminished. This is why each panelist needs to be conversant with at least two other areas of expertise, as in Figure 15.1.

We may now compare the working of the roundtable of Figure 15.1 to the virtual roundtable on the Internet, which solved the speeding neutrino problem described in Chapter 12. There, we saw how a complex scientific problem of unknown origins can be solved communally by experts – theorists, engineers, experimenters, mathematicians or even philosophers – in time of crisis.

The virtual roundtable on the internet involved in solving the neutrino problem consisted of a far larger number of experts than the twelve of Figure 15.1. But it led to the same basic result, namely that all aspects of the problem had been thoroughly vetted by the time the process had run its course, and that at least all key participants came to fully understand the problem that had arisen and how to fix it.

* * *

This has been a lengthy but necessary diversion in order to deal with the task raised at the outset of the present chapter – the maintenance of clarity of language to steadily advance astrophysics in the face of occasional erroneous or misunderstood information that can hinder day-to-day progress, distort the formulation of long-term communal plans, or thwart the interpretation of aging archived data.

The formal procedure I have outlined indicates how difficult scientific questions can be solved communally today by excellent people exchanging ideas. Just as spacecraft experts solve an instrumental problem far out in space, identical approaches can solve difficult theoretical problems of astrophysics. Panels of scientists, expert in their own discipline but also versed in the vocabulary of two or more others, quite evidently can readily communicate among themselves to sort out and evaluate new findings before they become incorporated into the core areas of physics or astrophysics accepted by the full community.

Clarity of language immunizes a community against error through the cross-checks it provides. When a new finding in one discipline is understood fully by another, and appears to violate its findings, a roundtable discussion can raise doubt, and curtail further dissemination of the conflicting information until the discrepancy is resolved. The virtual roundtable investigating the speed of neutrinos generated at CERN could point to countervailing astrophysical evidence. Neutrinos arriving at Earth from the 1987 supernova explosion in the Large Magellanic Cloud had certainly not exceeded the speed of light by anything like the speed claimed for the CERN neutrinos. The new measurements needed at the very least to be viewed with this in mind.

To a large extent the scientific experts sitting on the panels correspond to Fleck's thought collective. Their judgment, just like that of the spacecraft experts, needs to be carefully vetted before acceptance. Whether or not the spacecraft experts have properly understood a problem is always first tested by an attempt to fully resuscitate the spacecraft. If this succeeds, the experts are likely to have solved the problem – unless it soon recurs. Similarly, to ensure that the findings of a panel of astrophysicists indeed are correct, their conclusions must also be tested before they are widely accepted.

Understanding and Control

The process of subjecting new findings to panels of knowledgeable experts who then test their analyses experimentally is the alternative for avoiding errors discussed in the final section of Chapter 12. Aside from isolating different disciplines from one another through quarantine, a second means for avoiding the transmission of errors from one discipline to others is provided by traditional experimental testing.

The doctrine on which experimental science is based is that full understanding of a process requires the ability to repeat it, alter it in predictable ways, and account quantitatively for the resulting behavior. To understand is to control.

The questions we need to ask are, "If a network of disciplines is exposed to a harmful shock, is it possible to control its consequences? Can the instabilities that appear inherent to the ways we pursue astrophysics today be managed so that only meaningful communication between disciplines will take place? Are there more effective ways to organize the network of astrophysical communities to control their transactions better?"

Such questions may be answered by the work of Jean-Jaques Slotine, the French-born MIT researcher in robotics, nonlinear control, and learning systems; Albert-László Barabási, the Romanian-born, Hungarian-ethnic expert in network science at Northeastern University in Boston, Massachusetts; and Yang-Yu Liu, the Chinese-born postdoctoral research associate at Northeastern. Their investigation, 'Controllability of complex networks,' attempts to understand how complex networks may be effectively controlled.[1] The lead sentence of their seminal paper, a deep-rooted attempt to study the control of complex networks of arbitrary types – mechanical, electronic, biological, or social – states the basic credo, "A dynamical system is controllable if, with a suitable choice of inputs, it can be driven from any initial state to any desired final state within finite time."

They might have added "at finite cost"!

Two primary difficulties have been encountered to date in attempting to control complex networks. The first is the task of recognizing the architecture of the network to determine which components interact with each other; the second is to discern the dynamic rules determining the sequence of the interactions. These detailed considerations remain. However, several transcending traits common to all networks can simplify the search for effective control, and the work of Liu, Slotine, and Barabási – LSB – reveals at least some of these.

The networks that LSB considered all are *directed*. They consist of nodes, communication centers, from which information is directed to other nodes along the *edges* or links connecting the nodes, as in Figure 12.1. The edges transmit information from a *driver node* to a subordinate node requiring control. In the networks of LSB, however, information is prevented from flowing bidirectionally along an edge, meaning that arrows connecting any two nodes in Figure 12.1 should point only in one direction, as they do in Figures 15.1 and 15.3. Otherwise the controls fail. Nonetheless, an affected node may still transmit information back to the driver node along other edges, potentially via several intermediate nodes, as Figures 15.1 and 15.3 also illustrate.

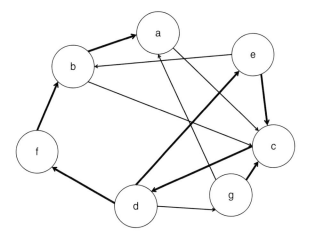

Figure 15.3. A network with seven nodes connected by links of varied coupling strengths, or *weights w*. But here, in contrast to Figure 12.1, each link or *edge* transmits information solely along a single direction. Such networks can be more reliably controlled. Stronger links among disciplines are indicated by darker arrows. Note that nodes *d,e,f*, and *g* each are controlled by a single *driver node*. Nodes *a* and *b* each are controlled by two drivers, while *c* is driven by four nodes, making it more difficult to control.

The state of a network consisting, as in Figure 15.3, of a number of nodes N, can be described by an enumeration $[x_1(t), x_2(t), \ldots, x_N(t)]_T$ of the states of each of its nodes at some specific time $t = T$. This state may, for example, denote the amount of information passing through each node at time T.

A network's evolution is described by two factors. The first is structural, depending solely on the network's architecture and its initial state. The second is the time-dependent control of at least some of the states of nodes controlled by an outside controller.

One of the simplest networks of this kind one can imagine is a pinball machine. A player exerts minimal control if he is merely able to launch the steel ball with a desired initial momentum but has no further interaction with the pinball machine. The initial state determined by the machine's internal architecture, and the ball's momentum at launch, then control the ball's trajectory and the sequence of states the ball and machine undergo. Clearly, a player able to exert additional control on the ball's trajectory after it is launched is better equipped to attain a higher pinball score.

Depending on a network's architecture, total control may be achieved through intervention at merely a few selected nodes, or may require control of a large majority of nodes. A minimum number of driver nodes required to gain full control of a network is determined by the total number of unshared paths. This means that if each driver node drives only one other node, it may suffice to control only the first node in a chain. Such a system may be thought of as equivalent to the pinball

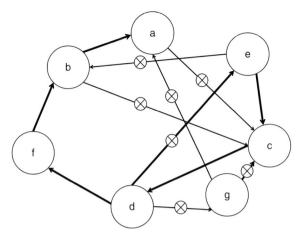

Figure 15.4. Control of the network in Figure 15.3 requires each node to have only one controlling superior node. If it has several incoming edges, all but one of these must be separately controlled, as must all but one of the outflowing edges. Symbols ⊗ denote externally controlled inputs. Note that driver node e can control nodes c, d, f, b, and a, provided controls are placed as shown along many of the other edges. But this arrangement of controls may not be the best strategy because it makes nodes c, d, and f critical, meaning that breakdown of one incapacitates others.

machine in which the player can solely control the ball's launch momentum. If a driver node drives two other nodes, control will have to be asserted not only over that driver node but also on at least one of the nodes it controls.

A second crucial factor is the distribution of *degrees*, the number of edges, k, providing incoming and outgoing information to each node. The quantity k is a measure of the connectance C that Gardner and Ashby considered in their Figure 12.3.[2] The direction in which information flows along the network edges of LSB may be changed, but appears to be relatively unimportant in determining the achievable control as long as the *degree distribution*, the probability distribution $P(k_{in}, k_{out})$ of the number of incoming and outgoing edges at the networks' nodes, respectively k_{in} and k_{out}, remains unchanged.

A question of importance is what happens under link failure, a common problem with large networks. Certain links are *critical*, in the sense that additional driver controls are needed if these links were to fail. Other links may be *redundant*, and can be removed without affecting the number or distribution of driver nodes. And still others are *ordinary* in that they are neither critical nor redundant; they perform a function but the network can still be controlled in their absence. In the arrangement of Figure 15.2 each member of the panel is *critical* in the terminology of Liu, Slotine, and Barabási, in that the absence of one panelist leads to dysfunction of the entire panel.[3] The configuration of Figure 15.1 shows each panelist to be redundant. The arrangement correspondingly is more robust.

Considering the roundtable of Figure 15.1 as a network, we may note that the paths along which the discussion transpires incorporates at least one of the requirements for full control, which Liu, Slotine, and Barabási emphasized.[4] Each of the links between nodes – panelists – conveys information along only one direction, although a longer return path to the initiating node is also included as a cross-check on reliability of transmission.

For small average degrees of connectivity $\langle k \rangle$, as in Figure 15.2, all links tend to be essential for control, so that the fraction of critical links approaches unity. Although one might think that the fraction of redundant links would always increase with increasing connectivity $\langle k \rangle$, it does not. This is because, as the connectivity increases, a giant component may emerge, as in Figure 9.8. The chances then diminish that a link could become nonessential for full control.

As LSB point out[5]

> To fully control a network, one has to … make sure every … node has is own 'superior'. If a node has no 'superiors', i.e., no nodes pointing to it, then obviously it is inaccessible and we lose control of it. If two or more nodes share one 'superior' … we cannot fully control the system. … If we control the central hub only, the system is uncontrollable because we cannot independently control … two subordinates if they share one superior.

These remarks make understandable why hubs generally will not serve as control nodes. Hubs, nevertheless, exert a major influence on network evolution, simply because they connect to and influence so many subordinate nodes, albeit in alarmingly indiscriminate ways if the aim is to drive the entire network expeditiously from any initial state to any desired final state.

LSB obtained some of their results analytically and others through insights gained from an examination of three dozen real-world networks ranging from regulatory networks in yeasts or the corporate world, food webs, power grids, electronic or neuronal circuits, scientific citations, the Internet and World Wide Web, e-mail exchanges, and intraorganizational manufacturing structures. Perhaps most surprisingly, they find that gene regulatory networks display a high fraction of driver nodes $n_D \sim 0.8$, meaning that about 80% of the nodes would have to be independently controlled to control behavior fully. This presumably accounts for the stability of biological species. It is difficult to make major changes when so many different nodes control the functioning of an organism. Mutations occur in small numbers and enter the main stream only if they prove advantageous in competition.

In contrast, several of the social networks studied by LSB are characterized by the smallest fraction of driver nodes, n_D, suggesting that a few individuals could in principle control the whole system. That this is likely to be true of the astrophysical network is shown by the mode of convincing colleagues I mentioned in Chapter 13, where I noted that a few exceptionally influential astrophysicists can greatly affect communal decisions.

This last finding of LSB is scary. What if these few individuals recognized as authorities are simply wrong, even if well-intentioned? It gives reason for pause. From the point of view of innovation, it is good that an individual may bring new approaches to the fore with relative ease. But, as we have seen throughout the many advances cited, the innovators in astrophysics are usually young and lack influence. The most influential astrophysicists often are no longer the most innovative. These are the individuals acting as hubs. They may not be able to control the system in the strict sense required by LSB, but their centrality in the network enables them to strongly affect the network's evolution through the cluster of intimately linked colleagues they influence.

A further worry in astrophysics may be that the nodes in the network of astrophysical disciplines appear to largely be linked by two-way rather than unidirectional edges – more like in Figure 12.1 than in Figure 15.3. Currently astrophysical information flows both ways among disciplines. This makes control of the network much more difficult and perhaps impossible in its present form. Shocks to the system may be unpreventable as long as the current network structure is maintained. A more readily controlled sharing of information among disciplines may need to be found to assure astrophysics greater stability.

One way of ensuring this would entail establishing a more workable link ordering among disciplines, in which a discipline establishing a new finding initially communicates this to a panel of experts representing the community of potentially affected disciplines. These experts constituting a roundtable, as in Figure 15.1, would first sift through the newly provided information to uncover and correct any misunderstandings or potential errors. Only after the panel released the clarified information for further transmission, would it be passed on to the linked disciplines. If the information provided then still appeared unclear or suspicious, a questioning discipline might seek further clarification from sources further up the chain, closer to the original source of information.

Such a scheme would be most in line with the findings of LSB that significant control can be established – here on the quality and clarity of transmitted data – only when each node, each receiving discipline, is fed information by only a single driver. The driver, here, would be the panel of experts who most readily grasp the vocabulary of both the discipline from which it received the information and the several disciplines to which the information is to be further conveyed.

The network of astronomical disciplines is just one of many social networks engaged today in sharing information. Fortunately, research in network theory continues to be vigorously pursued. This raises the prospects of finding progressively better ways of organizing exchanges of information within the astrophysical community to ensure it greater stability.

The findings of LSB are also partially in line with traditional ways of publishing, which reinsert unidirectional edges, leading from authors to journal referees and then to the journal-reading community. Here the control node resides in the journal's editorial offices. Bypassing this system through preprint distribution may transmit information more rapidly, but the propagated information may be less reliable and cause unintended confusion and harm.

What we need is an improved method of disseminating and sharing information, starting with methods of control similar to those currently practiced by journals, but specifically honed to act as barriers to misinformation shocks. Whether this is practicable is not yet clear and warrants further investigation.

Ludwik Fleck might warn us against trusting the thought collective, the proposed panel of experts, to such an extent and yielding to them such large measures of control. But our expectations from the panel are not that they control, merely that they sift through information to eliminate potential misunderstandings, and that their findings still be subject to objective testing wherever possible.

Informed Long-Term Planning

Every 10 years, the National Academy of Sciences' National Research Council assembles a panel of astrophysicists to recommend directions of research and the construction of new astronomical facilities in the decade ahead. These panels may be constituted in ways comparable to those required of experts dealing with puzzling interdisciplinary topics.

As in the resolution of spacecraft or interdisciplinary problems, panel members planning the future of astronomy and astrophysics should ideally be selected based not only on expertise in their own area of research, but also on intimate familiarity with two or more other astronomical disciplines. Whether this second criterion is ever given serious consideration in selecting Decadal Survey panelists is not clear.

An – admittedly subjective – assessment is that members of the community serving on the decadal reviews are most likely to be selected because they either are leading experts in their respective disciplines, or are generalists in theoretical astrophysics able to understand and mediate among the disciplinary representatives. This is undoubtedly a credible mix, but it does not appear to be quite as effective a group as required for the thorough roundtable investigation depicted in Figure 15.1. It ascertains only that the generalists understand the experts representing the individual disciplines, but not that the disciplinary experts fully understand each other or the generalists.

The same goes for participants at symposia convened from time to time to discuss particularly urgent problems of astrophysics to see how these might be successfully advanced. Where difficult problems need to be explored, language invariably plays a central role. Phrasing questions in ways that colleagues can understand fully is one of the most difficult tasks astrophysicists face when a previously unknown phenomenon surfaces and requires explanation. Often it is unclear whether the phenomenon really exists or merely reflects defective instrumentation. Frequently, an exciting new discovery fades away when checks reveal the observations to have been faulty. But even when a finding seems doubtlessly confirmed, the responsible causes may be so varied that it is not even clear where to begin. The discovery of dark matter and dark energy are two such findings.

Acquisition of a sufficiently large astrophysical vocabulary – comprising instrumental, astrophysical, mathematical, and computational concepts – is essential for any but the most superficial levels of understanding. And because few astrophysicists master vocabularies extending far beyond those required to operate successfully within their own research discipline, novel findings often require dedicated conferences to which experts in all apparently related and germane subjects are invited. Just like their engineering colleagues assembled around a table to deduce complexities of spacecraft operations, astrophysicists can be thought of as sitting around an imaginary roundtable as they attempt to tap all areas of expertise their community offers to come to grips with new phenomena.

Where a key element eludes the assembled experts, progress is thwarted and the field remains unstable. Abstract mathematicians, for example, are seldom invited to participate in debates conducted by the astrophysical community. Yet, we recall from Chapter 14 that this is precisely the advice that Paul Dirac would offer. These are aids to our understanding with which astrophysicists tend to be particularly unfamiliar. This was why the problem of black holes seemed intractable for nearly half a century after Schwarzschild's original formulation in 1916. It required David Finkelstein's realization that a new manifold needed to be envisioned in which physical processes were no longer symmetrical under time reversal.[6] The further insight of Roger Penrose that topological approaches could enrich our understanding of black holes then ushered in a new era in which the existence of black holes could be better understood.

The theoretical physics community has had a tradition, for a hundred years by now, of assembling at roughly three-year intervals its recognized leaders to discuss, and possibly point to ways of resolving, intractable problems of physics. Initiated by the Belgian chemist, industrialist, and philanthropist Ernest Gaston Joseph Solvay, in 1911, the Solvay Conferences were particularly influential during their early years, when the group of invited attendees was small and exchange of ideas was easy. As Werner Heisenberg wrote in 1974, "the historical influence of the Solvay Conferences on the development of physics was connected with the special style introduced by their founder: a small group of the most competent specialists from various countries discussing the unsolved problems of their field and thereby finding a basis for their solution."[7]

> • Small conferences of leading experts held to discuss new findings that seem at odds with the accepted lore of the astronomical disciplines they affect can be particularly useful aids to an orderly development of astrophysics. Initiated only when significant new problems arise and assembling only those experts who might provide penetrating new insight, such conferences could help in sorting out anomalies that otherwise could delay progress for many years.

Control and the Decision Favoring the Great Observatories

In Chapter 11 we noted the tumultuous three years that preceded the decision to construct the family of four space observatories in sufficiently rapid sequence to enable simultaneous or near-simultaneous observations of a large variety of astronomical phenomena at wavelengths covering most of the electromagnetic spectrum, from the γ-ray region to the far-infrared. Additional radio observations could then be conducted also with ground-based telescopes, which in the United States were under NSF control, whereas the space observatories came under the aegis of NASA.

Some of the difficulties Charlie Pellerin's Astrophysics Council faced in its attempts to persuade the administration to build the Great Observatories can be understood in terms of the control theory of Liu, Slotine, and Barabási (LSB) and the organizational chart of Figure 11.2 with its many bidirectional arrows flying in the face of a system optimized for control.

It is clear that the organizational structure reflected in Figure 11.2 can easily lead to chaotic decisions, not only through the many advisory bodies attempting to simultaneously influence NASA, but also through the multipronged initiatives introduced, respectively top-down by the President of the United States and the Congress, and bottom-up by the science community. Perhaps this is the nature of democratically reached decisions; but making the system workable requires the cooperation of the different factions, as well as courageous public servants willing to make tough decisions and stick by negotiated compromises.

In the end, several factors brought about a favorable decision on the Great Observatories. The Astrophysics Council worked out mutually acceptable ways, with Associate Administrator Burton Edelson's Space and Earth Science Advisory Committee, and with the National Academy of Science's Space Science Board, so that Edelson would hear more or less unanimous advice on the Great Observatories from all three committees. Charlie Pellerin also reached agreement on cooperation on a number of projects of common interest to him and Geoffrey Briggs, his counterpart in the Solar System Exploration division. In retrospect, Burton Edelson's insistence on unanimity was thereby at least significantly met.

Later, the arrival of Lennard Fisk, who replaced Edelson, introduced a structured list of long-term priorities within the Office of Space Science and Applications that was appreciated both at higher and lower levels within NASA; and several members of the Astrophysics Council were fortunate to raise the interest of the President's Science Advisor, Bill Graham, in the Great Observatories. Graham had direct access to the President concerning the framing of new scientific and technical initiatives, as indicated by the box at the upper left of Figure 11.2 marked "President's Science Advisor."

Finally also, members of the Astrophysics Council concentrated on periodically updating members of the Congressional staff about the anticipated power of the Great Observatories. They did this in part through their dedicated series of colloquia on the exciting new astrophysical and cosmological insights the Great Observatories would provide, and in part through Harvey Tananbaum's occasional visits to the offices of Congressional staff who carried out much of the work of the formal Congressional committees and exerted considerable influence.

By choosing cooperation across the entire space science community and within the governmental structure, the bidirectional arrows in Figure 11.2 effectively were converted into unidirectional arrows so that the advice channeled through each office reflected unanimity and agreement.

Although the organizational chart of Figure 11.2 continued to exhibit two-headed arrows, a measure of systemic control was achieved through decisiveness and persistence. Liu, Slotine, and Barabási's criteria for achieving control were more or less met, but only by way of a circuitous route of coordinated persuasion.

The Critical Role of Structured Archives

Nations go through cycles of prosperity and decline.

After half a century of unparalleled support in the post–World War II era, we have thought too little about how astronomy will cope when leaner times force delays and interruptions in our searches, possibly for decades or longer.

How can the astronomical community ensure that hard-won knowledge will not be forgotten during years of lag or inactivity, and that our understanding of the Universe will remain intact so that, after a long hiatus, we will be able to revive the field, starting, if we so choose, where an earlier generation left off?

Language and control are the key factors involved in the long-term preservation of information and data! We need to better ensure that our successors, in good times and poor, will correctly decipher the formats and understand the material we store for future recovery.

Our current archives are not designed to cope with decades-long interruptions. They are budgeted on the tacit assumption that every decade or two we will witness increasingly powerful observatories making earlier data obsolete. Our archives thus need to make data accessible only for a couple of decades. Over longer periods, with each passing generation, the information stored will no longer be fully recoverable as intimate knowledge of precisely how the observations were gathered is lost.

* * *

This is not an abstract problem. The construction of new observatories is currently becoming so expensive that an advance in observational astronomy along the broad front to which recent decades have accustomed us may die out with the present generation, unless we find ways to pace research differently. The expense of constructing the James Webb Space Telescope is already delaying construction of other highly recommended space telescopes. Specialists working in other wavelength ranges accessible only from space may have to wait a decade or two before work on their urgently needed space telescopes can begin. In the meantime these researchers will need to archive current data so they can be accessed again when activity in their wavelength ranges resumes, spearheaded by a younger generation.

The astrophysical community has still not adequately addressed this problem. Archival efforts to store data admittedly are already underway in what is now called the *Virtual Observatory*. But storage of astronomical data alone will not suffice without an equally meticulously archived set of specifications portraying the performance of the observatory that produced them, including the specification of each component, its characteristics, detailed records of all observatory software commands, and a dictionary of the vocabulary that observatory operators employed in the course of the observatory's operations, irrespective of whether the observatory was ground-based or operated in space. Finally, a diary of daily operations and sequences of failures and implemented fixes will need to be included, as will a record of the changing versions of software used in reducing observational data in the course of the months and years during which the observatory was active.

Regrettably, the data being stored in our archives today comes nowhere near attaining such completeness. When asked about it, archivists respond that available funding does not permit it. Yet it is difficult to see how an astronomer would be able to make full use of the archived information even as little as a couple of decades after the data were stored – not to mention a century later. Much of the success of using the archives today depends on the ability to tap the memories of people who did the original work.

Below, I sketch the complexities facing astronomers working on just one contemporary space observatory – complexities that, in one way or another, need to be reflected in any set of archives aiming to be complete.

Evolving Language and Long-Term Instability

In 2009, the European Space Agency launched Herschel, its second-generation billion-dollar infrared telescope.[a] Named for the eighteenth century astronomer, William Herschel, it was considerably larger than its immediate predecessor, the Infrared Space Observatory (ISO) and had a quite different set of instruments onboard.

Of particular significance was an elaborate data processing tool with which astronomers were expected to reduce their data. Called the Herschel Interactive Processing Environment (HIPE), it was sufficiently complex to require experienced astronomers to attend three-day workshops to learn how to use it. There, experts from the Herschel Science Center, who had created the HIPE language, gave talks on the reduction – interpretation – of data obtained with the three different instruments onboard. In part, the workshops were designed

---

[a] The Herschel mission was initiated, constructed, and funded by the European Space Agency, with significant participation by NASA. It operated successfully in space from 2009 to 2013.

to teach the new HIPE language; in part, they also sought to clarify minutiae compiled in book-length user manuals – one depicting spacecraft and telescope performance, three more to specify the characteristics of the three instruments onboard.

Once the Herschel observatory was launched into space and became operational, HIPE was continually upgraded as experience suggested potential improvements. By the end of space operations, roughly four years after launch, HIPE version 10 had come into use. Results published early in the life of the Herschel mission were reduced with early versions of HIPE. Those published in the mission's waning days were reduced with more advanced versions.

Observing procedures with the three instruments onboard Herschel similarly morphed throughout the mission as characteristic strengths and weaknesses emerged in different ways of operating the instruments. Thus maps or spectra of astronomical sources obtained early in the mission were assembled by means that significantly differed from those obtained later.

After Herschel operations had ceased, all data accumulated over the mission's lifetime were eventually to be reduced using a final version of HIPE and then archived for posterity. The basic observational data obtained throughout the succession of revised and improved instrumental operating procedures, however, could no longer be changed. Moreover, journal articles published early during mission operations were based on data obtained through early observing procedures and were reduced with early versions of HIPE, whereas observations conducted late in the mission were carried out with different procedures and reduced differently. These limitations were unavoidable, of course; performance can be improved only through experience gained in operating a spacecraft and reducing incoming data. However, careful archiving that may later provide astronomers with access to the raw data and potentially better means for analyzing them, should remain a priority. The richer and more complete the archives, the greater will be their potential!

* * *

By now, many leading observatories, both ground-based and those launched into space, have developed their own computer programs for operating telescopes and for handling data. Often these are based on distinct computer languages. Largely responsible is the bewildering speed at which different computer languages have been introduced and subsequently discarded. This problem is endemic and, at least in part, beyond control of scientific communities. The computer industry remains profitable only as long as it keeps offering improved performance while withdrawing support and maintenance of earlier generations of software and hardware. This automatically forces adoption of ever-changing computer languages and procedures, regardless of whether the long-term stability of science benefits or suffers.

* * *

With the departure of the experts who worked on a mission, memory of earlier procedures and vocabularies rapidly fades.

Thirty, fifty, or a hundred years from now, data obtained today may no longer be fully understandable. Each generation adapts language to its own needs so that archived material becomes increasingly difficult to interpret. But, although nobody has ever stopped language from morphing, a careful set of logs that spells out all procedures, step by step, paying attention to detail, can significantly prolong the useful life over which archives will remain instructive. With sufficient care, such records might remain useful considerably longer.

Archives Conceived as Commons

A new approach for preventing the degradation of knowledge over decades and possibly centuries has arisen in the past two decades and promises to help in the long-term maintenance of astronomical archives. This thrust, whose promise was originally revealed through the work of the sociologist and political economist Elinor Ostrom at the University of Indiana, illustrates how society manages communal resources whose value can be sustained only if the entire community is willing to pitch in so all may benefit.[8]

Sociologists have coined the word *commons* to designate terrain representing communal property that may be threatened by proprietary interests. The term originated in common grazing grounds – commons – that could not be maintained if any one herdsman attempted to graze more of his cattle than allowed. Violation led to overgrazing that hurt an entire village's common good.

In recent decades, sociologists and economists have been studying sensible ways of sustaining *knowledge* threatened by the cost of safe archiving, national security considerations, and other factors. A particularly extensive study on *knowledge* as a common good has been carried out by Charlotte Hess at Syracuse University and by Elinor Ostrom.[9] The issue of safely preserving knowledge over long periods, with the ability to recover and reap its benefits after long fallow periods, is a problem pursued by many communities. Knowledge, in the sense used by Hess and Ostrom, readily encompasses astronomical data and information.

The economics of the commons is a system that runs counter to the philosophy on which much of twentieth century capitalism was based. But, in a variety of settings, the approach has shown itself to work remarkably well. It can be understood in terms of everyday experience: Many of us, stumped when trying to get new software to work, turn to Google, to see whether someone else had the same problem, solved it, and was thoughtful enough to share the correct approach. This works only because Google, accessed by millions of people, can be used to establish a forum open and free of charge to anyone interested, a commons whose value and prospects depend only on the good will of a community interested in its maintenance.

For scholarly work, a well-known knowledge commons is *Wikipedia*. The success of the Wikipedia website can be traced to the many individuals willing to contribute essays to this wide-ranging encyclopedic resource and to the willingness of others to make donations enabling the administration of the site. A number of statistics provide a measure of what a well organized knowledge commons can achieve. In its annual report for the year from July 2010 to June 2011, Wikipedia's parent organization *Wikimedia* reported total revenues of about $25M raised through contributions from more than 570,000 individual donors. Only three donations exceeded $1M. The average donation was about $40. Eighty Wikimedia employees carried the workload of a site that created nearly 8400 new articles daily and was visited by 423 million individual visitors that year. Wikipedia is still expanding rapidly, but whether and how long this can or will continue is uncertain. Wikipedia still needs improvement, as any young venture does; but it covers a wide range of subjects, is readily accessible, and many of its offerings are carefully conceived, trustworthy, and useful.

* * *

For astronomy, one of the most difficult tasks is maintaining the utility of a data archive for periods of decades or longer, despite rapidly changing computer languages, scientific expressions, or shades of meaning of common words – all threatening to steadily erode the archive. A communally sustained effort to update the archive continually through periodic clarifications could appreciably extend the useful life of the archive at little added cost. An astronomer attempting, decades after the archive had been established, to retrieve stored data, might then find a useful explanation for their provenance through a targeted request for information on the meanings of specific expressions or data processing techniques that no longer are in common use.

At some point, to be sure, when we no longer master the language in which data were stored, we may need to repeat observations or accept some interpretation of them on faith. Unless observatories then become more powerful at more affordable cost, the rate at which we will be able to increase our understanding of the Universe may lag the rate at which previously gathered data become obsolete.

Notes

1. Controllability of complex networks, Yang-Yu Liu, Jean-Jacques Slotine, & Albert-László Barabási, *Nature*, *473*, 167–73, 2011.
2. Connectance of Large Dynamic (Cybernetic) Systems: Critical Values for Stability, Mark R. Gardner & W. Ross Ashby, *Nature*, *228*, 784, 1970.
3. Ibid., Controllability of complex networks, Liu, Slotine, & Barabási.
4. Ibid., Controllability of complex networks, Liu, Slotine, & Barabási.
5. Ibid., Controllability of complex networks, Liu, Slotine, & Barabási: See the online supplementary material to the *Nature* article of LSB, p. 13.
6. Past-Future Asymmetry of the Gravitational Field of a Point Particle, David Finkelstein, *Physical Review*, *110*, 965–67, 1958.

7. *The Solvay Conferences on Physics – Aspects of the Development of Physics since 1911*, Jagdish Mehra, with a foreword by Werner Heisenberg. Dordrecht, The Netherlands: D. Reidel, 1975, p. vi.
8. Introduction: An Overview of the Knowledge Commons, Charlotte Hess & Elinor Ostrom, in *Understanding Knowledge as a Commons – From Theory to Practice*, edited by Charlotte Hess & Elinor Ostrom. Cambridge, MA: The MIT Press, 2007, pp. 3–26.
9. A Framework for Analyzing the Knowledge Commons, by Elinor Ostrom and Charlotte Hess, in *Understanding Knowledge as a Commons – From Theory to Practice*, edited by Charlotte Hess and Elinor Ostrom. Cambridge, MA, 2007: MIT Press, pp. 41–81.

16

An Economically Viable Astronomical Program

A theme encountered throughout this book has been the resourceful approach of astronomers and astrophysicists of the twentieth century as they searched for new ways to advance our understanding of the Universe. By adopting the new atomic and ionic theories of Niels Bohr and Meghnad Saha, Henry Norris Russell and Cecilia Payne determined the abundances of the chemical elements in the Sun and stars. Willem de Sitter and Arthur Stanley Eddington taught us how Einstein's relativity theories could lead to new insights on an evolving Universe. And the nuclear theories of George Gamow, Hans Bethe, and Edwin Salpeter provided novel insight on the energy sources of stars and the origins of the chemical elements. Astronomical observers similarly adopted techniques developed by physicists and engineers to greatly expand the wavelength ranges across which the Cosmos might be studied. By adopting methods developed elsewhere, the cost of astronomy was kept low.

This enterprising spirit will stand us in good stead as we face a newer set of challenges posed by the global economic downturn of late 2008 that still persist today. Finding new ways to advance astronomy under these conditions will be important for solving the cosmological problems now to be overcome. Those efforts will undoubtedly take time, and will succeed best if we choose an appropriate economic approach aimed at assuring the long-term stability of our astronomical program.

Exploring these alternatives is the purpose of this final chapter.

* * *

Practical problems to which astronomers of the twentieth century devoted themselves tended to promise benefits mainly in the long term: studies of asteroids that could some day threaten catastrophic impacts on Earth, investigations of violent solar storms or extreme quiescence on the Sun, and other long-term evolutionary trends of potential consequence to a planet's climate and thus also to life on Earth.

The most gifted teachers astronomy has produced also attracted generations of youngsters to careers in science, showing them how thrilling the search to understand Nature can be, and how it can contribute to the welfare of their fellow citizens.

But, because most of astronomy's activities are aimed at more esoteric investigations, we realize that society pays for the conduct of astronomy not solely for its utility, but because humans have long wondered about their place in the Universe – a mixture of religion and awe that comes to mind mainly in leisure hours. Where economic hardship denies leisure, attention to astronomy wanes.

The times into which we are heading are likely to be difficult. Astronomy is approaching, or has already reached, the highest funding levels society can now afford. To productively move forward, we need to acknowledge this.

One early economic indicator is the shifting focus of high-energy physics: Once the pride of the U.S. physics community, the central activities of the field have now moved to the Large Hadron Collider at CERN in Switzerland. Another signal is the planned construction of the next large radio-astronomical facility, the Square Kilometer Array, partly in South Africa and partly in Australia. Both these moves are accompanied by the decommissioning of facilities in the United States that are no longer viable or are deemed unaffordable. Construction of powerful ground-based cosmic ray and γ-ray observatories in Argentina and Namibia are other signs of shifting priorities.

We see the same warning signs in NASA's Astrophysics Implementation Plan announced on December 20, 2012, which postpones many of the space missions the Decadal Survey of 2010 most highly recommended.[1] NASA's budget no longer is able to accommodate these, as anticipated costs of the James Webb Space Telescope have steadily mounted. The Wide Field Infrared Survey Telescope (WFIRST), which was conceived to study dark energy and exoplanets, among many other projects, is going to be greatly delayed. The Laser Interferometer Space Antenna (LISA), which was to be a mission jointly implemented with the European Space Agency, is now on hold – its future quite uncertain. It was designed in part to conduct precision tests of general relativity and potentially to detect low-frequency gravitational waves generated in black hole mergers.

Faced with a level budget, we will need to consider new ways of advancing our understanding of the Universe and maintaining astronomy a vibrant, exciting enterprise able to attract society's brightest, most talented young scientists. Without the best people, we will not succeed.

A level budget reduces much of the freedom of action our field has long enjoyed and come to take for granted during half a century of rapid expansion following World War II. We may now need to seek other, more innovative ways of continuing our quest into the origins and evolution of the Universe. A number of potential approaches for doing this will require taking stock.

The Need for an Evolving Economic Approach

Six factors demand a reevaluation:

• Budgets for the pursuit of astronomy, world wide, have reached a plateau and may, at least for the next decade, be expected to remain static or to partially decline.

• As astronomy's priorities turn to investigating the esoteric dark matter and dark energy, of limited practical interest because the prospects of mining these forms of energy seem remote, the field may no longer benefit from infrastructure provided free of charge by the military or industry. For Federal infrastructure support astronomy will increasingly have to turn to NASA and the National Science Foundation, both of whose resources are far more limited than those devoted to defense.

• The cost of astronomical observatories has rapidly risen as increasingly powerful capabilities have come into demand. In times of frozen budgets, the rate at which new observatories can be constructed will correspondingly have to decline.

• As currently customary with space observatories, the launch of new telescopes may require the closing down of existing facilities unless provision for continuing their operation at reduced cost can be found. Access to some wavelength ranges or forms of activity may then be curtailed for significant periods to provide for renewed vigor in other areas. Pursuit of astronomy along the broad front customary in recent decades may then become unaffordable unless we find new ways of conducting astronomy.

• If activities in any given area are thus curtailed, significantly improved archives will be required to maintain memory of all we have learned. These archives will be expensive because the rapid obsolescence of virtually all current technologies will require maintenance of far more detailed records than now customary to describe precisely the means with which data were gathered and how they were subsequently processed.

• Finally, some types of esoteric observations of little public interest but nonetheless of pressing astrophysical importance will require the support of private donors and volunteers. Where these ventures require significant expenditures or steady long-term support, steadfast communal backing will be essential.

The astronomical community's long-term plans will thus have to include not only nationally funded projects but also those for which campaigns for private funding will need to be maintained. This will mean pacing progress based both on instrumentation and capabilities that industry and governments may continue to develop or are planning to utilize, while also targeting exploration of significant topics with no known societal application or governmental support.

An Evolving Scientific Rationale

In the United States, policies for Federally funded research advocated by 'Science – The Endless Frontier' depicted an explicit dependence on nationally mandated ways of conducting research.[2] They required conformance to Presidential and Congressional directives based on the nation's economy and other political exigencies. The landscapes of ambient activities sketched in Figure 9.4 now influenced the conduct of astronomy and astrophysics as never before. As we saw in Chapter 11, major projects could be started and then delayed or abandoned as national priorities changed. And, as noted in Chapter 7, U.S. collaboration on international projects could be curtailed if doubts were raised about whether sensitive information might somehow be revealed.

Although such concerns are understandable, and often can be equitably resolved, a more difficult question they raise is this:

> • Do the means made available to us today in our quest to understand the workings of the Universe predetermine what we will find? Are we selectively restricted to uncovering just those cosmic features that the nation's form of support and its accompanying constraints permit?

A more mundane, but still important and more readily answered question is "Do national security considerations, the projects they are likely to fund, and the facilities they make available to astronomers facilitate only certain types of astronomical work while discouraging other investigations?"

The answer to this second question is undoubtedly "Yes." U.S. astronomers would never have had their successes in infrared, X-ray, and γ-ray astronomy without the use of instrumentation originally developed for defense at an expense dwarfing, by orders of magnitude, any conceivable, purely astronomical outlays. The use of powerful launch facilities to transport into space the large telescopes astronomers required was equally important. Where practical applications appeared to be lacking, as in instruments for the detection of gravitational radiation, dark matter, or dark energy, funding has been harder to come by and progress has been slow. Direct detection of any of these forms of matter or energy still eludes us.

If the U.S. government had found urgent reasons, after the end of World War II, to stress the detection of neutrinos and gravitational rather than electromagnetic waves, astronomers today might be fully engaged in studying a cosmos appearing markedly different from the one we now envision – quite likely no more complete, but certainly different. The world in which we live, portrayed in the textbooks we would now be writing, would hardly resemble the depictions our students are taught today. Our research efforts would be directed at clarifying questions we currently have no reason to ask.

Some may argue that sooner or later the tools to pursue dark matter and dark energy directly will surface, and that some researchers already are engaged in this effort. But if the requisite funding, to come to grips with these forms of matter and energy, should keep rising without any hint of progress, these efforts might be abandoned as unaffordable. We may then have to face the question:

- What are the chances that we may have to stop investigating the origin of the Universe and related questions, because their effective pursuit would cost more than society will ever be able to afford?

I plan to show how we might deal with such contingencies if, or as, they arise, and argue that they will require modifying our current economic approach to forge a stronger contract with private donors and volunteers that diverges from the utilitarian vision of 'Science – The Endless Frontier,' the working model to which we have become accustomed. It will not be as convenient, will undoubtedly require far greater efforts to raise and husband funding, but might nevertheless be the only workable means for the pursuit of certain critical questions.

Consideration of such factors suggest a number of ways we may need to proceed:

- As the conduct of astronomy progressively abstracts itself from conceivable societal applications, we will need to maintain our search into the origins of the Universe vibrant to attract the best young scientists.

A vibrant field requires ambitious goals. Without them, astrophysics will lose our best and brightest colleagues:

- Even with limited budgets, retaining focus on grand, ambitious goals, but pacing and budgeting our advances at levels society can afford, should remain the highest priority.

Sight of these longer-term goals should never be abandoned because where important problems are emphasized unexpected solutions often surprise us. We need only recall the many decades during which Einstein's gravitational redshift was unsuccessfully sought in the spectra of white dwarf stars. Then, as we saw in Chapter 3, the discovery of the Mössbauer effect suddenly demonstrated the redshift with an elegant laboratory set-up that a professor and his graduate student had assembled.

The Economics of an Affordably Paced Program

The question we now need to ask is, how might ambitious astrophysical and cosmological research programs best be conducted in the foreseeable future with affordable means?

One certainty is that service to society should remain a primary astronomical goal.

  • Freedom to pursue esoteric scientific problems is contingent on astronomers serving humanity in practical ways wherever we can. This is our prevailing tacit contract with society.

Work on practical problems often goes hand in hand with far more abstract investigations. Recall Einstein's remark in Chapter 13 that he had found working on practical problems at the Swiss Patent Office actually to be a blessing as he privately also worked on deep-rooted problems of physics that could not be solved rapidly. Trying to understand the everyday world often yields insight on more obscure problems.

A second successful approach, which we should similarly continue to pursue, is the maintenance of strong ties to other scientific and engineering communities. Astronomy is a far smaller field than physics, chemistry, biology, or engineering. And, because funding for astronomy tends to be limited:

  • We should continue to capitalize on the good fortune we have had in importing both theoretical and instrumental techniques developed by engineers, physicists, chemists, mathematicians, and biologists. Many of these tools have played central roles in advancing astronomy.

As we noted in Chapter 9, these benefactors to astronomy frequently not only provided new tools, but also helped out with expertise and manpower our discipline was lacking.

As we have seen throughout, most of the successes of twentieth century astrophysics have come from importing observing techniques developed by the military and communications industries. Imported theories developed by mathematicians and theoretical physicists have similarly elucidated the nature of the Universe governed, as far as we can tell, by general relativity and particle physics. Our understanding of stellar interiors correspondingly is based on the laws of thermodynamics, relativity, and nuclear physics.

Along with such techniques, astrophysics has also benefitted from the inventors of new tools, some of whom, having introduced a powerful new technique, chose to join astrophysics for the remainder of their careers, establishing areas of research that were totally novel and the basis for further progress. Their contributions suggest that:

  • Astronomy should aim at accommodating innovators whose contributions will supplement or surpass those that regularly trained astrophysicists may provide. The next wave of talented individuals imported into astrophysics might come from the fields of chemistry or biology, whose perspectives could have a direction-changing impact on our field. Astrophysicists should encourage and welcome these colleagues. At the same time we should beware indiscriminate growth of our field.

Nobody gains from an enlarged cadre of workers when, as we saw in Chapter 14, requests for observing time at large observatories and opportunities for astrophysical research funding are already oversubscribed. Competition among astronomers for these resources can bring out the best; but excessive competition is wasteful.

Fortunately, the talented individuals from other fields whom we invite to join our efforts often come with their own funding from organizations that support experimental physics, chemistry, biology, or instrumentation, so that any incremental funding that astronomical budgets might need to procure may be easier to raise.

None of this is a major departure from current practice. For decades, astronomy has advanced thanks to detection systems and computers provided by the government, by industry, and by colleagues from other fields. But the flow of these subsidies will almost certainly decline as our interests in the investigation of gravitational radiation, dark matter, and dark energy at least partially disengage from more practical priorities. We will then have to find other sources willing to fund the development of the required tools.

- As astronomy ventures into new realms increasingly removed from current national and industrial priorities, ambitious goals will require long-term investments in infrastructure to develop the required tools. With limited budgets, the astronomical community will need to carefully weigh the appropriate balance between two competing approaches: One will yield small but steady gains achievable through systematic enhancement of existing tools; the other, promising major advances along entirely new directions, will entail longer-term investments in novel infrastructure, and yield success only after the requisite tools are at hand.

Astronomical Planning in a Changing Economy

The organization of the U.S. astronomical enterprise is complex, as Figure 14.1 shows. New research directions may be initiated at virtually any level, starting with the President of the United States or the Congress, and ranging down, through the decadal surveys of astronomy and astrophysics assembled by the National Academy of Sciences, to individual astronomers or astrophysicists proposing imaginative observations or computations. No single agent or agency fully controls the activities that emerge because, counter to the dictates of Liu, Slotine, and Barabási in Chapter 15, and Figure 15.3, the number of two-headed arrows in the organizational chart, and the number of offices subjected to multiple inputs, break with the requirements for full network control. Only when the President, the Congress, and all levels of a funding agency agree that an activity can go forward, thereby rendering the two-headed arrows moot, can even a top-down initiative go forward.

A major initiative proposed at lower levels requires even greater steadfastness in navigating the hazards of rapidly alternating political pressures. As we saw in Chapters 11 and 15, the thrust to launch the Great Observatories succeeded only after agreement had been obtained through all the advisory bodies making recommendations to NASA, and after the Astrophysics Division and Solar System Exploration Division had reached agreement on missions of mutual interest. These conditions again made the two-headed arrows in NASA's organization chart moot, as agreement among the various participants all aligned to point toward a favorable outcome.

In selecting a balance between efforts aimed at major projects, such as launching the Great Observatories, as contrasted to the many smaller steps that might more easily be taken, we should remember that undertaking the most significant projects often require us to cross a substantial threshold:

> • The astronomical community will need to identify and keep in mind questions we still fail to understand and, whenever new opportunities present themselves, grasp them to gradually answer these questions. This does not mean that we should disregard favorable opportunities that technological advances or improved economic climates may from time to time provide to rapidly advance our understanding of lesser questions. But such successes should not distract us from long-term goals of addressing major problems without whose resolution more substantive advances will continue to elude us.

Long-Term Efforts

We may have to reconsider whether or not the present system of decadal surveys and 10-year plans will continue to be useful. It emphasizes problems that, for the moment at least, can still be affordably solved within about a decade. It does not favor programs, which could be so costly, or require observations spanning so many decades, that activities and expenditures would have to be spread over 50 or 100 years, or more.

For the immediate future, this may not be a problem. But the most costly missions, even now, take well over a decade to build and may not be sufficiently utilized if they are soon cast aside. Most space observatories constructed today are designed to operate for no more than a decade or two, if that. If they continue to operate reliably that long, they frequently are retired early to make room for a next generation of missions, which otherwise would not be affordable. Even routine mission operations following the successful launch of a spacecraft are costly. The staff handling the observations to be carried out, the engineers maintaining the health of the entire system, the scientists carrying out investigations, all need to be budgeted for and paid.

> • As we continue to solve the easier questions and are confronted by ever fainter, less frequently occurring observational phenomena, we will probably have to launch larger, more sensitive, more complex, longer-lived,

and thus more expensive missions whose costs may have to be spread over many decades or even centuries. This will require longer-term planning and steadfastness of purpose virtually unprecedented in the astrophysical community. Decadal surveys may then be insufficiently matched to plan, guide, and control longer-enduring projects.

Concentrating on Modest Advances on a Broad Front

Some of the more expensive astrophysical space observatories constructed in recent decades have been a tremendous success, both scientifically and from a viewpoint of managing cost. Others have not, and we should heed their lessons.

The Spitzer Infrared Telescope Facility was exceptionally well managed. An initially proposed version would have exceeded available funding. A series of cost-cutting measures combined with an imaginative redesign then led to an affordable mission that, nevertheless, proved immensely powerful.[3] The history of this infrared telescope may have been a painful, 20-year-long, pre-launch haul for the team of scientists and engineers who conceived, launched, and then operated the mission. But it illustrates how a tightly managed well-run mission can advance astronomy at a high but sustainable cost.

In contrast, two other infrared astronomical missions repeatedly suffered cost overruns threatening cancellation but received reprieves, perhaps partly because they were international projects that cannot be as readily pared back and redesigned once bilateral or multilateral agreements have been signed.

The Stratospheric Observatory for Infrared Astronomy (SOFIA) is currently beginning operations. The James Webb Space Telescope (JWST) is still under construction. Both missions should be a warning to the astrophysics community that our cost estimates are unacceptably inadequate and need to be drastically overhauled.

This is not a problem that can be entirely blamed on managers and administrators.

> • Cost estimates predictably become unreliable when proposed technological steps are too far ahead of their times. Understandably, we astronomers want to plan for the greatest possible returns from the observatories we build. And industry, eager to build them, tries to show how the observatories can be constructed at low cost. The unrealistic expectations on both sides have so consistently led to hubristic bids and painful cost overruns that we urgently need to adopt more stringent methods of assessing and controlling cost.

Unless the astrophysics community tempers its appetites and comes up with more realistic proposals, we will find ourselves predictably facing setbacks.

• A lesson the planning for the Great Observatories taught us is that the various astrophysical disciplines must content themselves with realistically paced advances so all might benefit. If one discipline aspires to advance too fast at the expense of others, the entire system breaks down.

This is the strategy that Charlie Pellerin, Director of NASA's Astrophysics Division, so clearly advocated in the mid-1980s. Recall his musing, which I paraphrase here for clearer relevance:

• The argument for advancing one mission at the expense of all others might be worth having if it supplanted the need for the others. However, each mission actually increases the necessity for the others. We need to create a fresh, compelling strategy that will mobilize the astronomical community behind all important missions, including those already underway.

This greater sense for astronomy's balanced needs should remain a firm guide.

• Astronomy today remains most successful when it finds ways of studying the Universe across all accessible energy ranges, complemented with a healthy program of theoretical work. Pursuing this objective through international collaborations suggests that multination or international investigations be undertaken only with explicit agreement that cost control must be stringent, and may require mutual agreement on moderating ambitions.

When a single mission becomes so expensive that the queue of all others suffers long delays the entire field is set back.

Budgeting Major Searches Requiring Private Funding

From time to time, problems emerge whose solution is critical for gaining essential cosmological insight but lacks any foreseeable practical application. To address such tasks, the astronomical community will increasingly need to raise private funding.

Philanthropic construction of large ground-based telescopes is nothing new in astronomy. Earlier we saw how the Hooker telescope on Mount Wilson and the Hale telescope on Mount Palomar had been constructed with privately raised funds in the first half of the twentieth century. In 1985 Howard B. Keck of the W. M. Keck Foundation similarly provided $70 million to fund the design and construction of the first of two Keck 10-meter-aperture telescopes on Mauna Kea, Hawaii. The Keck Foundation established by William Myron Keck, founder of the Superior Oil Company, completed this telescope in 1992 and went on to complete an identical second telescope in 1996. Around the turn of the millennium, contributions of the order of $100M toward the construction of large ground-based telescopes were thus rare but conceivable.

Two related questions, however, are less clear. First, is it possible to raise substantial private funding to answer astronomical questions that are largely curiosity driven and by no means certain to yield significant results? Second, can philanthropic organizations be expected to construct a space telescope far from Earth, an all but invisible monument to a philanthropist's munificence? Philanthropic organizations usually require visibility and recognition to survive.

* * *

The first of these two question may be best answered by the current Search for Extraterrestrial Intelligence (SETI). SETI has been supported through philanthropy and the willingness of interested volunteers to pitch in with donations of their time.

The Allen Telescope Array (ATA), constructed for SETI through the generosity of Microsoft co-founder Paul Allen, has been a prime pillar of support. Allen contributed an initial $30M to an eventual total of about $50M required to build the current 42-dish radio array, which began science operations in 2008. The original intention was to expand the array with further funding. The international financial crisis that started in late 2008, however, has delayed those plans.

The SETI Institute has attracted a high-quality professional staff paid largely through federally sponsored research grants won for investigations of astrobiology and extrasolar planetary systems, studies ranking high among NASA and NSF research priorities.

Separate funding for the search for other technologically advanced civilizations able to communicate across interstellar space has also been provided by NASA and the NSF, at varying levels at different times. Additional moneys raised through donations, again mainly through philanthropy and contributions from industry, and smaller donations from amateur astronomers and enthusiasts, have also been a steady source of support, as have been services provided free of charge. Averaged over the years since its incorporation as a nonprofit organization, the SETI Institute has operated at annual levels in the $8M–$12M realm – measured in fiscal year 2000 dollars – with roughly half of this funding dedicated to the search for extraterrestrial civilizations.

Day-to-day and year-to-year operating costs for the search for advanced technological civilizations have had to be covered largely by private means, difficult to raise since the economic downturn, which has curtailed philanthropic contributions and reduced funding for the extraterrestrial search to a level around $1M per year.[4]

Figures like this provide a rough measure of how much support a largely curiosity-driven cosmological effort might reasonably expect to receive from philanthropic institutions and an appreciative public. It makes clear that whatever sizable program of this kind may be envisioned, its efforts may have to be spread across many carefully budgeted decades, potentially ranging up to or exceeding a century. In turn this will demand a steadfastness, a stable observing platform on which data will be gathered, and a similarly stabilized archive for data storage maintained over a corresponding time span. Archived material will need to be documented far more carefully than is currently customary. It will require clear, traceable language, so that several generations of workers will be able to precisely conduct, and correctly interpret and evaluate, the long-drawn-out observations.

- The price for investigating pressing cosmological problems using private means when necessary will be the premium placed on discipline and steadfastness of purpose. Should such a privately funded mission entail expenditures of the order of $1B in current dollars, based on operating costs budgeted at levels of about $10M a year, it would likely entail an operating life spanning a century, and correspondingly the professional lives of at least three generations of researchers.

Long-Term Maintenance and Stability of Archived Data

If we have to proceed more gradually to keep astronomy affordable, we will need increasingly to ensure the long-term stability of data. Otherwise steady progress will remain beyond reach.

- In view of apparently inevitable evolutionary trends in the computer industry, astronomers and astrophysicists will need to devise fault-free methods for frequently transcribing records to maintain ready access to decade- or century-old data, including detailed descriptions of how those data were collected and reduced. This will be costly but necessary to prevent the irretrievable erosion of data and insight that earlier generations had garnered at great cost.

The alternative would be to periodically launch new missions to check what might have already been established decades or centuries earlier. Those original missions may have been expensive. Perhaps, decades later, similar observations could be repeated less expensively, albeit with quite different, by then more recent, technologies that might be more affordable. A wealthy society might agree to launch such missions. A population beset by other priorities would not.

Funding never rises indefinitely. At some point leaner times inevitably arrive. All this suggests the need for greater concern for how knowledge might be better preserved and made understandable for longer periods.

Information will increasingly be stored in higher-capacity storage devices. But the stored information will be successfully accessed only if we know how to ask the right questions, and this will require appreciation of changing vocabularies, evolving language, and the nature and history of obsolescent technologies.

- Improved archiving of data and knowledge is becoming increasingly urgent as we come to recognize that we are under continuous threat of losing what we at one time fully understood.

Loss of memory is terribly costly. In science it can lead to folklore, where we accept what has been handed down to us by earlier generations because we lack the means – linguistic, technological, or financial – to verify whether or not the knowledge they gathered is still valid.

Rapidly Executed Philanthropic Projects

The most ambitious space mission proposed by a private foundation, to date, is a search for asteroids that might someday collide with Earth. In 2011, the B612 Foundation began to seek philanthropic funding aimed at building and launching a space telescope intended to survey the Solar System to identify and track at least 90% of all asteroids with diameters larger than 140 m that might predictably come close to Earth.[a] Called "Sentinel," the mission would also track many asteroids as small as 50 meters that could impact Earth. The cost of such a mission lasting about 6.5 years is in the $500M range, which the B612 Foundation hopes to raise sufficiently early to support a launch of Sentinel around 2018. The total number of asteroids catalogued by this mission should range around 500,000. Construction of the spacecraft will be carried out by a commercial firm.

Such a mission is urgent. A 140-m asteroid colliding with Earth would release an amount of energy equivalent to that of the largest hydrogen bombs ever exploded, bombs roughly in the 20 to 100 Megaton range. The damage would be horrendous. On December 20, 2005, the U.S. Congress passed Public Law 109–155, whose Section 321, the George E. Brown, Jr. Near-Earth Object Survey Act, instructed the NASA Administrator to "plan, develop, and implement a ... program to detect, track, catalogue, and characterize the physical characteristics of near-Earth objects equal to or greater than 140 meters in diameter in order to assess the threat of such near-Earth objects to Earth" [6]

Although NASA never initiated this survey, the agency has agreed to provide the B612 Foundation the use of NASA's Deep Space Network for receiving telemetry from the Sentinel spacecraft and tracking its orbit. In addition, NASA personnel will participate on Sentinel's technical reviews. The arrangement will permit the scientific community to benefit from the B612 effort, which will make its mission's data freely available.[7]

- The Sentinel mission has obvious practical benefits and requires little infrastructure development. This may make it more likely to garner philanthropic support. But any successful start on a mission of this magnitude

---

[a] The B612 Foundation's name refers to the fictional asteroid B-612 from which Antoine de Saint-Exupéry's Little Prince had arrived.[5]

may bode well for the promise of other large-scale philanthropically supported space missions of more esoteric cosmological interest that might be expedited without governmental support.

Maintenance of Reserves and Liquidity on Options

In the experimental sciences, new findings can usually be rapidly verified – often through several independent approaches that rule out potential misunderstandings. In astrophysics, verification is considerably harder to attain. Many critical phenomena can be observed only on rare and unpredictable occasions and may be short-lived, lasting only seconds or hours. Comprehensive studies may then be thwarted because critical tools cannot be marshaled in time. Often the tools required to provide an independent confirmation no longer exist. They may have been disposed of to permit the funding of other facilities.

- Observatories launched into space are generally kept operational for only a limited number of years. This deliberate culling releases funding for other missions in line for launch. An essential observatory may then no longer exist to record the characteristics of a rare reappearing phenomenon. A way to cope with such operational concerns could be to build at least some space missions so they may be placed into a hibernation mode for extensive periods and then revived for further use when needed.

This may not be possible for all missions, in particular those that require cryogenic coolants or other expendables generally consumed within a few years. But if even a limited number of space observatories could be commanded into a safe hibernation standby mode, from which they could be awakened from time to time for especially important observations, considerable gains could be expected at incremental cost.

Where hibernation is not a viable option, alternatives meeting the same goals may be available. In the infrared and submillimeter domains, the Stratospheric Observatory for Infrared Astronomy (SOFIA) will carry out observations that previously were solely accessible with the Herschel space telescope. With its estimated 20-year life span, the SOFIA telescope aboard its Boeing 747 aircraft enables observations on a wide range of infrared or submillimeter phenomena of exceptional importance, whenever called upon, even on short notice. SOFIA's sensitivity may not match that of its predecessor, the Herschel mission, but still may be adequate for many interesting observations that otherwise would be entirely beyond reach.

A potential way of prolonging the lives of deserving observatories is exemplified by the lifeline thrown to the GALEX mission. In 2012, NASA loaned its Galaxy Evolution Explorer (GALEX) mission to Caltech for continuing operation and management supported by private funding. Extending the operation of this ultraviolet astronomical observatory, which had been scheduled for retirement after operating in space for nine years, has enabled an international collaboration of astronomers to conduct observations previously not attempted with GALEX in part because they make use of instrumental capabilities for which the telescope had not been explicitly designed.[8]

- Where possible, national facilities scheduled for retirement should first be offered for use by private groups. A privately constituted group often can operate an observatory more affordably by designing an observing schedule that minimizes costs of operation by sacrificing flexibility. Because such a group is generally quite small and working primarily on a more limited range of research topics, it can often operate more economically than a national agency handling proposals from the entire astronomical community.

The Economics of the Commons

The last three sections may appear to recommend that astronomical efforts that cannot be governmentally funded revert to means pursued before World War II, when wealthy benefactors often underwrote major projects. This is somewhat misleading. The scope of the more modern thrust is considerably broader. The economics of the commons, introduced in Chapter 15, illustrates how the astronomical community can maintain the long-term use of communal resources, whose value will decline unless the entire community, including its amateur membership, invests in its upkeep so that all may benefit.[9]

The feature that brought this powerful economic model to the fore in the 1990s was the emergence of the Internet. An astronomical archive is a communal property – a *commons* open to any astronomer. A particular commons intended largely for the benefit of the astrophysics community is the *arXiv* site, in particular *arXiv astro-ph*, to which many active researchers in the field contribute. For 2013, the maintenance for this archive is provided through a single large gift from the *arXiv* home institution, Cornell University, a second substantial annual gift from the Simons Foundation, and by several hundred research institutions paying in the range of $1500 to $3000 annually, based on the number of articles they download. Together, these meet the anticipated annual operating costs of *arXiv* running in the range of ~$825k for the year.[10] The primary value of *arXiv*, however, resides in the papers annually written by thousands of astronomers world-wide and uploaded to the website, free of charge by or to their authors and to interested readers.

Important sets of archives maintained by various institutions are similarly a set of commons dedicated to the preservation of astronomical data gathered by ground-based and space observatories over periods of years. These archives can be communally sustained for long periods only if users encountering a problem in the archives' use and successfully solving it are encouraged to record their notes on how the problem can be overcome.

As language gradually morphs, and technical expressions lose clear meaning, a user who first encounters an ambiguity and provides advice on how it is to be read in more modern terms helps a next generation of astronomers encountering the same problem. The gradual evolution of language and of technical usage can thus be continually remedied to prevent, or at least postpone, otherwise inevitable loss. The lifetime of archives can thus be communally extended. But this requires the full cooperation of the archives' users. Every user who finds a creative solution to a problem must responsibly cooperate and record the new insight so the vitality and use of the archives can be maintained. Those who benefit from clarifications and explanations their predecessors noted, without similarly recording their own updates, gradually contribute to the archives' decline.

The difference between such modern efforts and help that might have been provided by volunteer efforts earlier in the twentieth century is that earlier volunteers usually had to be local. Through the Internet, global volunteerism has emerged to aid the common good. A global net increases the prospects that an expert on an arcane subject can solve a problem that otherwise would be intractable or at least difficult or expensive to solve. A global net also increases the chances that not only individuals but also motivated industries can contribute to, and in turn derive benefit from, significant astronomical advances.

The full range of prospects of an economic model based on common interests has only just begun to be explored. For the pursuit of astronomical problems that governments may not be inclined to fund, the commons approach may prove a particularly useful way forward.

Notes

1. Astrophysics Implementation Plan, National Aeronautics and Space Administration: NASA Headquarters, Science Mission Directorate, Astrophysics Division, December 20, 2012.
2. *Science – The Endless Frontier*, Vannevar Bush, reprinted by the National Science Foundation on its 40th Anniversary 1950–1990, National Science Foundation, 1990.
3. *Making the invisible visible: A history of the Spitzer Infrared Telescope Facility (1971–2003)*, Renee M. Rottner & Christine M. Beckman, Monographs in aerospace history, NASA-SP 4547, 2012.
4. I thank Dr. Jill Tarter who, until her recent retirement, headed the SETI Institute ever since its inception. She provided me with several different budget estimates from which I culled representative figures presented here. Her estimates are contained in two e-mail compilations she kindly sent me, both dated June 16, 2012, and a clarifying e-mail dated February 3, 2013.
5. *The Little Prince*, Antoine de Saint-Exupéry, translated from the French by Katherine Woods. NewYork: Harcourt Brace, 1943.
6. The B612 Foundation Sentinel Space telescope, E. T. Lu, H. Reitsema, J. Troeltzsch, & S. Hubbard, *NewSpace*, January 2013. See also http://www.gpo.gov/fdsys/pkg/PLAW-109publ155/pdf/PLAW-109publ155.pdf

7. I am indebted to the B612 foundation and to Dr. Harold J. Reitsema, the Sentinel Mission Director, for the cited information.

8. http://features.caltech.edu/features/372

9. *Understanding Knowledge as a Commons – From Theory to Practice*, edited by Charlotte Hess & Elinor Ostrom. Cambridge, MA: The MIT Press, 2007.

10. http://arxiv.org/help/support.faq33c

Epilogue

In the course of the twentieth century science gained insights on the nature of the Cosmos that might have seemed unimaginable even as late as the year 1900. New words and concepts, novel ways of conceiving space and time, and massive efforts to understand the structure of matter were needed to describe the world we were encountering. I embarked on writing this book to gain clearer insight on the factors that had most effectively contributed to all these advances.

Astronomy is a small field whose tools were largely imported from physics and engineering. Early in the century, theoretical physicists provided us with new perspectives for devising and interpreting novel observations.

Then, in mid-century the United States adopted a deliberate policy of coupling basic research to practical national priorities. Other countries followed suit, leading to an explosive expansion of all of the sciences and engineering. Scientific progress led to novel engineering ventures. These, in turn, offered the sciences increasingly powerful research tools. The coupling could not have been more fruitful. For astronomy the benefits were immense.

Today, we may be entering a new era, in which this bond may loosen between astronomy and engineering. Astronomers are turning their efforts to studies of dark matter and dark energy, whose potential utility to society is not at all apparent. In view of competing national priorities governments may find themselves unable to support the development of the novel tools required for conducting such esoteric cosmological searches. Astronomers may then have to find alternative ways of extending their investigations of the true Universe.

I use the word "true" to convey that the Cosmos may have a structure that could elude us unless we search for it with highly specific tools hard to imagine or come by. Lacking these tools the Cosmos we unveil could then be largely deceptive.

To realize our goal of establishing how the Universe began, transformed itself, and over the eons gave rise to galaxies, stars, and living matter, we will need to keep firmly in mind that tools are central to shaping our understanding of the Cosmos, and that the cost of developing the right tools will play a critical role in search of the true Universe.

Appendix: Symbols, Glossary, Units, and Their Ranges

Symbols

~	Denotes that a particular numerical value is approximate.
″	Denotes an angle measured in seconds of arc.
′	Denotes an angle measured in minutes of arc.
°	Denotes an angle measured in degrees.
α	Greek letter "alpha." See also *alpha particle*.
Å	See *Angstrom unit*.
β	Greek letter "beta." See also *beta particle* and *electron, e^-*.
c	The speed of light in vacuum, 2.998×10^{10} cm/sec.
γ	Greek letter "gamma." See *gamma rays*.
δ	Greek letter "delta."
e	The electron charge, 4.803×10^{-10} electrostatic units.
e^-	Electron. See also *electron, e^-*.
e^+	Positron. See also *positron, e^+*.
ζ	Greek letter "zeta."
G	The gravitational constant, 6.674×10^{-8} cm^{3}g^{-1}sec^{-2}.
h	Plank's constant, 6.626×10^{-27} erg sec.
$\hbar$	Planck's constant divided by 2π, $\hbar = h/2\pi = 1.055 \times 10^{-27}$ erg sec.
k	Boltzmann's constant, 1.381×10^{-16} erg/K.
K	Degree Kelvin. See also *Kelvin temperature, K*.
$L_\odot$	The solar luminosity 4×10^{33} erg/sec.
λ	Greek letter "lambda." See also *wavelength, λ*.
Λ	Greek letter "Lambda." See also *cosmological constant*.
$M_\odot$	The solar mass 2×10^{33} g.
μ	Greek letter "mu" for "micro," one part in a million.
n	Neutron.
ν	Greek letter "nu." See both *neutrino, ν* and *frequency, ν*.
$\bar{\nu}$	Designation for an antineutrino. See *antineutrino, $\bar{\nu}$*.
p	Proton.

Glossary

absolute magnitude: See *magnitude of a star or galaxy.*

adiabatic expansion: A gas expanding without input or loss of energy is said to expand adiabatically. An expansion of the Universe without infusion or extraction of energy is similarly said to be adiabatic.

alpha particle, α: The highly energetic nucleus of a helium atom consisting of two protons and two neutrons.

Angstrom unit, Å: A measure of length equaling 10^{-8} cm.

annihilation of matter: The destruction of matter on encountering antimatter, with an accompanying liberation of energy and the formation of pairs of photons or particles and their antiparticles. See also *antimatter.*

antimatter: Matter consisting of antiparticles.

antineutrino, v̄: Antiparticle of the neutrino. See *neutrino.*

antiparticle: See also *annihilation of matter.* Matter consists of atoms that contain neutrons, protons, and electrons. Corresponding to each of these three particles there exists an antiparticle with identical mass, but opposite charge, if any. Neutrinos, which are neutral particles, are distinguished from antineutrinos by the direction of their spin. Particles and antiparticles annihilate on encounter.

arc second, ": An angle that subtends 1/3600 of a degree.

asteroid: A small planet, often barely large enough to be detected with a telescope. The smallest observed asteroids have diameters of 50 to 100 meters.

atomic mass unit, amu: A mass equivalent to one twelfth the rest mass of an isolated atom of carbon ^{12}C in its ground electronic and nuclear states.

AXAF: Advanced X-ray Astrophysics Facility, a powerful X-ray astronomical space observatory renamed *Chandra* after launch in July 1999.

beta particle: An electron or positron emitted from a nucleus, usually at high energy.

billion: 10^9.

binary stars: Two gravitationally bound stars orbiting a common center of mass.

bit: A contraction for "binary digit," usually designated by symbols 0 or 1; a unit of information.

black hole: A highly compact massive object, whose gravitational attraction is so strong that no matter or radiation can escape. See also *Schwarzschild radius.*

blue shift: A shift of light to higher frequencies.

Boltzmann constant, k: A constant of nature enabling the conversion of an equilibrium temperature T into units of energy. For an ideal gas the product of pressure P and the volume v occupied by particles, that is, atoms or molecules of the gas, provides a measure of the mean energy of the individual particles, $Pv = kT$, with $k = 1.381 \times 10^{-16}$ erg/K.

brown dwarf: A body of mass lower than that of stars. Stars are sufficiently massive to convert hydrogen into helium in their interior. Brown dwarfs are incapable of doing this. But even low mass brown dwarfs can derive energy from fusing deuterium nuclei, an activity that planets are not sufficiently massive to exhibit.

The distinctions between the least massive brown dwarfs and the most massive planets are not always well defined.

carrier of information: Any particle or wave that transmits information from a source.

cascade: A succession of processes, each of which triggers the next.

Cepheid variable: A bright yellow variable star that pulses regularly with a period as short as 2 days for some cepheids and as long as 40 days for others. Cepheids are sufficiently luminous to be detected in nearby galaxies. Because their periods are directly related to their luminosities, Cepheids can serve as distance indicators.

cluster of galaxies: A grouping of galaxies that may contain as many as several thousand individual or interacting galaxies. A small cluster usually is called a group of galaxies.

comet: A solar system body that disintegrates on approaching the Sun, leaving a trail of debris and often a long tail of ionized gas pointing away from the Sun.

cosmic background radiation: See *microwave background radiation*.

cosmic maser: A monochromatic source of radiation whose high surface brightness is produced through the same process active in man-made masers and lasers. See also *maser*.

cosmic ray: A highly energetic particle that travels at a speed close to the speed of light. Electrons, positrons, protons, and nuclei of atoms have all been found in the rain of cosmic ray particles that continually impacts Earth's upper atmosphere from as-yet-unspecified cosmic realms.

cosmological constant, Λ: A uniform vacuum energy density pervading the Universe.

Cosmos: The Universe; all that we can survey.

cross section for an interaction: A significant interaction of two particles takes place only if the encounter is sufficiently close. The particles act as though they had a cross section for interaction whose diameter exceeds the center-to-center distance between the two particles whenever they interact.

curvature of space: In many cosmological models, light does not propagate along straight lines, but rather along curved trajectories. The curvature of these trajectories is the curvature of the space.

dark energy: This is a hypothetical energy permeating the Universe and tending to accelerate cosmic expansion. Little is known about its physical origins. It might take the form of a cosmological constant Λ, an energy content of the vacuum that constantly regenerates itself as the vacuum expands. But it could equally well represent other energy distributions, which might vary with time and place.

dark matter: This hypothetical form of matter appears to dominate the gravitational forces exerted on stars and interstellar matter within galaxies. It also dominates the gravitational interactions of galaxies within clusters. Its nature is not understood.

diffraction: The spreading of a light beam around a body that blocks part of the beam. Also the spreading of different wavelengths of light along different directions.

disk: A thin, circular aggregate of gas, dust, or stars, gravitationally attracted to and circling about a massive central body.

Doppler shift: The systematic shift of an entire spectrum of radiation toward longer wavelengths – lower frequencies – when the source of radiation rapidly recedes from the observer, and toward shorter wavelengths – higher frequencies – when the source approaches.

dust: Fine grains of solid matter. In interstellar space dust appears aggregated in dark irregular clouds.

dwarf star: A low-luminosity main sequence star.

eclipsing binary: A pair of mutually orbiting stars in which one star passes in front of the other and blocks its light.

electromagnetic radiation: A class of radiation comprising radio waves, infrared rays, visible light, ultraviolet radiation, X-rays, and γ-rays, differing from each other only in their wavelengths, which also determine the frequency and energy of each quantum of radiation.

electromagnetic theory: A mathematical depiction of electrical and magnetic processes.

electromagnetic waves: See *electromagnetic radiation.*

electron, e^-: A negatively charged particle often found orbiting an atomic nucleus, but able to travel through space by itself if removed from the atom.

electron Volt, eV: A unit of energy. The energy carried by a quantum of yellow light is about 2 eV. An X-ray photon has an energy of several thousand electron Volts. See Table A.2.

equation of state: The relationship between the density and pressure of a substance as its temperature changes.

erg: A unit of energy. See Table A.2. A mass of one gram moving with a velocity of 1 centimeter per second has an energy of 0.5 erg.

eruptive variable: A variable star that suddenly changes its output of light, usually from a normally low to a far higher eruption level. See also *nova.*

eV: See *electron Volt* and Table A.2.

event horizon A surface separating events that can be observed from others that disappeared when the recession velocity of the objects involved exceeded the speed of light.

evolved star: See *main sequence.* A star that has already converted its available store of hydrogen into helium and no longer is found on the main sequence.

extreme ultraviolet: Ultraviolet rays of very short wavelength and X-rays of very long wavelength define the upper and lower bounds of this wavelength range.

Faraday rotation: A plane polarized wave passing through an ionized gas along the direction defined by a local magnetic field experiences a rotation of its direction of polarization. This rotation is largest at long radio wavelengths and decreases monotonically at shorter wavelengths.

field theory: A theory in which the forces between particles is transmitted by a *field* permeating the space encompassing the particles. The field continually generates and destroys intermediary particles transmitting the force between particles. The particles on which the field acts can be considered to be concentrations in the spatial energy distribution of the field.

flare star: A star whose luminosity can increase by a substantial factor in a matter of minutes, with a subsequent, slower decline to normal levels. In some rare stars the luminosity increases by more than a factor of 100. In most it increases only by a factor of 2 to 5.

flat space: Cosmic space is said to be flat if the Universe is infinite and its geometry obeys Euclid's postulates, in particular, that parallel straight lines, no matter how far they are extended, never intersect. In Einstein's general relativity this happens only when the mass–energy density of the Universe has a critical value ρ_{crit} related to the Hubble constant H, and the gravitational constant G, by $\rho_{crit} = 3H^2/8\pi G$. See *Hubble constant, H.*

flux: The energy passing through unit area each second within the measured range of frequencies.

frequency, ν: The number of crests of a wave, for example, an electromagnetic wave, passing an observer during an interval lasting one second. See also *Hertz (Hz).*

galactic plane: The disk-shaped aggregation of stars and gas in the central plane of a spiral galaxy.

galaxy: A galaxy is an isolated grouping of 10^9 to 10^{11} stars and associated interstellar matter, mutually bound by gravity. The *Galaxy* – where the word *galaxy* is capitalized – is the particular galaxy in which the Solar System resides. It contains $\sim 10^{11}$ stars, measures a hundred thousand light years across, and contains some stars older than 10^{10} years. The Galaxy is sometimes called the Milky Way because, seen from the Solar System, which lies $\sim 25,000$ light years from the Galaxy's center, it looks like a white diffuse band stretching across the night sky.

gamma ray, γ: An electromagnetic wave whose wavelength is less than 10^{-10} cm and whose energy is higher than 10^5 eV. See Table A.2.

gamma-ray bursts: Periodic burst of gamma rays reaching Earth from different parts of the sky, at unpredictable times, often lasting no longer than a few seconds. Most arrive from distant galaxies and appear to be generated in hypernova explosions or by the merger of two neutron stars. See *hypernova.*

gauss: A unit of magnetic field strength. Measured on Earth magnetic field strengths are around half a gauss but vary with latitude. In magnetic stars the field strength can exceed 10^4 gauss. In interstellar space the field strength tends to be less than 10^{-5} gauss.

GeV: See Table A.2.

GHz, Gigahertz: 10^9 Hz. See Table A.2. See also *Hz.*

giant star: Any highly luminous star. See *red giant.*

globular cluster: A closely packed spherical aggregate of 10^5 to 10^7 stars bound by gravity. The Galaxy contains several hundred of these clusters, all formed about 10^{10} years ago.

Gpc, Gigaparsec: See Table A.1.

gram, g: A unit of mass. A cubic centimeter of water, approximately a thimbleful, has a mass of one gram. See Table A.3 and Figure A.2 for the relations between different units of mass.

gravitational radiation: A form of radiation expected to be liberated by massive, accelerating bodies. To date, these gravitational waves have not been directly observed, but their existence is inferred from gradual changes in the orbits of binary neutron stars. The quantum of gravitational radiation is sometimes called a graviton.

gravitational waves: See *gravitational radiation*.

Heisenberg uncertainty principle: See *uncertainty principle*.

Hertz, Hz: A measure of frequency named after the nineteenth-century German physicist Heinrich Hertz. 1 Hz equals 1 cycle per second. See Table A.2.

Hertzsprung–Russell diagram: A drawing that plots the luminosity of stars as a function of their spectral type. See Figure 4.1.

homogeneity: The property of having a uniform consistency throughout.

Hubble constant, H:, The expansion rate of the Universe, generally expressed in kilometers per second, per Megaparsec, km/sec-Mpc. See Table A.1.

hypernova: An extremely powerful supernova.

Hz: See *Hertz (Hz)*.

ideal gas: A gas idealized as consisting of non-interacting point-like particles in random motion. The concept is useful in analyzing the behavior of highly dilute gases.

information: A quantitative measure of data content. See also *bit*.

infrared galaxy: A galaxy that emits most of its energy at infrared wavelengths.

infrared radiation: Radiation of wavelength in the 0.7 micrometer to 1 millimeter range. See also Table A.1.

infrared star: A star that emits most of its energy as infrared radiation.

intensity of radiation: The energy content of a beam of radiation; a measure of brightness.

interferometer: Apparatus used for interferometry. See *interferometry*.

interferometry: The use of interference between superposed beams of electro-magnetic radiation to measure the angular or spectral structure of a radiating source.

interplanetary space: The region between the planets orbiting the Sun.

interstellar cloud: A cloud of gas or dust in the space between stars.

ion: An atom charged through addition or (more often) removal of electrons orbiting the nucleus.

ionized hydrogen region: A domain of interstellar space containing hydrogen ionized through removal of electrons from their parent nuclei. It largely contains freely moving electrons, protons, and smaller concentrations of other ions and atoms.

isotropy: Independence of orientation; having identical characteristics along any direction.

Kelvin temperature, K: A temperature measured on a scale with units identical to those of the conventional Celsius (centigrade) scale, but with its zero point at absolute zero, where the energy of all matter attains an absolute minimum. The zero point on the Kelvin scale corresponds to −273.15 on the Celsius scale.

keV: See Table A.2.

kHz: See Table A.2.

limb: The edge of the projected disk of the Sun or Moon, or stars and planets, as viewed by an observer.

logarithm: The logarithm of a number N, measured to base 10, is n, if $10^n = N$. We write $\log_{10} N = n$.

logarithmic scale: A scale on which numbers that differ by a constant ratio, say a factor of 10, are plotted a constant distance apart. See, for example, this glossary's Figure A.2.

luminosity: The energy emitted in unit time by a radiant source.

Mach's principle: The hypothesis that the aggregate of distant matter in the Universe defines a local state of unaccelerated motion everywhere, and determines the inertia of a body – its resistance to acceleration. Some versions of the principle advanced at different times also proposed that the constants of Nature might similarly be determined, but this now appears unlikely.

magnetic field: A field that exerts a force on moving charges and magnetized bodies. See also *gauss*.

magnetic variable: A variable star having an extraordinarily high magnetic dipole field, often exhibiting a magnetic field thousands of times higher than the Sun's.

magnitude of a star or galaxy: The brightness of a star or a galaxy. The magnitude of a star or galaxy, as seen from Earth, is called its apparent magnitude. Seen from a standard distance, chosen as 10 parsecs – about 30 light years – the magnitude of the star or galaxy is called its absolute magnitude. Two stars or galaxies that differ in luminosity by one magnitude, differ by a factor of ~2.5 in brightness. Five magnitudes amount to a brightness difference of 100. Magnitudes measured in the visual wavelength band, are called visual magnitudes.

main sequence: See *Hertzsprung–Russell diagram*. On a plot of luminosity of stars as a function of color or temperature, more than 90% of all stars fall along a band that stretches from bright, hot, blue stars to faint, cool, red stars. This band marks the main sequence.

main sequence star: A star that falls anywhere along the main sequence in a Hertzsprung-Russell diagram. See *main sequence*. See also *Hertzsprung-Russell diagram*.

manifold: A manifold in general relativity is a space which, near each point resembles a four-dimensional Euclidean space. Over larger distances, however, the space can be curved depending on the distributions of mass–energy within the space.

maser: An intense source of electromagnetic radiation emitted in an avalanche of photons all of which have identical wavelength, polarization, and direction of travel.

mass defect: The difference between the mass of an atom and the sum of the masses of its constituent particles when not bound to each other. The more tightly the constituent protons and neutrons are bound in an atomic nucleus, the higher is its mass defect.

Megahertz, MHz: A million Hertz = 10^6 cycles per second. See Table A.2.

meteor: A grain of interplanetary matter that burns and disintegrates as it enters the upper atmosphere at high speed and produces a streak of light along its trajectory. Colloquially, meteors are often called shooting stars.

meteorite: See *meteor.* A sizable meteor, much of which survives passage through the atmosphere to hit Earth as a solid chunk of matter.

metric: A metric in four–dimensional relativistic space-time is a mathematical expression for the distance between two elements in the space–time.

MHz: See Megahertz.

micrometer: 1 micrometer = 10^{-4} cm = 10^{-6} m. See also Table A.1.

microwave background radiation: Isotropically arriving microwave radiation from the Universe. The intensity of this radiation is roughly equivalent to that which would be found inside a cavity whose walls were kept at a temperature of 2.73 degrees Kelvin.

microwave radiation: Electromagnetic radiation in the wavelength range from approximately 1 mm out to beyond 10 cm.

moon: A planet-like body gravitationally bound to, and orbiting, a larger parent planet. Earth has only one moon; Jupiter has well over a dozen. See *planet.*

neutrino, ν: A subatomic particle with zero charge, a spin of 1/2, a rest mass close to zero, and a speed usually close to the speed of light. The neutrino's antiparticle is the antineutrino. See also *antineutrino.*

neutron, n: A neutrally charged particle with a mass somewhat in excess of a hydrogen atom. In all atomic nuclei, except for ordinary hydrogen, it is found bound to protons and other neutrons. When isolated from a nucleus, a neutron decays with a mean life of 885 seconds, giving rise to an electron, a proton, and an antineutrino.

neutron star: A collapsed, compact star whose core consists largely of neutrons.

noise: Spurious signal registered by a detector.

nonthermal radiation: Radiation emitted by a gas in excess of the radiation expected from its thermal glow alone.

nova: A star that erupts in a matter of hours or days to increase in luminosity by a factor of $\sim 10^5$, returning to its original luminosity, roughly comparable to the Sun's, in a matter of weeks. Novae are binaries consisting of a cool red giant and a compact, hotter companion, which tidally strips the giant's outer layers, gathering this material until a critical surface mass triggers a thermonuclear explosion. Such cycles periodically repeat.

nucleon: Generic name for protons and neutrons in atomic nuclei.

observation: A passive form of study in which the observer has no way of, or refrains from, stimulating the system under investigation.

occultation: The extinction of light from a celestial source by a body that passes between the source and the observer.

old star: A star that has been evolving for more than $\sim 10^8$ years. A young star, in contrast, may have been formed less than 10^7 to 10^8 years ago.

orbital velocity: The speed with which an orbiting body moves along its trajectory.

orthogonal dimensions: Two or more lines that are perpendicular to each other are said to be orthogonal. Lengths measured along such lines specify the dimensions of a physical system.

parallax: The change of apparent source position in the sky as an observer moves from one location to another. In judging the distances of stars, the motion of Earth is taken to be equivalent to one radial distance between Earth and the Sun, 1.5×10^{13} cm, along a base line perpendicular to the line of sight to the star.

parameter: A trait that can be quantified and whose value describes the physical state of a system.

parsec, pc: The distance of a star at which its parallax is one second of arc. This distance, 3×10^{18} cm, roughly corresponding to three light years, nearly 20 trillion miles.

perihelion precession: The perihelion point of a planet orbiting the Sun in an elliptic orbit is its point of closest approach to the Sun. This point can slowly but persistently circle the Sun in a motion called a precession.

phase: A stage of development of an evolving system.

phase of matter: Atomic and molecular matter can exist in a number of different states, called phases. Major distinctions in phase are the ionized, gaseous, liquid, or solid phases. But other phases, such as different crystalline states of a solid, can also exist.

phenomenon: A class of objects or patterns of events that drastically differs from all other classes.

photometry: A low-spectral-resolution brightness measurement that generally encompasses a bandwidth comparable to the mean frequency of the observed radiation.

photon: A quantum of radiation.

Planck's constant, h: A constant whose value is $\sim 6.6 \times 10^{-27}$ erg-sec. When multiplied by the frequency ν of a photon, Planck's constant gives the photon's energy, $E = h\nu$.

planet: The Sun is orbited by eight major bodies called "planets," six of which in turn are orbited by moons, and at least three of which exhibit rings. In order of increasing distance from the Sun, the planets are Mercury, Venus, Earth, Mars, Jupiter, Saturn, Uranus, Neptune.

planetary nebula: A cloud of gas ejected from an evolved star and ionized by the ultraviolet radiation emitted by the star's white dwarf remnant.

planetary system: A grouping of planets gravitationally bound to a star about which they orbit. The Solar System is a planetary system, but by now a sizable fraction of other stars are known to also exhibit planetary systems.

plasma: An ionized gas. See also *ionized hydrogen region*.

positron, e^+: Antiparticle of the electron and identical to it in all ways except that it carries a positive, instead of a negative, electrical charge.

proton, p: Nucleus of a hydrogen atom.

pulsar: A source of sharp radio or γ-ray pulses emitted at regular intervals. In some pulsars the intervals can be as short as milliseconds; in others as long as several seconds. Most pulsars are known to be rotating, highly magnetized neutron stars that are remnants of supernova explosions.

pulsating variables: A star whose radius and brightness vary regularly or semi-regularly.

quasar: A compact source of radiation occupying the nucleus of a galaxy. Quasars harbor giant black holes with masses ranging up to $\sim 10^9 M_\odot$. They often are highly red shifted and exhibit irregular variations in brightness.

radar: A technique that transmits radio pulses to distant objects and measures that distance by the delay in arrival of the returning reflected pulse.

radio star: A star that emits radio waves, generally a rare star that becomes a radio emitter when its luminosity suddenly increases.

radio wave: An electromagnetic wave whose wavelength exceeds a millimeter.

recession velocity: Speed of recession along an observer's line of sight.

red giant: A luminous red star whose luminosity and surface temperature place it on the red giant branch in a Hertzsprung–Russell diagram. See Figure 4.1.

redshift: A move of an entire spectrum toward longer wavelengths, lower frequencies. A galaxy from which the observed arriving wavelength, say of a particular hydrogen spectral line, is λ_1, whereas the same spectral line emitted by hydrogen locally is λ_0 is said to have a redshift $z = [(\lambda_0 - \lambda_1)/\lambda_1]$. See also *blueshift*.

relativistic: An adjective indicating that a body's motion is significantly governed by the laws of relativity.

relativistic particles: Subatomic particles that move at velocities close to the speed of light.

resolving power: Spectral resolving power, R_s, is the ratio of the wavelength, λ, at which an observation is carried out, to the wavelength difference, $\Delta\lambda$, which can just he resolved:

$$R_s = \lambda/\Delta\lambda.$$

If an angle $\Delta\theta$ radians can just barely be resolved, the angular resolving power – also called the spatial resolving power – is:

$$R_\theta = 1/\Delta\theta.$$

rest mass: A body's mass at rest is at a minimum; at high velocities its mass increases significantly.

satellite: There are two kinds of satellites: natural and artificial. Natural satellites of planets are moons. Artificial satellites are devices placed in orbit around a planet or moon, generally for scientific, communications, or military purposes. See also *moon*.

scalar quantities: A scalar quantity can be described by a single magnitude, such as temperature, mass, density, or speed, none of which specify a spatial direction.

scalar particles, fields or perturbations: A scalar particle is a hypothetical particle having no directional properties. Thus, it has no spin. Scalar fields are similarly hypothetical and are the fields generated by scalar particles. In principle, these fields can perturb the density distribution in the space to produce scalar perturbations.

Schwarzschild radius: The radius of a sphere constituting the surface of a black hole through which no outward directed radiation or matter can escape. See *black hole*.

sensitivity: The capacity of an instrument to detect weak signals.

SETI: A project to Search for Extra-Terrestrial Intelligence, primarily by analysis of radio signals that could be reaching us from nearby stars with planetary systems.

Seyfert galaxy: A spiral galaxy with a compact nucleus in which gas velocities of several thousand kilometers per second are observed. These compact nuclei are believed to generally harbor supermassive black holes.

shock: A sudden impact that changes the state of a system.

solar mass, $M_\odot$ The mass of the Sun. A unit of mass equaling the mass of the Sun. See Table A.3.

Solar System: The Sun and the system of planets, moons, asteroids, comets, and all other matter orbiting the Sun. See also *planet, moon, asteroid, comet*.

spatial resolution: See *resolving power*.

spectral energy distribution: The distribution of radiated energy across the range of wavelengths or frequencies at which a star or a galaxy radiates.

spectral line: A narrow, dark or bright feature in a spectral display of light due to an excess or a lack of radiation at one particular color or wavelength.

spectral resolution: See *resolving power*.

spectral type: The spectral type of a star, determined by its spectrum, largely depends on the star's surface temperature. Surface chemistry can also be a factor. Stars whose spectra exhibit similar features are said to be of the same spectral type.

spectrometer: Apparatus used in spectroscopy. See *spectroscopy*.

spectroscopic binary: A binary in which the presence of the two stars is discerned by virtue of two sets of superposed spectra, one spectrum corresponding to each star. The two spectra exhibit opposite Doppler shifts that vary with time, as each star orbits a common center of mass, alternately approaching and receding from the observer.

spectroscopy: The separation of light into its wavelength or color components.

spectrum: The display of the different color components in light, or wavelength components in other types of electromagnetic radiation, the intensity of each component being separately displayed.

speed of light, c: The speed of light is about 3×10^{10} cm per second or, equivalently, 300,000 kilometers per second.

spin: Every fundamental particle is characterized by a spin angular momentum. For electrons, protons, neutrons, and neutrinos that spin has a value $(1/2)\hbar$. For light quanta the spin is $\hbar$. For gravitational waves the spin is believed to be $2\hbar$, where $\hbar$ is Planck's constant, h, divided by 2π. Some subnuclear particles also have zero spin.

spiral galaxy: A galaxy that exhibits stars, gas, and dust arranged in lanes or segments of lanes that stretch outward from the galaxy's center in a spiral pattern. Barred spirals are galaxies in which spiral arms appear at the ends of an elongated bar-shaped aggregate of stars at the galaxy's center.

stability: The ability to withstand small disturbances and return to equilibrium.

star: A gravitationally bound compact mass containing between 10^{32} and 10^{35} grams of matter. It can keep shining as long as nuclear or gravitational energy keeps being released by activity in the star's highly compressed central regions. See also *binary star, brown dwarf, evolved star, flare star, giant star, main sequence star, old star, red giant, variable star, young star.*

statistical reasoning: Reasoning based on the probabilities of randomly occurring events in a system of known structure.

superluminal source: A source whose components are expanding at a rate appearing to be faster than the speed of light.

superluminal velocities: Speeds that appear to be greater than the speed of light.

supermassive black holes: Black holes in the nuclei of galaxies whose masses may reach $\sim 10^9 M_\odot$.

supernova: A star whose luminosity can increase by a factor of $\sim 10^8$, over a period of hours or days as the star explodes. The most luminous supernovae, termed hypernovae, are the most luminous individual stars known. The supernova bright phase declines over a period of months. See *hypernova.*

surface brightness: The energy emanating from each unit of area in a second.

thermonuclear conversion: The conversion of hydrogen into helium, or more generally of any element into another, at temperatures sufficiently high for nuclear reactions to take place. The nuclear reactions in a star's interior liberate energy that ultimately is emitted at the star's surface as starlight.

time dilatation: The slowing down of the pace at which a clock will run. Time dilation occurs in bodies that move at velocities close to the speed of light and in bodies placed in the close proximity of massive objects.

topology: The properties of an object or of a space that are preserved under continuous deformation, including stretching and bending without tearing. A sphere has the same topology as a cube, but not as a doughnut. A tea cup with a curved handle attached near the top and bottom of the cup does have the same topology as a doughnut.

ultraviolet radiation: Electromagnetic radiation in a range of wavelengths shorter than those of violet light and ranging in wavelength from ~4000 Å down to 100 Å. The human eye does not see (sense) ultraviolet radiation.

uncertainty principle: The uncertainty principle due to Werner Heisenberg states that certain complementary pairs of properties of matter, or radiation, cannot be simultaneously measured with arbitrary accuracy. Thus, the frequency and time of arrival of a particle or photon cannot both be determined beyond a certain level of precision.

Universe: The entire world in which we live; all we can survey. See also *Cosmos.*

variable star: See *Cepheid variable, eruptive variable, flare star, pulsating variable.*

virial theorem: The theorem states that a system of gravitationally coupled particles in orbital motion about a common center has a long-term mean kinetic energy $\langle T \rangle$ equal to half the absolute magnitude of its long-term mean potential energy $\langle V \rangle$, that is, $\langle T \rangle = -\langle V \rangle / 2$.

visual binary: A system in which two mutually orbiting stars are sufficiently far apart to be spatially resolved when viewed through a telescope.

Watt, W: A unit of power equal to 10^7 erg per second.

wavelength, λ: The distance between successive crests of a wave.

white dwarf: A compact star that has gravitationally contracted to its present size, after exhausting all its sources of thermonuclear energy. Its size is comparable to that of Earth. Its mass is comparable to the Sun's.

X-ray: A photon whose energy is 10^3 to 10^5 times that of visible light.

X-ray background radiation: Diffuse X-radiation from the sky, largely due to X-ray emission from supermassive black holes in galaxy nuclei. Its wavelengths range from ~1000 Å down to 0.1 Å.

X-ray galaxy: A galaxy that emits an appreciable fraction of its energy in the X-ray wavelength range.

X-ray star: A star that emits an appreciable fraction of its energy in the X-ray wavelength range.

young star: See *old star.*

Units and Their Ranges

Very large or very small numbers are expressed in powers of ten. Thus 10^6 stands for 1,000,000, that is, a million, where the exponent 6 indicates the number of zeroes following the number 1. Similarly 10^{-6}, standing for 0.000,001, represents *one millionth*, or one part in a million; and 7×10^{-6} is seven parts per million.

Table A.1. *Relation between units of length*

Unit	Length in Centimeters	Length in Meters
Angstrom unit, Å	10^{-8} cm	10^{-10} m
Micrometer, μm	10^{-4}	10^{-6}
Millimeter, mm	10^{-1}	10^{-3}
Centimeter, cm	1	10^{-2}
Meter, m	10^{2}	1
Kilometer, km	10^{5}	10^{3}
Light year, ly	9×10^{17}	9×10^{15}
Parsec, pc	3×10^{18}	3×10^{16}
Kiloparsec, kpc	3×10^{21}	3×10^{19}
Megaparsec, Mpc	3×10^{24}	3×10^{22}
Gigaparsec, Gpc	3×10^{27}	3×10^{25}

Table A.2. *Relation between units of photon energy, wavelength, and frequency*

Unit		Energy in ergs	Photon Wavelength	Photon Frequency
1 electron Volt (eV)	=	1.6×10^{-12} erg	1.2×10^{-4} cm	2.5×10^{14} Hz
1 keV = 10^{3} eV	=	1.6×10^{-9}	1.2×10^{-7}	2.5×10^{17}
1 MeV = 10^{6} eV	=	1.6×10^{-6}	1.2×10^{-10}	2.5×10^{20}
1 GeV = 10^{9} eV	=	1.6×10^{-3}	1.2×10^{-10}	2.5×10^{23}

(Wavelength, λ) $\times$ (Frequency, ν) = (Speed of Light, c): $\lambda\nu = c = 3 \times 10^{10}$ cm/sec

(Energy, E) = (Planck's constant, h) $\times$ (Frequency ν).

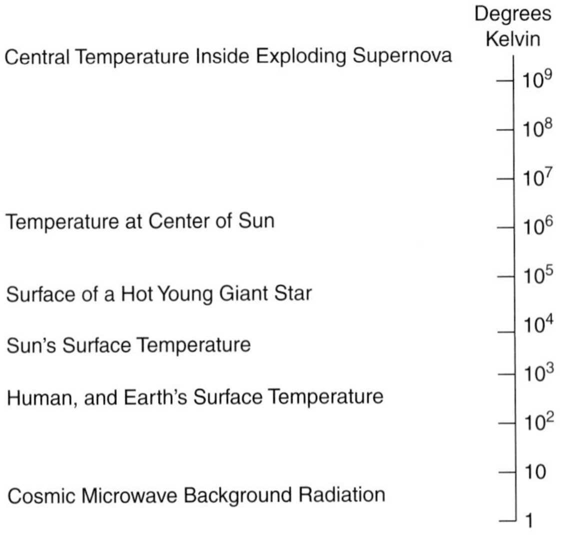

Figure A.1. The range of temperatures encountered in the Cosmos.

Table A.3. *Relation between units of mass*

Unit	Mass in Grams
Microgram, μg	10^{-6} g
Milligram, mg	10^{-3}
Gram, g	1
Kilogram, kg	10^3
Metric ton	10^6
Solar Mass, $M_\odot$	2×10^{33}

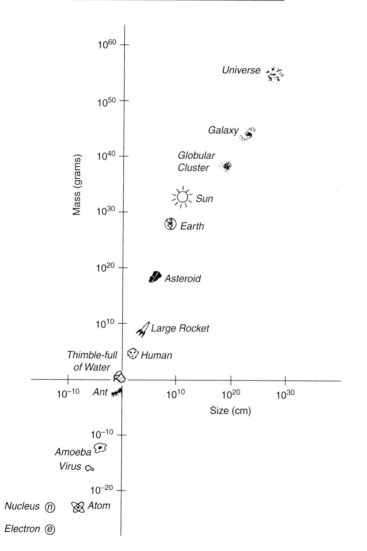

Figure A.2. The masses and sizes of different bodies shown on a logarithmic scale, expressed in the powers of 10 by which they exceed that of a cubic centimeter, roughly a thimbleful, of water. The range stretches from the electron, whose mass is ~10^{-27} g and whose radius is ~10^{-13} cm, to the visible part of the Universe with mass ~10^{55} g and size ~10^{28} cm.

Index

Susskind, Leonard, 264
Sutherland, Gordon Brims Black McIvor, 139
symbols, 369
symmetry breaking
   spontaneous, 218
symmetry properties, 19, 20

Tammann, Gustav Andreas, 267
Tananbaum, Harvey, 235, 244, 248, 344
Tarter, Jill Cornell, 360
Tate, John Torrence, 112
Taylor, Joseph Hooton, 136
Teller, Edward, 105, 238
temperature, *see* Kelvin temperature scale
The Great Observatories, 230–53, 310, 343, *see*
   Great Observatories, 237
The Structure of Scientific Revolutions, 18,
   200, 299
Thomson, Joseph John, 3, 58
Thonnard, Norbert, 294
't Hooft Gerardus, 221
Thorne, Kip Stephen, 48, 93, 110, 142, 202,
   204
Thorne-Żytkow objects, 110
thought
   collective, 18, 188
   convention, 16, 18, 188, 298
see also Fleck, Ludwik
thumbtacks, 20
Tolman, Richard Chance, 91
tools
   craftsmen of theoretical, 191
   mathematical, ix, 11
   observational, 11
   versus new ideas, 196
topology, 203, 208
Tousey, Richard, 137
Trimble, Virginia, 41
Triple-alpha process, 163
Truman, Harry S, President, 123
Turkevich, Anthony Leonid, 161
Turner, Edwin L., 260
Turner, Michael S., 264
Tye, Sze-Hoi Henry, 222, 260

U.S. Air Force, 148
U.S. Astronomical Community
   organization of, 290, 356
U.S. House of Representatives,
   234
U.S. Senate, 234
understanding and control, 335
units, 369
Universe
   age of, 5, 9
   birth of, 10
   curvature of, 40
   early nucleogenesis, 104
   Euclidean, 40
   expansion of, 9

static, 40
   understanding of, 12
Unsöld, Albrecht, 70
Urey, Harold Clayton, 101

vacuum
   false, 219
   fluctuations, 323
   phase transitions, 219
van Vleck, John Hasbrouck, 113
Vela Project, 141, 180
Virtual Observatory, 344
vocabularies
   importance of shared, 290, 329
Vogel, Hermann Carl, 4
Volkoff, George Michael, 91
von Braun, Wernher, 137

Wagoner, Robert V., 259
Wali, Kameshwar C., 87
Walker, Russell G., 132
Washington Conference on Theoretical
   Physics, 105
Waterston, John James, 32
Watts, Duncan J., 182, 262, 276
   global cascades, 198, 300
Weaver, Harold Francis, 135
Webster, B. Louise, 269
Weinberg, Erick J., 218
Weinreb Sander, 135
Weisberg, Joel M., 136
Weiss, Rainer (Rai), 235
Weizsäcker, Carl Friedrich (von), 102, 142, 194
   heavy element production, 102
Wheeler, John Archibald, 142, 202, 209
Whipple, Fred Lawrence, 133
white dwarfs, 79
   limiting mass, 82
   polytropic models, 83
Wikipedia, 348
Wilkinson Anisotropy Probe, WMAP, 323
Wilkinson, David Todd, 135
Wilson, Robert Woodrow, 135, 303
Wolszczan, Aleksander, 268
World War II, 123, 129, 131

X-ray
   galaxies, 138
   stars, 138
      binaries, 218, 270
   telescopes, 3
X-rays, discovery of, 1

Yang, Chen-Ning, 48, 195

Zel'dovich, Yakov Borisovich, 142, 215, 221,
   224, 323
Znajek, Roman L., 216
Zwicky, Fritz, 113, 294
Żytkow Anna N., 110